AF605830

Visual Vestiges

SUNY series, Studies in Technical Communication

Miles A. Kimball, Derek G. Ross, and Hilary A. Sarat-St. Peter, editors

Visual Vestiges

Time, Rhetoric, and Information Design

CHARLES KOSTELNICK

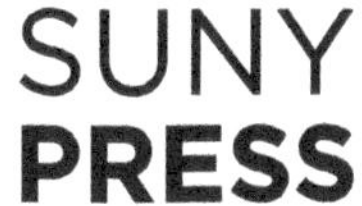

Cover credit: Shutterstock

Published by State University of New York Press, Albany

Printed in the United States of America

EU GPSR Authorised Representative:
Logos Europe, 9 rue Nicolas Poussin, 17000, La Rochelle, France
contact@logoseurope.eu

For information, contact State University of New York Press, Albany, NY
www.sunypress.edu

Library of Congress Cataloging-in-Publication Data

Name: Kostelnick, Charles, author.
Title: Visual vestiges : time, rhetoric, and information design / Charles Kostelnick.
Description: Albany : State University of New York Press, [2026]. | Series: SUNY series, Studies in technical communication | Includes bibliographical references and index.
Identifiers: LCCN 2025034736 | ISBN 9798855805963 (hardcover : alk. paper) | ISBN 9798855805987 (PDF) | ISBN 9798855807097 (epub)
Subjects: LCSH: Visual communication. | Information visualization.
Classification: LCC P93.5 .K68 2026
LC record available at https://lccn.loc.gov/2025034736

To all of my family, who have been with me since the start
and continue to support and encourage me,
and especially to my grandchildren—
that they will fully value the past as they explore the future.

Contents

Illustrations

Acknowledgments

To my family, especially my wife Clare, I owe my gratitude for their patience and support as I worked on this book over the course of a few years. To Miles Kimball, coeditor of the State University of New York Press series in Rhetoric and Technical Communication, I appreciate the opportunity to move forward with this book. His encouragement and his understanding of both the complexities of studying visual communication, as well as its endless possibilities, made this project seem both feasible and worthwhile. I also greatly appreciate the guidance of Richard Carlin, acquisitions editor at Routledge, through the early stages of the book's development and through the review process. I truly appreciate his patience and encouragement as this project slowly found its way to completion, often behind schedule. I am truly grateful for the suggestions of several anonymous reviewers who provided valuable and detailed feedback that pointed me to additional sources, opened new avenues for exploration, and guided me toward a more cogent and cohesive analysis.

Here at Iowa State University, my professional home for over four decades, I thank the English Department and the College of Liberal Arts and Sciences for their continuous support of my research. I also thank my many colleagues, past and present, for their insights and encouragement over the years. I'm especially grateful to Andrew Sutton for his technical support as I prepared the many figures in the book. I also wish to thank Olivia Garrison, Amy Bishop, and other staff members at the Special Collections Department at the Iowa State University Library who gave me ready access to historical materials, both print and digital, that enabled me to examine a wide range of artifacts. Additionally, I thank Jason Carpenter at Interlibrary Loan, Iowa State University Library, for obtaining some of the historical figures for the book.

I also want to express my gratitude to the Library of Congress for the many historical examples that I used in the book; their expansive digital collections make possible the work of scholars like me. I also thank the many other libraries that generously shared their materials with me for this project, as well as the National Gallery of Art in Washington, DC, for their open access to paintings and drawings that allowed me to use materials from their collection. Finally, I thank the many businesses, non-profits, government agencies, and other organizations that kindly granted me permission to reproduce their images, all of which enabled me to illustrate and analyze the many ways in which time and rhetoric coalesce.

I also thank Jennifer Bennett from the State University of New York Press for her guidance on preparing the images and final manuscript for publication. She patiently and skillfully guided me through the publication process and made it both productive and enjoyable. I also greatly appreciate the timely and generous support of Caitlin Bean, associate production editor for the State University of New York Press, and her production team as they transformed my manuscript into a published book. I especially thank John Raymond for his thorough copyediting of my manuscript by making suggestions and edits that greatly enhanced its clarity and consistency. I'm also grateful to Ryan Morris, editorial, design, and production manager for the State University of New York Press, for consulting with me as she developed the vivid cover design for the book.

Finally, I sincerely thank my many graduate students for their encouragement and insights as this project unfolded over several years. Their discussions with me about many of the issues that surfaced during the writing of this book helped me greatly to sharpen and articulate my ideas and to consider fresh perspectives along the way.

Introduction

In any given workspace we occupy or visit, we experience the visual presence of practical communications, both print and digital. Their immediate materiality places them within our line of sight and our physical grasp: letters and reports; newsletters and annual reports with pictures, charts, and graphs; business cards in the desk drawer; large and small screens filled with emails, web pages, interactive displays, and social media; instructions for devices we use, warnings on equipment. Many of these and other communications also pervade our private spaces: bills and brochures, calendars and timetables, owner's manuals, bank and investment statements, do-it-yourself videos, and a host of other digital images and messages.

Simply put, we are inundated with text and images designed to help us make sense of the world and to get things done. As audiences, our encounters with these visual artifacts are mostly, if not entirely, *in the moment*. Although some of them are old (business cards proffered at last year's conference) and some of them new (emails that just popped up on the screen), these communications don't demand that we think much, if at all, about time: where their visual language came from, its genealogy and provenance, whether or not it's fresh and novel, or popular and conventional, or what it might look like in the future after languishing on a digital desktop or in the back of a drawer.

Still, time dominates our perception of our workplace lives: We record things we accomplished, meetings we attended, observations we've made, places we visited, people we met. However, when we think of time, we usually think of time *past* (or time *future*), the historical sensibility that everyone possesses in one form or another, and usually as a series of unique and indelible dots on a timeline. One dot follows another, like an interminable string of pearls, in a discrete succession of events. In this ongoing temporal world, the one that Heraclitus imagined where we "cannot step twice into the same river" (Wheelwright 1959, 29), each moment claims its place along a timeline, never to be revisited. This linear concept of time creates the impression that things past become perpetually more distant, like stars and galaxies ever expanding outward into the universe. Time rushes inexorably forward, and no yearning or exhortation can coax it back—had we even the intention to do so.

However much we consciously acknowledge it, we also experience time in a very different way, as a dynamic and recursive force that continues to have a visible presence in our lives. Buildings, landmarks, public spaces, and statues; fine arts objects like paintings, sculptures, and drawings; and on a more personal level, books, furniture, houses, and domestic articles like dishes and silverware—these artifacts, and many others, connect us with the past but also have immediate functions in the present. So too with newsletters, websites, annual reports, warnings, and a host of other information designs that occupy our personal and professional spaces—all have a constant presence in our daily lives, though their pedigrees might be less perceptible than those of other visual artifacts we encounter.

We mostly take it as axiomatic, then, that the past influences the present. Who would argue with that? The question rather is *how*, and to what degree, and to what rhetorical ends? Moreover, how *aware* are we of its effect on our everyday acts of design and interpretation, and furthermore, how mindful of our resistance to its relentless presence?

Waxing and Waning of Forms, Both Fine Arts and Practical

Our ability to measure and experience time through visual artifacts, however, varies between practical, everyday designs and works of fine art. Paintings like Leonardo da Vinci's *Mona Lisa* (1503–1519) and Sandro Botticelli's *The Birth of Venus* (1485), buildings like the Parthenon and Notre-Dame Cathedral, and statues like Michelangelo's *David* (1501–1504) or Auguste Rodin's *The Thinker* (1903) are sure-footed milestones in art and cultural history, their status regenerating from one era to the next with varying levels of intensity. Works of fine art with a regional identity like Grant Wood's *American Gothic* (1930) also achieve long-term status, though more locally and with periodic surges in popular culture. The fortunes of most works of fine art wax or wane over time depending on their perceived aesthetic, cultural, social, political, or commercial value.

However, even obscure paintings or sculptures always have the potential for regeneration, to be rediscovered by a critic or collector and catapulted into the limelight.

In contrast, most practical forms of information design are highly ephemeral and vulnerable: As material objects, they get *used*, they deteriorate—they get marked on, folded, torn, stained with coffee, and gather dust. And then they disappear, sometimes slowly, sometimes abruptly. And unlike Victorian-style courthouses, public monuments, and family heirlooms, hardly anyone notices their absence. Of course, *any* material object these days can defy the passage of time when it becomes "collectable"—practical artifacts like technical and scientific books, maps and charts, promotional pieces, and commercial signs—but their valuations fluctuate with the tastes and resources of collectors. Unless they were connected to famous people, hardly anyone today collects handwritten business letters, instructional manuals, technical drawings, typewritten reports, or train timetables; more often than not, these artifacts become fodder for shredders, recycle bins, and landfills—or the bottomless trash can lurking on the desktop.

Because of their vulnerability to the degradation of time, then, practical designs require continual use to sustain their currencies and remain relevant. Unlike fine arts images or buildings whose reputations can extend for centuries, information designs must earn their keep through constant use, primarily through the well-worn paths of visual conventions that guide text design, data displays, pictures, and other forms of visual language that designers and their audiences learn to deploy and interpret (Kostelnick and Hassett 2003). Although their materiality is highly susceptible to the ravages of time, information designs embody conventional forms that are rhetorically potent over long stretches of time, bridging past, present, and future.

Unless they are truly novel, most designs are propagated through visual conventions—*all* of them vestiges with varying genealogies. In this way, the workings of the past often remain imperceptible, hidden in plain sight and often flying under our rhetorical radar. However, conventions are constantly in flux, living and perishing in the hands of their users: They develop, evolve, gain currency, and sometimes mutate, as designers reshape them for their own purposes and pass them along until audiences no longer find them useful. Ultimately, then, conventions are vulnerable to the effects of entropy and degeneration, they devolve into obscurity or face extinction, and they occasionally experience revivals. The forces regulating these changes are exerted from several sources: cultural shifts in aesthetics or values, social changes that redefine how we relate to one another rhetorically, and breakthroughs in technology that transform our tools for visualizing and interpreting information.

Some visual conventions that lose their currency manage to survive narrowly, mutate into hybrid forms, or are rhetorically repurposed. These vestigial forms—typewriter fonts, ornate Victorian images, early modernist icons—are testimonials to conventions long retired, their glory days in the rear-view mirror. When they reappear today, they explicitly connect us to the past, prodding memory, nostalgia, and wonder or perhaps skepticism or even revulsion. These vestiges aren't mere historical curiosities, however; they have a more explicit rhetorical presence that captures our attention. They remind us of the fragility of visual language but also the rhetorical power of the past, the "historical consciousness," as Richard Weaver put it (1970, 161–62), that's central to rhetoric and its ability to persuade. Time and visual language supply the abundant raw materials for this enterprise, elastically stretching backward and forward with the information designs we create and encounter.

However, we largely ignore the presence or influence of the past. Its opacity results partly from the powerful influence of modernism on the ways we deploy and interpret visual language. Acknowledging the dynamic life cycle of visual language runs counter to the age of modernism and its quest for timeless, universal design forms that resonate across generations (and geography) and thereby seem to escape both the cultural baggage and the corrosive effects of time. The quest to sever design from its past occurred radically in the fine arts with the Fauve and Cubist painters and with the Futurist designers and theorists, and later in the applied arts with the Bauhaus and other modernist schools of design. However, even the modernist minimalism that springs from these, with its high-contrast geometrical forms and dynamic international style, showed its wear in the late twentieth century. Its visual codes began to age, postmodern aesthetics brought renewed attention to historical context (Venturi 1966), and postcolonial perspectives challenged modernist tenets on a global stage.

However much visual vestiges are ignored, acknowledged, or denied, they interfuse past, present, and future. This, then, will be my main line of inquiry in this book: how the past lives in the present, and the present in the past, and both extend into the future—in short, how past, present, and future intermingle.[1]

So how does rhetoric figure in this past-in-the-present-in-the-future formulation? To answer that question, I will explore how conventional forms guide information design, often within the structure of genres. Time provides both resilience and ballast in generating and sustaining these forms, in recovering and reviving them in their residual traces, and in gauging their immediate impact (kairos). I'll study these phenomena through the key rhetorical purposes they serve—visualizing epideictic praise and blame; appealing to the emotions with sentiment and memory; narrating stories with text, pictures, and charts; and building trust, both individually and collectively. The visual vestiges all around us will provide the data, the raw materials, for these rhetorical explorations. These and the many other visual artifacts we encounter warrant our scrutiny—part of the larger domain of visual rhetoric that, as Sonja Foss (2004) contends, widens the scope of rhetorical studies.

Reasons for Studying Vestigial Forms

So why should we seek out and study vestigial forms, whatever their state of preservation, neglect, devolution, or denial? What scholarly or practical value do they have? Why not just wish them good riddance? Several reasons can be offered. First, vestigial forms help us understand the *processes* of history: They give us clues about why these forms flourished in their earlier lives, matured into conventions, why they later devolved or atrophied, and conversely, why they still have value today or have even experienced a revival or been reinvented digitally. Why any living entity devolves (or faces extinction) poses a compelling question that demands an answer not only for itself but for all other living things, which are inevitably vulnerable. Like its biological counterparts, visual language is shaped by several interacting factors—cultural, social, rhetorical, and technological. Observing the synergies among these (and other) factors externalizes the processes that govern visual language over time.

Second, vestigial forms keep us constantly connected to the past, their historical traces immediately visible and tangible. As Gerda Lerner (1997) contends in *Why History Matters*, "History is the means whereby we assert the continuity of human life" (116). Vestigial forms foster this "continuity," saturating our visible world in a variety of incarnations—visual conventions, evolving genres, retro and repurposed designs, and makeovers and mutations afforded by digital technology. Moreover, the social, cultural, and rhetorical forces that shaped—and continue to reshape—vestigial designs and all forms of visual language resonate all about us, giving us insights into how old forms blend with the new. By keeping the past alive in the present, vestigial forms of information design enable us to achieve this "continuity" with the past, to contextualize history and the rhetorical and cultural forces that shape it. Some vestigial designs form tenacious links in the temporal chain. The popular telephone icon, for example, still envisions a low-tech rotary model, technology that scarcely exists anymore and with which most millennials have no experience, yet its residual form persists as a functional symbol. So we always stand on the precipice of what's lost, or about to be lost, and on the threshold of what's new.

Third, vestigial forms constantly replenish the skills and knowledge, the techne, that enable us to design and interpret them: skills required to deploy conventions embedded in information designs but also skills required to create legacy forms that have largely been lost but still survive today. How many design practices still reside quietly among us that are *thought* to have been lost? And how often do we even recognize those vestigial practices and the artifacts that embody them? Chisel and stone, pen and ink, the printing press, typewriters, digital interfaces, projection systems, and various copying technologies—each has its own affordances and limitations for making information visible. In the seventeenth and eighteenth centuries, for example, accomplished writing masters taught penmanship as a practical art, skills that eroded as handwriting became increasingly obsolete but that many calligraphers still possess today. Still, even though these skills have been diminished by technology (typewriters, laser printers, digital media), technology has made their proxies more accessible: Script typefaces on the desktop can be deployed by virtually anyone, albeit without the dexterity and expressiveness of the human hand.

Fourth, and perhaps most importantly, visual vestiges provide insights into applied visual rhetoric—that is, the processes by which visual language is deployed to achieve certain practical ends with a given audience. Studying visual vestiges enables us to make rhetorical connections across time and to observe how strategies for deploying visual language have evolved from one era to another to make practical information persuasive, accessible, credible, and useful to audiences. For example, business cards have been around for two centuries, and even though their ethos-building function remains intact, changes in the social and commercial purposes they serve have altered the rhetorical dynamics between rhetor and audience. In these and many other ways the development of modern rhetoric is profoundly intertwined with the history of *visual* rhetoric, with the visual and the verbal melding conceptually and pragmatically as part of the same larger enterprise. Visual vestiges provide both a gateway and avenue for exploring these relationships over time.

Examining time and visual communication, then, enables us to peer through both ends of the telescope: Scrutinizing the past provides a potent lens for understanding the present, and the present supplies a lens for understanding the past. Like a baseball team that wears jerseys from another era, the past visibly resonates in the present—and the present reshapes our perception of the past, those old-time jerseys enabling us to witness the game played by the phantoms in the grainy black-and-white photos. Studying time also enables us to define the methods and value of historical analysis: the challenges of recovering ephemeral artifacts, the ebb and flow of visual conventions, and the tenacious rhetorical power of vestigial designs. Studying time also alerts us to the omissions and cleansings of design forms, exemplified by twentieth-century modernism and its search for an international design language that eschewed the cultural associations of the past and their continual fluctuations. By clarifying these tensions and affinities, studying vestigial designs has the reciprocal benefit of giving us insights into present practices and, by extension, a window into the future.

Scholarly Foundation for Analyzing Time and Visual Language

The study of rhetorical time, and the forces that shape it, can hardly occur in a vacuum, given the rich and varied body of existing scholarship in visual rhetoric, design history, memory studies, information design, and professional communication—as well as historiography more generally. Although the deterministic notion of historical progress lies outside the domain of my inquiry, my approach correlates with Georg Wilhelm Friedrich Hegel's ([1837] 1901) organic concept of history guided by larger social and cultural forces that are inextricably connected to one another. Indeed, the rhetorical processes and design artifacts that I survey here can be understood only by examining the social and cultural—as well as intellectual—contexts in which they evolved. In other words, like other designed objects and spaces—buildings, parks, furniture, jewelry—practical artifacts marinate in the historical milieus in which they function rhetorically.

More recently, the conceptual basis for bridging past and present (and future) has been explored by the study of "haunting," or hauntology, which scholars have applied to a wide realm of disciplines ranging from literature, film, and anthropology to visual design, architecture, and performance arts.[2] In the realm of rhetoric, Michelle Ballif (2013) contends that the hauntology approach to history recognizes "that linear, historical time is 'out of joint,' and that the borders between the past and the future remain—not only permeable, but—impossible" (141). In short, advocates of this approach regard "the dead as undead, the *revenant* as the *arrivant*" (Ballif 140). In contemporary communications, such "hauntings" occur, for example, in social media where famous authors are given voice in spaces like Facebook and Twitter and thereby retain an authorial presence among us (Baggott 2013; Best 2020).[3] The fungible temporality of this theoretical approach, which mixes past, present, and future in innovative and surprising ways, clearly aligns with my own, though my study of practical communication has taken me down a more independent and pragmatic path.

That path has been illuminated by a large body of scholarship in the fields of rhetoric, information design, and technical and professional communication. Among the most expansive of these scholarly studies is Walter Ong's *Orality and Literacy: The Technologizing of the Word* (1982), in which he analyzes the cultural and rhetorical shift that accompanied the emergence of print and the interplay between past and present during that transformation. The change in modality from hearing to seeing had a profound effect on how audiences interpreted text, as well as how authors produced, designed, and claimed ownership of it. Concurrently, Ong argues, the long legacy of orality, and of handwritten text, continued well into the age of print, with technology intermingling old and new practices. Accordingly, the switch from one modality to another didn't occur overnight; rather, the transition to print was a long incremental process in which rhetors and their audiences continually negotiated past and emergent forms.

Since Ong's study, numerous others have examined strands of visual rhetoric in historical context. Combining history, aesthetics, rhetoric, and practice, Victor Margolin's *Design Discourse: History, Theory, Criticism* (1989) gathers together many of the seminal articles from the journal *Design Issues*, ranging from the history of page composition (Williamson 1986), the origins of Isotype design (Lupton 1986) and classical tropes as invention tools (Ehses 1984), to the visual rhetoric of product design

(Buchanan 1985) and the rhetorical nature of modernist aesthetics (Kinross 1985). In the broader realm of visual rhetoric, Charles Hill and Marguerite Helmers' *Defining Visual Rhetorics* (2004) also contains historical studies of visual language in several media—paintings, photographs, films, advertising, data displays, architecture, needlework—using a variety of rhetorical and ideological frameworks to interpret these artifacts. In *Nostalgic Design: Rhetoric, Memory, and Democratizing Technology* (2018) William Kurlinkus explores the roles of memory and nostalgia in the uses and design of technology, and in *Writing in the Clouds* (2022) John Logie analyzes contemporary communication technologies in the rhetorical and historical context of their predecessors. All of these studies, with their analyses of visual artifacts in multiple media, provide a broad background for my exploration.

Examining how we *remember* the past also informs my study because it reveals the ways in which public audiences constantly refresh and reinterpret the past. A growing body of scholarship has emerged that examines the rhetoric of memory—of how artifacts like buildings, statues, and memorials rhetorically filter and articulate time by engaging audiences with past events, people, and achievements. These studies focus, for example, on the rhetoric of museums and public monuments (Dickinson, Blair, and Ott 2010), on fine arts and painting (Helmers 2004), on how the shared experiences of landscape and travel foster national identification (Clark 2004), and on cultural memory more broadly (Erll and Nünning 2008; Phillips 2004; Prelli 2006). Several of these studies illustrate the socializing effects of rhetoric that bind communities together as the past is contemplated and experienced by public audiences.

More specifically germane to my own project, the visual history of technical and professional communication has been studied in a wide variety of forms, including genre (Brasseur 2003; Kostelnick 2019, "Pervasive"; Kostelnick 2016, "Mosaics") mapping (Barton and Barton 1993, "Ideology"; Kimball 2006), and text design (Tebeaux 1991, 1993, 1997, 2014; S. Walker 2003, 2014). Historical studies have examined the visual rhetoric of pictures and, more specifically, of human forms (David 2001; Kostelnick 2019, *Humanizing*), as well as the evolution of visual language in engineering (Ferguson 1993; Fox 2009), science (Gross and Harmon 2014; Gigante 2018), medicine (Zhang 2016), technology (Brockman 2002), and practical comics (Yu 2015; Eisner 2008; McCloud and the Google Chrome Team 2009). Historical studies of text, picture, and data design also appear in *Information Design: Research and Practice* (Black et al. 2017).

Data design has also been the focus of numerous historical studies (Wainer 1997, 2005; Kimball and Kostelnick 2016; Navarro 2022), especially those focused on the formative period of the nineteenth century (Funkhouser 1937; Friendly 2008) and the rhetorical strategies embodied in those emergent displays (Brasseur 2005; Kostelnick 2004, 2016, "Re-Emergence"). Michael Friendly and Daniel Denis (2001) comprehensively document the history of data design in their interactive website Milestones in the History of Thematic Cartography, Statistical Graphics, and Data Visualization. The history of charts, graphs, and other forms of data design has also been studied in an organizational and management context (Yates 1985, 1989; Robles 2018) as well as in relation to science, statistics, health, education, and crime (Friendly and Wainer 2021). Howard Wainer's books (1997, 2005) provide detailed and insightful accounts of seminal achievements of data designers along with historical narratives of visual problem-solving. All of these studies—among others, far too many to mention here—give us insights into how visual language functioned at various historical moments, with several of these studies doing so within a rhetorical framework.

In addition to these scholarly contributions, many accomplished practitioners have enlisted historical examples to contextualize their work and the work of others, deepening our understanding of how visual language evolved into its present forms. Ellen Lupton's *Thinking with Type* (2010), for instance, traces the origins of typefaces like Palatino in the "humanist" handwriting styles of the Renaissance (15). History also figures importantly in the work of several other graphic designers. In *The Visual History of Type*, for example, Paul McNeil (2017) documents the cultural, aesthetic, and commercial origins of 320 of the most popular and influential typefaces from the fifteenth century to the present. In this case, and others like it, history and practice are deeply connected, implying that practitioners who know the origins of typographic language will design more intentionally and effectively. Data designers also integrate history extensively into their work to illustrate the ingenuity and lasting influence of pioneers in the field. Edward Tufte's *The Visual Display of Quantitative Information* (1983) showcases data designs from the past, including those of innovators like William Playfair, John Snow, and Charles Joseph Minard, whose chart of Napoleon's Russian campaign Tufte regards as one of the best data displays ever created (40). Collectively, Tufte's assemblage of historical examples plays a key role in transforming the design sensibility of his readers (10). Most current data designers who have a public presence via publications, websites, or social media provide some account of past discoveries to situate contemporary design in a larger historical framework, to create epideictic moments celebrating these achievements, and to explain their own strategies and approaches (see, for example, Yau 2013).

Collectively, all of these contributions to design history, visual rhetoric, and related lines of inquiry provide a broad foundation for my own exploration, which extends and differs from them in a few important ways. First, I plan to focus on the rhetorical dynamics of time—not only on how it operates in a given moment, fixed and complete, but also its fluid motion, both forward and backward, as design forms emerge and evolve, gain currency, mutate, and decline. In these ways, information design moves in cycles rather than in a straight line. Second, I'll examine how practical forms of communication are fragile and ephemeral, even when widely deployed and interpreted. Although this transient materiality obscures their provenance, the visual conventions embodied within these forms give them vestigial resilience, connecting them with the past and future. Third, I'll explore how design forms acquire vestigial lives even after they lose their currency—regenerating in new and sometimes obscure roles or experiencing a revival of their original forms. Fourth, I'll explain how the visual rhetoric of information design bridges past and present. Using Western concepts like chronos, kairos, pathos, ethos, and epideictic rhetoric as touchstones, I'll examine how current uses of visual language constantly reshape and refocus past practices in diverse modes and genres. Finally, I'll show how macro-level forces—culture, aesthetics, moral values, technology—shape, sustain, and reshape information design diachronically over long stretches of time.

Toward these ends, throughout the book I'll address several key questions: what key rhetorical strategies are embodied in information designs of the past—for example, emotional appeals, storytelling, trust building, praise and blame—and what kinds of visual language were deployed to implement those strategies? How do contemporary print and digital designs echo, revive, or repurpose visual language to enact (or redefine) these rhetorical strategies? How do genres and their visual conventions extend the visual language of the past? What social, cultural, and other historical forces have guided and transformed these rhetorical processes? Answers to these questions will further enable scholars, teachers, and practitioners to understand how visual language functions rhetorically across time in the documents they study, interpret, and design.

Contents and Organization of This Temporal Exploration

Because visual design inundates practical communication, this book will encompass several genres, disciplines, and historical periods. Although I'll focus primarily on Western concepts and traditions and on artifacts from North America and Europe, I'll remain mindful of the global implications and international context of contemporary information design. I'll draw from examples as early as the sixteenth century, though I'll emphasize designs from the nineteenth century to the present—Victorian, modern, postmodern—a historical swath of time that most immediately shapes and informs current practices in the U.S. and other Western countries as well as our ways of thinking, feeling about, and remembering design, culture, and rhetoric. The design artifacts that I'll analyze encompass business, technical, engineering, and scientific documents that appear in print and digital forms, including

- Instructions for completing tasks in the workplace or at home.
- Letters, business cards, newsletters, and annual reports for clients and employees.
- Drawings that illustrate machines, buildings, and other objects and spaces.
- Schedules, timelines, decision trees, and comic-style narratives.
- Warnings, safety signs, and interactive maps intended to protect people.
- Charts and graphs that visualize data about economics, population, and health.
- Promotional materials that market products, events, and nonprofits.

Information designs like these are time stamped, bearing the marks of their origins and development, their waxing and waning currency, their resilience and fragility, and their future trajectory. Rhetoric provides a lens for examining this temporal flux by defining information designs as interactions with audiences that are aimed at achieving practical goals. Using time as a constant reference point, I'll draw on strategies and concepts from classical to contemporary rhetoric to examine these interactions in their social and cultural contexts.

In chapter 1 ("Timely Design: Chronos, Kairos, and the Resilience of Visual Language") I examine how design forms age—gradually over decades and centuries and quickly during their kairotic moments. They reach their zenith and then experience a process of degeneration until they become rhetorical curiosities or collectables—or simply become irrelevant. This ebb and flow, which occurs over long historical periods on the macro level and in immediate interactions on the micro level, parallels ancient Greek conceptions of time: chronos and kairos. On the one hand, then, we can survey the visual rhetoric of information designs over long stretches of time (*chronos*) to see how they evolve, gain and lose their currency, or find vestigial lives that connect us to our past; and we can look at specific moments of time (*kairos*) when a given design is optimally deployed, achieving its full rhetorical force in a particular situation.

In chapter 2 ("Epideictic Design: The Visual Rhetorics of Praise and Blame") I examine how epideictic forms of rhetoric since the classical age have been associated with amplification, the ability to magnify praise (or blame) in order to create a powerful impression on an audience. Visual amplification has long been achieved through the skillful dexterity of the designer—for example, flourishes in handwriting, pictorial and graphical comparatives, and vivid description (*enargeia*). The visual epideictic also materializes in designs associated with ceremonial occasions, such as invitations, awards, and diplomas. Visual language that praises or celebrates, moreover, has a socializing function by promoting the values that underpin the communities in which it's deployed and interpreted. In contrast, the visual epideictic of "blame" serves to caution its audiences: in warning signs that proscribe certain activities, data displays that tell disastrous stories or that map crime and disease, and pictures showing distress resulting from harmful behaviors—each of these serving as a form of risk communication.

In chapter 3 ("The Pathos of Memory: Sentiment, Nostalgia, and Rhetro Design") I explain how memory becomes a catalyst for expressing emotion in visual design, which had its roots in the eighteenth century and early Romanticism with a passion for ruins and picturesque landscape, the revival of the Gothic, and the emergence of sentimental morals and aesthetics. Following in the wake of these developments, the expression of sentimental emotion permeated many areas of practical design in the nineteenth century, including commercial designs promoting agricultural machinery, professional services, and domestic products. Pathos appeals today invoke memory and nostalgia through "rhetro" design—the rhetoric of evoking the past with vestigial elements like handwritten text, simulations of old-school technology, and traditional typefaces, images, and color. Conversely, modernist design principles seek to escape sentimental appeals through lean universal forms that bridge cultures and eschew the past.

In chapter 4 ("Visual Storytelling: Temporal Elasticity and the Interplay Between Micro- and Macro-Narratives") I explore how information design genres tell stories in temporal increments ranging from minutes to millennia, from simple how-to tasks to synoptic, panoramic views of the past or future. These designs envision the past, present, and future in textual, graphical, and pictorial modes. Text designs envision temporal relationships in schedules, decision trees, and flow charts; data displays visualize time in charts and graphs about population, health, climate, and economics; pictures and comics narrate stories about workplace procedures and how-to tasks. All of these narratives invoke the past by deploying visual conventions embodied in designs that preceded them and, moreover, are often embedded in larger social and cultural narratives—about Manifest Destiny, civil rights, economic prosperity, climate change, and pandemics—that provide interpretive lenses for audiences that share their values.

In chapter 5 ("The Myriad and Mutable Faces of Ethos: Visualizing Trust with Text and Image") I analyze how visual language has been deployed to foster ethos, both individual and collective, in the near and distant past. Building ethos has traditionally taken two forms: initial (preexisting) and acquired (invented)—the former resulting from the position and status of the rhetor and the latter fostered by the rhetor's communication skill and virtuosity. These two methods of generating ethos visually have evolved along with social and cultural conditions and innovations in print and digital technology. Visualizing initial ethos has been achieved with letterheads and nameplates and by pictorial associations with people, places, myths, and historical events. Acquired ethos has been cultivated by deploying the visual conventions of a given genre (a business card, newsletter) or aesthetic style (Victorian, modern) or by creating visual identification with an audience through images of people and visual branding. Visual novelty and misdirection can also play a role in building ethos as well as degrading it.

The historical traces in these five chapters can inspire and direct effective practice in the present—not just for an elite group of designers who publicly showcase historical examples but anyone responsible for composing text and images on pages or screens, creating logos or illustrations, designing charts and graphs, or managing an organization's newsletter or social media. In this respect, information design parallels other professions—architecture, graphic design, engineering, science—engaged with visual forms: Practitioners glean from the past to enrich and contextualize the present. At the conclusions of each of the first five chapters, I'll briefly provide some insights about how the past provides a resource for rhetorical design—for example, in effectively deploying conventions, developing emotional appeals, using design technology, and counteracting the corrosive effects of time.

In chapter 6 ("Making Visual History: Applications for Teaching, Learning, and Discovery") I explore how the past can enrich pedagogy and scholarship. Teaching historical aspects of information design enables students to understand the nature and origins of visual conventions, define the social and cultural influences that shape design, compare and contrast current values with those embedded in historical examples, and experience the power of visual storytelling. Studying the past also offers plentiful opportunities for scholarly inquiry, prompting students to discover hidden artifacts, to establish conceptual frameworks for analyzing them, and to connect those artifacts to larger social, political, and cultural issues.

Given that visual rhetoric is a holistic activity whereby its elements combine to animate information designs, the rhetorical concepts I address in these six chapters necessarily overlap and interweave with each other. Moreover, culture, technology, and visual conventions are constantly interacting diachronically, however transparent or opaque these processes might appear. Although these chapters offer multiple explorations of these affinities between time, rhetoric, and information design, many other methods and approaches might offer alternative avenues. This book, then, will provide both a prelude and roadmap for that more expansive enterprise.

Chapter 1

Timely Design

Chronos, Kairos, and the Resilience of Visual Language

Design forms have varying lifespans, but eventually they all age and degrade, a process often subjected to forces outside the designer's purview or control. Designs mature and deteriorate gradually, even glacially, over decades and centuries, and they also do so with blazing speed in the hours and days following their kairotic moments—those especially propitious situations in which they're deployed and interpreted. Temporally divergent, designs are, as metaphysical poet John Donne ([1663] 2007) put it enigmatically, "the rags of time." However long they survive, they develop a following among their audiences, reach their zenith, and then experience a process of degradation—slow or rapid—until they become rhetorical curiosities or collectables, or simply irrelevant, their audiences as vapid and feckless as ghosts. This rhetorical ebb and flow operates on two different levels: on the macro level over long stretches of time, as far as the eye can see; and on the micro level within the temporal confines of immediate interactions.

These two levels parallel divergent conceptions of time held by the ancient Greeks: chronos and kairos.[1] According to John Smith (1969), chronos "expresses the fundamental conception of time as measure, the *quantity* of duration, the length of periodicity, the age of an object or artifact," while kairos, on the other hand, "points to a *qualitative* character of time" that fits a particular moment, "to a time that marks an opportunity which may not recur" (1; see also Smith 2002). As Debra Hawhee puts it, "*chronos* marks duration while *kairos* marks force" (2002, 18). Chronos rolls leisurely across the temporal landscape as far as the eye can see, while kairos is as potent and savvy as the mythological figure of kairos himself, nimbly ready for action when the occasion arises (Hawhee 2002, 19–21).

On the one hand, then, we can look at the visual rhetoric of information design over long stretches of time (*chronos*) to see how design elements evolved, gained currency, and then disappeared or found obscure residual lives, were revived in retro form, or reinvented with digital technology. And we can look at specific moments of time (*kairos*) when an information design was optimally deployed, when it achieved full rhetorical "force" in a particular situation as well as when that moment passed and it became powerless and irrelevant.

Practical forms of visual communication require continuous use to sustain their currencies and to maintain their visual presence among us. Systematically, over extended periods of time—years, decades, even centuries or millennia—they achieve this longevity through visual conventions. Initial letters, tables, organizational charts, line graphs, mechanical drawings, and many other recurring forms have had long, rhetorically robust lives because they continually serve useful purposes for audiences who interpret them and designers who deploy them. And when that's no longer the case, their currency dwindles, sometimes in fits and starts, depending on rhetors' perceptions of their value, audience expectations, available technology, and social and cultural conditions. Nonetheless, many design conventions survive, wholly intact or as echoes of their past iterations, visual vestiges from a previous time that connect us to our past—and likely our future as well.[2]

On the other end of the temporal spectrum, visual forms of practical communication also function on the more immediate temporal level of kairos, on their ability to resonate the moment an audience interacts with them. A concept originating in classical rhetoric, kairos is defined by James Kinneavy as "the right or opportune time to do something, or right measure in doing something" (Kinneavy 2002, 58). As such, visual designs all around us—newsletters, warning signs, emails, text messages—are constantly waxing and waning in their ability to engage us, to entice us to interact with them. This ephemeral, micro-level temporality defines the scope of their rhetorical power.

In this chapter, then, I will explore the effects of time on information design through the macro and micro lenses of chronos and kairos. I will examine both the sustaining powers and the ravages of time over the long term, I'll illustrate its residual traces in the contemporary world, and I'll examine what lessons we can learn from these about the past, present, and future. I will also examine the more immediate, micro-level effects of time and how it regulates the ways we visualize and interpret practical information. At the end of the chapter, I'll provide some insights and guidelines about how chronos and kairos can be applied in everyday communications.

Chronos and Visual Language: Toward a Theory of Residual Design

When viewed through the lens of time, practical forms of visual language—text design, data displays, pictures, and other forms of information design—present something of a paradox. On the one hand, they are in constant flux as they emerge, develop, and mutate, with their currencies growing and shrinking over time; on the other hand, they evolve into stable, reliable conventions that designers and their audiences learn to deploy and interpret. Stout as boulders in a fast-moving stream, those conventions have tenacious staying power (Kostelnick and Hassett 2003, 163–96). Even when innovations in technology occur, designers continue to accommodate design conventions, using the momentum of the past as rhetorical ballast. However, over time designers reshape those conventions for their own ends and pass them along to their successors, in this way extending the vestigial life of conventions until designers no longer find them useful. The fast-moving stream slowly dwindles to a trickle, and the bedrock boulders lie dormant. Acknowledging these dynamic rhythms of visual language runs counter to the age of modernism and its quest for universal design forms that resonate across space and time—across geography and generations—and thereby seem to escape the ravages of time. But even modernist minimalism with its geometrical forms and perceptual efficiency has been diachronically shaped by the ebbs and flows of time.

However, most vestigial forms we encounter every day seem unworthy of our scrutiny. When we consider the temporality of visual design, we typically mark *beginnings* rather than middles or endings, the nascent rather than the vestigial: the first painting or building in a given style, the first bridges, skyscrapers, printed books, photographs, and more specific to information design, the first bar charts, technical drawings, typewritten letters, websites, and so on. Like explorers searching for the source of the Nile River, we long to find the headwaters, the progenitors that spawned succeeding versions.[3] In contrast, rarely do we take notice when design forms become normalized, let alone mark their endings: landmark buildings torn down, fonts removed from desktops, organizational logos retired. Even when we do take notice, we're content to watch the lights wink out, momentarily mourn (or celebrate) the loss, and move on. Not surprisingly, then, the life cycle of design conventions, both their vestigial forms and rhetorical footprints, mostly escape our attention. However, we might gain some insights into design by studying both nascent *and* vestigial forms. In the natural world, the plights of endangered species also tell us a great deal about other living things that share the same environment. Likewise, whether thriving or endangered, visual language is both stable and robust as well as mutable and fragile, evolving along with the social and cultural values that underpin it, the technologies used to design and circulate it, and the rhetorical situations in which it's deployed.

Residual design practices usually have a low profile, however, and when they've run their course slip silently into the deep, a process that unfolds in our personal and professional lives with familiar objects and documents that have weathered the corrosive effects of time. For example, an orange office chair from the 1970s—when such bright primary colors were popular—finally finds its way to the dumpster, never to be seen again, leaving only a sigh of relief in its wake. Some departures are more widespread and collectively systemic. Many years ago when I began at my university, memos sent to faculty were routinely printed on green paper. Green signaled a memo's important status and, especially if it was sent by an administrator, usually meant that it warranted careful scrutiny and retention for future reference. At some point (probably in the late 1990s), the long-standing convention of the green-colored memo disappeared unceremoniously from people's mailboxes, its visual presence relegated thereafter to file cabinets and recycling bins.

The green memo convention could be easily implemented today with digital technology, but by the time this software was widely available, the practice had nearly vanished. Still, despite the degradation of that convention, some of those green memos still languish in my file cabinet, echoes of that long-ago practice and begging the question of what caused its demise. Did green paper become too scarce (or expensive), or was it too cumbersome to be fed manually into a laser printer, the new technology that swept through offices during that time? Had the rhetorical situation or the organizational culture changed, whereby the need to differentiate such messages visually was no longer deemed necessary? Countless other organizations, disciplines, and cultures have similar visual codes and designs that thrived for a time and then vanished, leaving only traces and prompting us to wonder what induced their demise.

To be sure, the green memo was just a fleeting local convention that required minimal training or skill to implement or interpret. More pervasive and durable conventions under duress appear everywhere in our midst, some of which retain rhetorical power. Phone books, for example, have long been divided into color-coded sections of white and yellow (and sometimes blue). Although the convention is fading because of the dwindling number of landlines and ever skinnier phone books, these color codes still prevail in everyday discourse. Likewise, due to environmental concerns and digital technology, cabinets stuffed with file folders are disappearing from offices. Nonetheless, even if we don't physically handle folders, the visible traces of the originals appear ubiquitously on the desktop. Similarly, although printed bookmarks are increasingly becoming obsolete, bookmarks still exist in a variety of digital forms. Of course, many other visual practices and conventions have forfeited their currency and become extinct, or nearly so. Boustrophedon text that runs right and then left (back and forth, as the ox plows), rubrication (different colors for sections of a manuscript), ligatures (æ), and various handwriting styles for personal and professional documents have largely become historical curiosities, their remnants preserved in archives and special collections.

Conventions as Time Capsules: Texts, Data Displays, Pictures

Visual language is constantly evolving, and observing the fortunes of its conventions reveals how and why this process unfolds. As visual conventions settle in as commonplace forms, each iteration by their users constitutes a vestigial reoccurrence that relies on its lineage for its rhetorical power. In this way, conventions are vestigial by their very nature, though we rarely recognize them as such because we become so accustomed to them that their genealogy remains invisible. However, even long-standing conventions eventually show their age, their currency begins to contract, and (rather ironically) their vestigial presence becomes more transparent as they transition from the visible mainstream to old-school anachronism. At this stage, however, vestigial designs don't fade away completely; often they hang on, reappearing in various forms, sometimes under the radar. Many of them make a comeback, materialize as retro designs, or are reinvented or repurposed by new technology; some merely languish as obscure remnants. In all of these instances, their rhetorical effects on audiences change, ranging from indifference or curiosity, to discomfort or aversion, to sentimentality and nostalgia.

In the next few sections, I'll examine how these temporal-rhetorical processes shape practical communications in three different modes: text design, data design, and pictures. First, I'll examine handwriting practices—their elegant forbears, their demise, and their current uses, both typographically and rhetorically—as well as explore the typewritten text that superseded handwriting. I'll then examine vestigial forms of data design that originated in the late nineteenth century when charts and graphs experienced a profusion of innovations for visualizing demographic and economic data. Finally, I'll show how visual conventions for practical pictures have ebbed and flowed rhetorically, especially in the fields of engineering, architecture, and information design. Along the way, I'll also briefly examine the role of technology in developing and deploying conventions.

Practical Handwriting: Waxing, Waning, Sustaining

Design practices wax and wane, depending on how strongly they're endorsed by certain groups of users, often referred to as the discourse communities (see, for example, Swales 1990, 21–32). Design practices also rely on how often the rhetorical situations in which they're used recur, along with a host of other factors, ranging from the social and cultural conditions that shape them to the tools and technology that enable rhetors to implement them. This evolutionary process of decline and transformation can take many twists and turns as it unfolds, with vestigial forms morphing into new forms, operating in new modalities, or idling in obscure enclaves.

All three of these occur with the practice of writing and transcribing documents by hand. Handwriting has played a long and critical role in the visual history of practical communication, with hundreds of "copybooks" appearing between the sixteenth and nineteenth centuries across Europe and North America instructing writers on how to shape text for various purposes: business, legal, civil, royal, and religious. Conventional handwriting included several styles or "hands": secretary, round, running, and bastard, among others. Figure 1.1 appeared in the collection of plates (*Recueil de Planches*, Diderot and d'Alembert 1762–1772) that accompanied the compre-

Figure 1.1. Handwriting tools and flourishes illustrated in the *Recueil de Planches* (Diderot and d'Alembert 1762–1772, vol. 2, "L'Art d'Ecrire," plate III, bottom), which accompanied the eighteenth-century French *Encyclopédie*. Courtesy of Iowa State University Library Special Collections and University Archives.

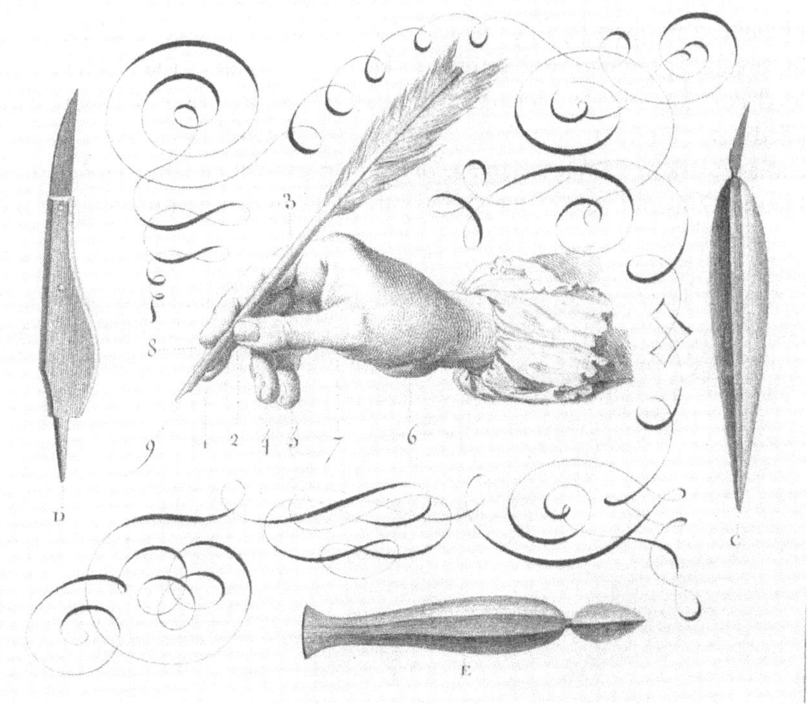

hensive and renowned eighteenth-century French *Encyclopédie* (Diderot and d'Alembert 1751–1765). The eleven volumes of plates (*Planches*) contained thousands of illustrations of practical knowledge about manufacturing, skilled trades, and the processes used to produce everyday things, including handwritten text. Figure 1.1 shows the handwriting tools of the period, including a quill pen and knives for sharpening the nib. Also shown are swirling embellishments called "flourishing," which require "command of hand": the skillful use of the arm, hand, and pen to execute thick and thin strokes for letterforms, borders, and decorative elements. Penwork provided the conventional workaday mode for producing visual language, for both personal and professional text, including correspondence, business and legal transactions, calling cards, accounting ledgers, and civil and religious records, among others. Writing in a “fair” and competent “hand” carried rhetorical weight—heightening clarity, signaling the type of communication, and enhancing the writer’s ethos by revealing the writer’s social class, level of education, and moral character.

Handwriting continued to evolve throughout the nineteenth century, influenced by increases in literacy, burgeoning commerce, and the need for greater speed and productivity to keep up with all of the business and legal correspondence. Herman Melville’s (1853) scrivener Bartleby, who plied his trade in a New York law office, epitomized the exhausting and endless work in the Victorian age of copying text with a fair, legible hand.[4] Figure 1.2 from the late 1880s shows the flowing handwriting style of the

Figure 1.2. Handwriting instructions from Thomas Hill’s *Manual of Social and Business Forms* (1888) showing extreme variations between the thick and thin strokes of the pen (plate V). Courtesy of Iowa State University Library Special Collections and University Archives.

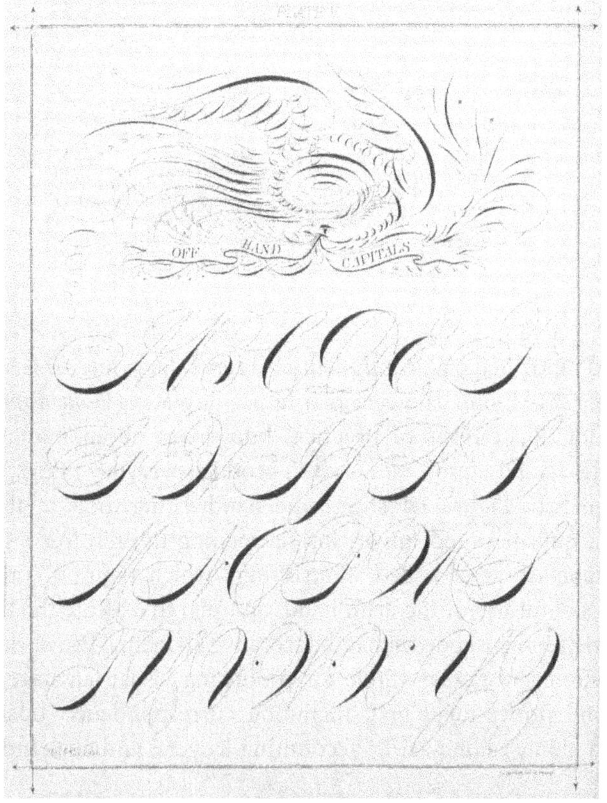

era: highly slanted and round, with extreme variations between thick and thin strokes. As the typewriter soon made its way into the workplace, challenging handwriting practices, methods for increasing the speed and efficiency of handwriting became even more urgent. In *Palmer's Guide to Business Writing*, which appeared the following decade (1894), handwriting expert A. N. Palmer argued that "speed" should supplant "beauty" with his lean, unembellished style (fig. 1.3).

Handwritten business messages continued into the early twentieth century, with business communication books still acknowledging their value well into the 1920s. As the mechanics and operating speed of the typewriter improved, however, it catapulted ahead of handwriting, displacing practices that had dominated the design of practical text for centuries. Typewriting vastly diminished the role of handwriting in professional communication, and by the middle of the twentieth century relegated it largely to fill-in forms, accounting ledgers, and notetaking by health professionals and law enforcement, most of which continue today. During the hiring process, moreover, some companies still use graphology to analyze

Figure 1.3. Business writing sample from A. N. Palmer's *Guide to Business Writing* (1894) that illustrates a faster, more simplified writing style (39). Courtesy of the Library of Congress [11026563].

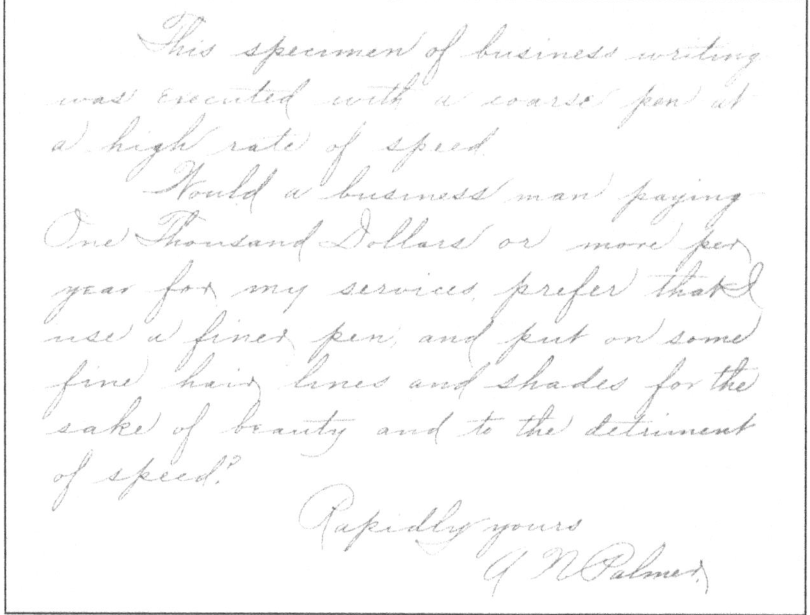

This specimen of business writing was Executed with a coarse pen at a high rate of speed.

Would a business man paying One Thousand Dollars or more per year for my services, prefer that I use a finer pen, and put on some fine hair lines and shades for the sake of beauty and to the detriment of speed?

Rapidly yours

A N Palmer

the handwriting of their potential employees to determine psychological traits. Handwritten signatures also continue to have a place in business, law, and civic life. For example, on many legal documents (marriage licenses, court proceedings), the handwritten signature is still required to authenticate and legitimize identity,[5] as it is on paper tax returns, checks, and ballots and for some in-store credit transactions. For the most part, though, handwriting has withdrawn to the domains of the personal and educational: correspondence with friends and relatives, private journals, elementary classrooms, comments on student essays, and the like.

As an indication of its marginal status today, handwriting is now deemed to be so passé that most schools no longer teach it, even though students who practice handwriting enhance their cognitive and expressive abilities (Chemin 2014). Nonetheless, its residual influence still pervades professional communication, though in new modalities. As Ellen Lupton (2010) observes, "Words originated as gestures of the body" (13), and from the beginning of print, typefaces were modeled on Renaissance handwriting (Lupton 2010, 13–15). Although this genealogy remains mostly unacknowledged today, contemporary rhetors and their audiences still find appealing the explicit look and feel of handwritten text, re-created digitally with script typefaces—Edwardian Script, Brush, Park Avenue, Zaph Chancery—for formal documents like wedding invitations, graduation announcements, certificates, awards, and other epideictic designs. These typefaces simulate a "fair," highly skilled hand capable of executing these strokes with pen and ink. Rhetors also have at their disposal an arsenal of other script typefaces like Comic Sans, Chalkboard, and Mistral to lighten the tone and personalize the text. In figure 1.4 I've illustrated two samples, Edwardian Script and Bradley Hand, one very formal and traditional and the other very casual and conversational. Some typefaces replicate handwriting for specific disciplines and purposes: for example, fonts that mimic the conventional block lettering that architects use on building plans, elevations, and construction details; and comics typefaces that simulate the casual, hand-drawn words that appear in pictorial stories, instructions, and graphic novels. Although few professionals still handwrite messages for their audiences, the multitude of handwriting typefaces available on the desktop testifies to their staying power and rhetorical resilience.

Despite their ability to simulate the look and feel of handwriting, however, script typefaces are intrinsically artificial because they are programmed to have a consistent design, with every letterform appearing exactly the same from one word and sentence to the next, betraying their

Figure 1.4. Script typefaces with different levels of formality: Edwardian Script (top) and Bradley Hand (bottom).

You are cordially invited to the marriage of
John Smith and Ann Jones.

Once you've assembled all of your tools and materials, you're ready to build your birdhouse. First, you'll need a smooth, hard surface on which to assemble the pieces. A driveway, garage, or basement floor works well.

lack of authenticity. In short, every rhetor deploys the same contrived typeface as everyone else does, which is why we can spot a fake signature in a heartbeat. However, with advances in AI, it may soon be possible to compensate for this uniformity by personalizing fonts for individual writers based on their unique handwriting styles or on what they *want* them to look like. Ironically, then, *more* technology may be needed to restore handwriting's personal connection with an audience.

Of course, writers can opt to handwrite their letters and other messages if they are sufficiently motivated and they have the patience and skill to do so. Occasionally we hear about an executive or someone in political office penning a note to someone expressing gratitude, congratulations, or condolences, or making a persuasive appeal for support. As part of college recruiting efforts, sometimes students and faculty handwrite notes to prospective students encouraging them to choose a certain institution. The rhetorical power of these personalized messages can hardly be exaggerated.

Perhaps this residual allure of handwriting explains why so many online businesses now offer handwriting services—text handwritten by an actual human being on behalf of the author (Letter Friend 2025), by AI-generated writing (Scribeless 2025), or by robots that are programmed to hold a pen and simulate the strokes of a human agent pressing pen to paper (Handwrytten 2025; Postaglia Ink 2025; Simply Noted 2025). These handwritten messages intend to convince audiences that they are authentic—that the writer took the time to personalize the message by hand, or at least hired a surrogate to do so. The act of *hand* writing these messages reveals our continuing value for things that are *hand* crafted, as opposed to those that are mechanically reproduced—a vestigial sentiment from the Victorian age and the Arts and Crafts movement that championed handmade artifacts, a subject I'll return to in chapter 4. Indeed, handwriting services claim that handwritten messages are significantly more effective than type; according to one service (Letter Friend 2025), audiences are likely to engage with handwritten messages, their audiences are more responsive and loyal, and they produce high referral rates.

Overall, then, handwriting for professional purposes has not yet completely disappeared, not by a long shot, even though we generally regard it as so because of the widespread availability of so many other (and presumably more efficient) forms of textual media.[6] The many residual forms of handwriting—script typefaces, handwritten letters, proxy handwriting, personal signatures—are hidden before us in plain sight. Whether these vestigial forms will resonate in the future remains to be seen, given the widening distance between handwriting as a practical, everyday activity and the digital experiences of contemporary audiences designing and interpreting text.

Typewritten Text: Its Rise and (Not Entire) Fall

The long process of abandoning handwriting in the workplace began with the rise of the typewriter in the late nineteenth century. Initially an expensive and novel technology, the writing machine rapidly became commonplace in offices across the U.S. and Europe—for commercial correspondence, legal documents, invoices, reports, and a variety of quotidian messages. Manuals appeared on how to use this new technology, including Isaac Pitman's *A Manual of the Typewriter* (1893) in England and O. R. Palmer's *Type-Writing and Business Correspondence* (1893) in the United States. Palmer's manual included all kinds of practical tips on how to use the writing machine effectively, visualizing typewritten business documents as well as designs for borders and embellishments.

The transition from handwritten to typewritten text in everyday life was exemplified by Western Union telegrams, which in the late nineteenth century were sent via Morse code and then transcribed by hand and delivered to their destinations. By the time the Wright brothers first flew at Kitty Hawk in December 1903, however, the typewriter was firmly ensconced in the workplace. Figure 1.5 shows the typewritten telegram that Orville Wright sent to his father Bishop Wright in Dayton, Ohio. This sparse typewritten message was sent via Morse code to Dayton, where it was transcribed on a typewriter and delivered to Wright's father. This curt, mechanically typed message—with its no frills design, odd spacing, and faint blue letters—must be one the most understated documents ever to record a historical milestone. The technology used to generate this message, with its mechanical one-size-fits-all design, was utterly incapable of capturing the excitement or emotion of the feat or the imminent global fame it would bring its sender. This visual oxymoron lacked the rhetorical elasticity the situation demanded.

The impersonal visual language of typewritten text posed a formidable break from the expressive capabilities of handwritten text, creating a rhetorical vacuum for writers and their audiences. However, the typewriter was hardly an a-rhetorical medium, if such a medium could exist. In the early twentieth century, as typewritten text caught on quickly in the business world, users of this new technology tenaciously retained the traditional visual structure of the handwritten letter (inside address,

Figure 1.5. Telegraph from Orville Wright to his father about successful flights at Kitty Hawk, North Carolina, in December 1903 (Wright). Courtesy of the Library of Congress, Wright Brothers Papers, Manuscript Division [LC-MSS-46706-5].

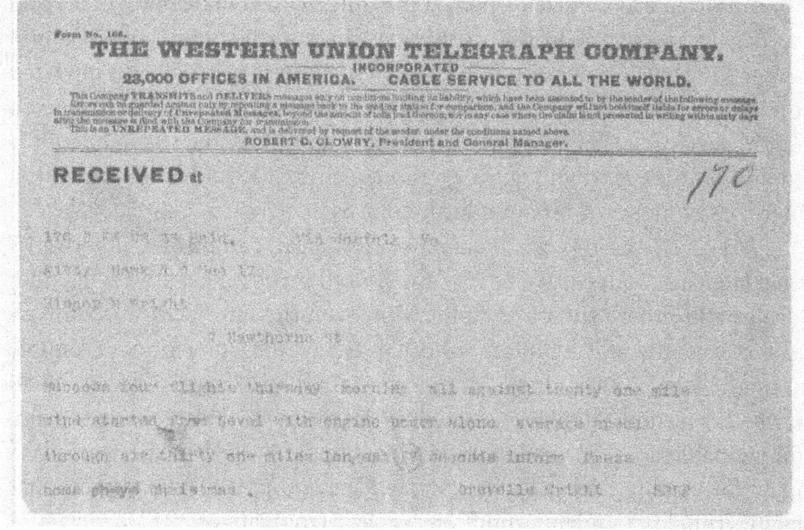

THE WESTERN UNION TELEGRAPH COMPANY.

INCORPORATED

23,000 OFFICES IN AMERICA. CABLE SERVICE TO ALL THE WORLD.

ROBERT C. CLOWRY, President and General Manager.

RECEIVED at 170

7 Hawthorne St

salutation, spacing), creating perceptual and rhetorical continuity for audiences long familiar with these conventions (see S. Walker 1984, 103–6; Kostelnick 1994, 100). Here again (like the larger shift from handwritten manuscripts to print), new technology conformed to traditional visual practices to accommodate its audiences. Eventually, however, despite the constraints of the new technology, as typewritten text became the workplace standard, a modest set of new visual conventions emerged (S. Walker, 2018; see also Logie 2022, 12–27). Today we look at typewriting practices such as centered headings in all caps and underlining text (S. Walker 2018, 145–47) as archaic remnants from a long-ago era, practices whose functional efficiency no longer serves an army of users. The impersonal visual language of the typewritten page, and its residual fonts readily available on our desktops, now engender nostalgia rather than tepid and forbearing compliance.

As technology intervened again in the 1990s with the appearance of dot matrix and laser printers, typewriters rapidly became obsolete in the workplace. However, even then the look of typewritten text and its popular Courier typeface continued to dominate the visual language of practical communications. The new desktop publishing software included this conventional typeface (and others like it) to accommodate audiences that were long accustomed to typewritten text; indeed, for a while at least, a memo composed in Times New Roman looked odd and pretentious compared to proverbial Courier. Nothing came close to looking as credible and comfortable as Courier, which at the time seemed to be perpetually and irrevocably ensconced in the workplace. During this period of transition to laser printing and desktop publishing, typewriters gradually disappeared from workplace desks, dorm rooms, and home offices and were relegated to storage closets, basements, and landfills. And along with that mass exodus, Courier gradually gave way to a rich new visual language with a much larger menu of typographical choices.

However, the typewriter did not disappear completely, nor did its visual language. Indeed, even today, most word processing software includes Courier and other typewriter fonts. Moreover, the interfaces for composing text digitally still maintain a host of typewriter conventions: "double spacing," "tabs," "indenting," "margins," "tables," "all caps," "bullets"—all of which we now take for granted as intrinsic to digital design. And, unimaginable as it may seem, typewriters are still used in the workplace today: for example, by government agencies and police departments that type documents requiring duplicate copies (Goldenberg 2012), as well as in accounting, law, banking, the military, and in funeral homes, oftentimes for the purpose of filling out official forms (Greenbaum 2019). Typewriters are also used by security agencies wanting to protect sensitive information from being traced digitally, as well as for personal, artistic, and expressive purposes (Polt 2019).[7]

Most of these activities are invisible to users of word processors, where text often never leaves the screen, and when it does it is printed by a free-standing mechanical device that separates "typing" from the physical act of printing. This separation from the printed word foregrounds one of the values of typewritten text: its physical, tactile nature. With the impact of keys on paper, blank spaces on forms can be filled in, duplicate copies can be made simultaneously, and envelopes can be directly addressed. Perhaps more importantly, words struck onto paper make them more amenable to their physical control because there's no electronic version that can be copied digitally, hacked, or circulated instantly—and no record of the keystrokes that created the artifact. Of course, copies (print and digital) of a typewritten text can be made, but that would require access to the single original paper source. So in our hyperconnected digital age, one of the former burdens of typewritten text, its lack of portability, has also become one of its chief benefits, at least for the few specialized users who require a high degree of confidentiality. Recognizing these vestigial forms of design clarifies why typewritten text declined precipitously (thankfully, for most of us), how it still inundates the digital tools we use today, and why it's worth preserving, if only marginally.

Data Design: Near Extinctions, Revivals, and Residual Forms

As we're starting to see, many vestigial design forms echo all around us but go unnoticed, either because we take them for granted as long-standing conventions or because they've lost their currency and their past lives lie hidden and buried. These vestigial processes also occur in the realm of data design, where hundreds of genres and subgenres have emerged in the last two and half centuries, most of which have been documented in Robert Harris's expansive collection, *Information Graphics: A Comprehensive Illustrated Reference* (1996), with legions of new digital designs emerging since then. Some of these genres have had dramatic and spectacular lives: They burst onto the scene, have their moment in the sun, and then dwell in the shadows, sometimes for decades, barely surviving with minimal, sporadic use.

These dramatic changes of fortune were experienced by mosaic area charts, which in the late nineteenth century appeared prominently in the *Statistical Atlases of the United States* (F. Walker and the U.S. Census Office 1874; Hewes and Gannett 1883; Gannett and the U.S. Census Office 1898) and the French *Albums de Statistique Graphique* (Ministère des Travaux Publics 1878–96). Figure 1.6 from the first *Statistical Atlas* (F. Walker and the U.S. Census Office 1874) shows an example of this novel, emerging genre. In this ingenious mosaic, large rectangles show the relative size of the population of each U.S. state, arranged from smallest (Nevada, upper left) to largest (New York, lower right) in boustrophedon style, back and forth as the ox plows. The smaller rectangles on the right show the number of people who emigrated elsewhere (probably out West). Each of the rectangles for each state is broken down into subcategories: the large rectangles by race and residency—whites born inside the state, in another

Figure 1.6. Mosaic chart the first *Statistical Atlas of the United States* (1874) showing the composition of the population of each U.S. state (F. Walker and the U.S. Census Office, plate XX). Courtesy of the Library of Congress, Geography and Map Division [05019329].

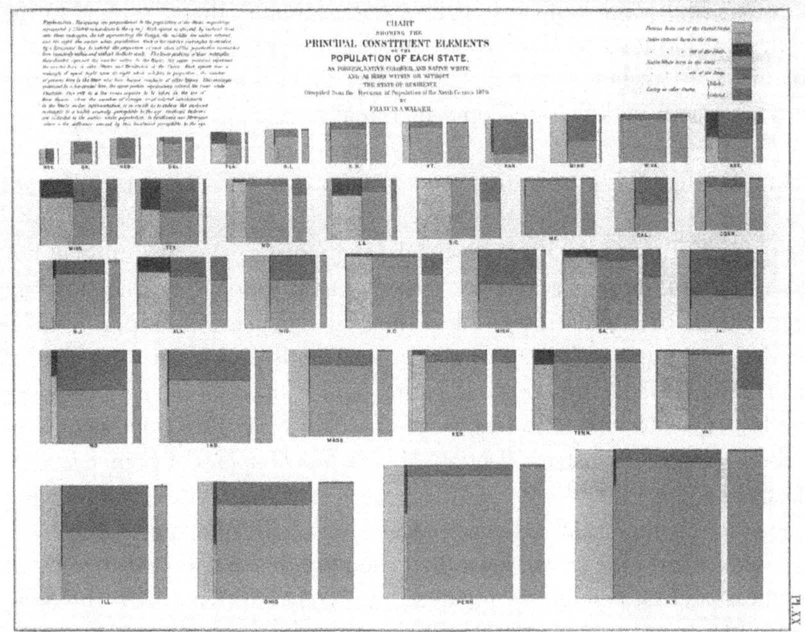

state, or outside the country; African Americans born inside or outside the state—and the smaller rectangles on the right by race alone. Audiences at the time must have been fascinated by this complex, data-rich chart and the sheer labor needed to construct it by hand, which may have contributed to the dwindling popularity of mosaics in the early twentieth century when they gradually lost their momentum as modernism simplified design forms, handmade artifacts lost some of their appeal, and the perceptual deficiencies of interpreting data displayed with areas were discovered.

Recently, however, with the affordances of digital technology, mosaics have experienced a strong, widespread resurgence in the form of treemaps and other rectilinear area charts (Kostelnick 2016, "Mosaics," 205–6; Friendly 2002)—contemporary reincarnations far exceeding the currency of handmade mosaics. The chart in figure 1.7 from the website StockCharts.com shows one of these re-creations, an interactive treemap chart that visualizes the S&P 500 on a March day. The rectangular area chart is divided into sectors of industries, within which each individual rectangle stands for a single company. Shades of green or red indicate the intensity of the gain or loss of a stock on that day, with green largely prevailing here, and stocks in some sectors faring better than those in other sectors. This chart genre, which thrived in the late nineteenth century and was largely

Figure 1.7. StockCharts.com interactive chart showing the values of S&P 500 stocks by sectors (2025, "MarketCarpet"). Copyright StockCharts.com. Chart courtesy of StockCharts.com.

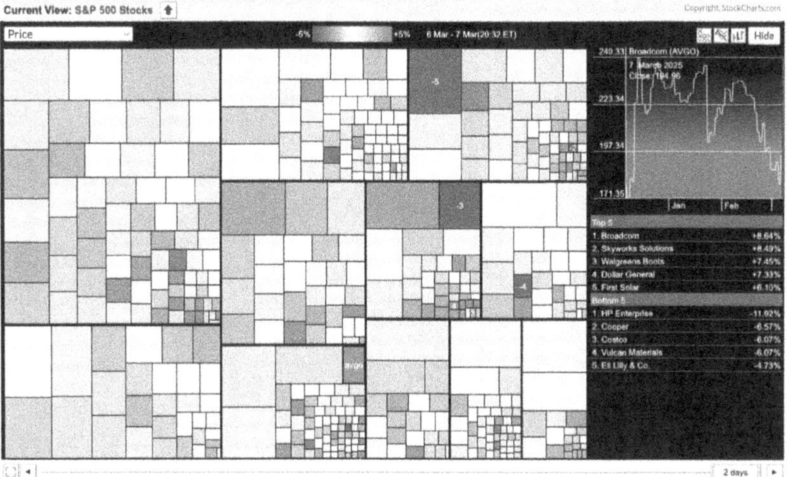

dismissed in the twentieth, suddenly flourished with the technology that afforded interactive data visualization (see Kostelnick 2016, "Mosaics"). The vestigial descendant of its forebears in the nineteenth century, this interactive treemap (and many others like it) has eclipsed the success of its earlier incarnations, a genealogy that's as obscure to audiences today as nameless ancestors in the old country.

Many other data design practices that rippled through the nineteenth century still have traces today. For example, in the innovative period of data design during its "golden age" (Funkhouser 1937, 330; Friendly 2008) in the late nineteenth century, pie charts with a dozen or more sectors (slices) visualized data about population, economics, and health. Figure 1.8 shows one of these "super" pie charts, which visualizes sixteen church denominations

Figure 1.8. Super pie chart from the 1898 *Statistical Atlas of the United States* that visualizes membership in various religious groups (Gannett and the U.S. Census Office, plate 33). Courtesy of the Library of Congress, Geography and Map Division [07019233].

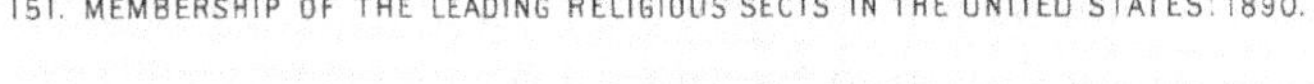

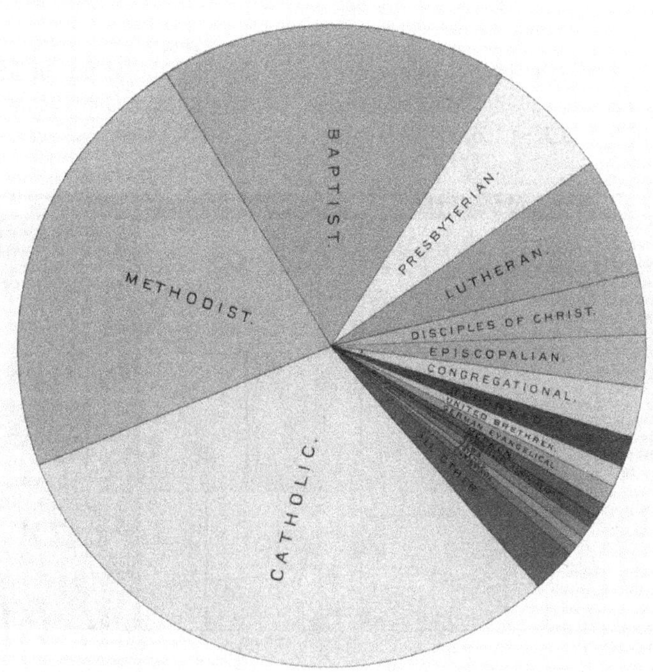

along with an "Other" category. Some of the sectors are large and easily labeled, others quite miniscule, with tight labeling. Gradually, given the challenges of labeling and deciphering so many odd-shaped areas, along with the influence of modernist minimalism, pie charts became simpler and more streamlined, with only a handful of sectors and simpler labeling. Today pie charts mostly adhere to the convention of eight or fewer sectors, though some still push the limits, which look overwrought because we've grown so accustomed to their simpler versions (Kostelnick 2019, "Pervasive").

Thus, the vestigial lives of data design conventions unfold in two ways. First and foremost, they embody the original foundational elements of the genre—for example, circles divided into sectors—which we adhere to today, though their distant ancestors remain virtually invisible to us. And second, those conventions embody the remnants of incipient practices—for example, circles with a dozen or more sectors—which, ironically, appear anomalous today but are also embedded in the genre's genealogy.

Super pie charts weren't alone in pushing the limits of conventional genres during the "golden age" of data design. Graphic inventions enabled line graphs and bar charts, for example, to accommodate data that exceeded the available area within the plot frame. These creative approaches are illustrated in *Scribner's Statistical Atlas of the United States* (Hewes and Gannett 1883). For line graphs visualizing highly variable data, the data lines simply continue outside the plot frame, breaking through the borders like wandering mustangs.[8] Figure 1.9 shows one of several charts in *Scribner's Statistical Atlas* with this effect—here displaying the cost of sugar and molasses, which like other goods spiked during and right after the Civil War. Because they extend outside of the plot frame, these abnormal data appear as graphical and rhetorical anomalies—outside the *normal* price range for these commodities and by today's standards outside the conventions for visualizing them.

Bar charts provided another novel method of visualizing extreme variations in data: If they exceeded the width of the plot frame, they were simply snaked back and forth until they were all accounted for. An example of this practice appears in figure 1.10, which shows the composition of the population of major U.S. cities, with the long, snaking bars showing the percentage of the city population born in their respective home states—for example, the percentage of New York City residents born in the state of New York, Philadelphia residents born in Pennsylvania, and Boston residents born in Massachusetts. These bars represent numbers disproportionate to the other numbers in the data set, which fit easily within the plot frame

Figure 1.9. Graph from Scribner's *Statistical Atlas of the United States* that extends lines outside of the plot frame (Hewes and Gannett 1883, plate 108). Courtesy of the Library of Congress, Geography and Map Division [a40001834].

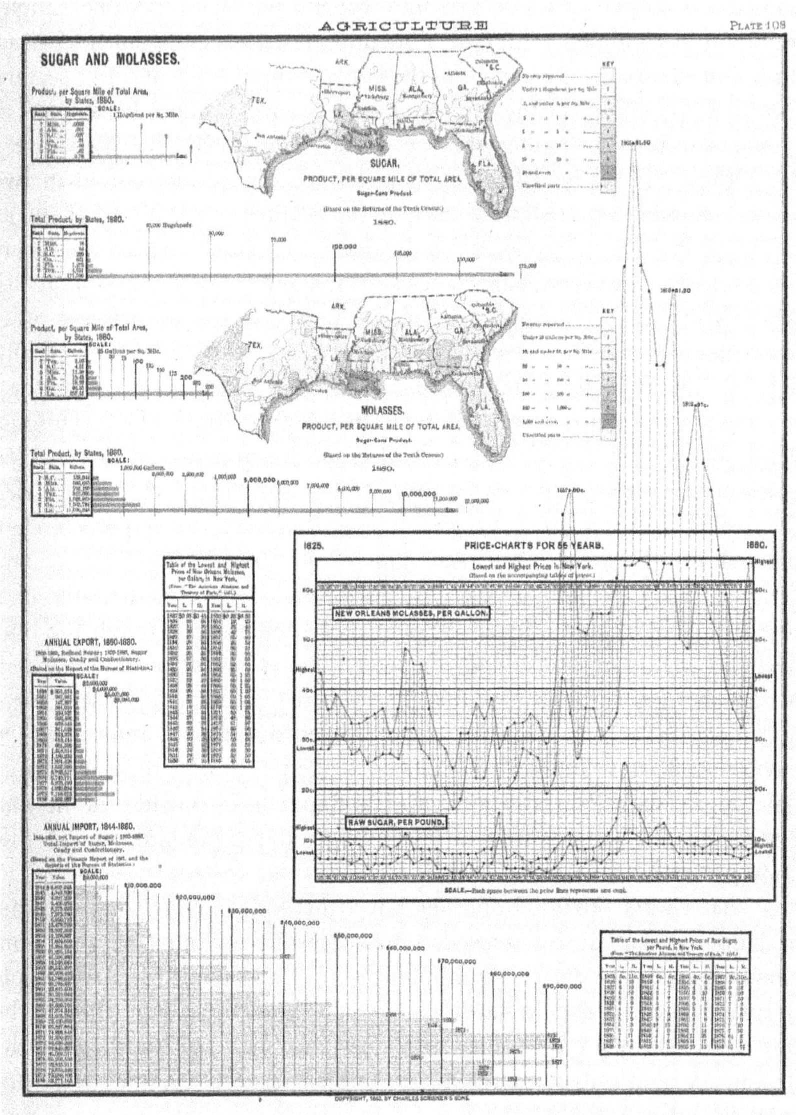

(except for Irish Bostonians, who poke through the border). The serpentine bars also appear in several other charts in *Scribner's Statistical Atlas* where large variations in the data require graphical adaptation.

Figure 1.10. Bar charts from *Scribner's Statistical Atlas of the United States* that cleverly display data to fit inside the plot frame (Hewes and Gannett 1883, plate 31). Courtesy of the Library of Congress, Geography and Map Division [a40001834].

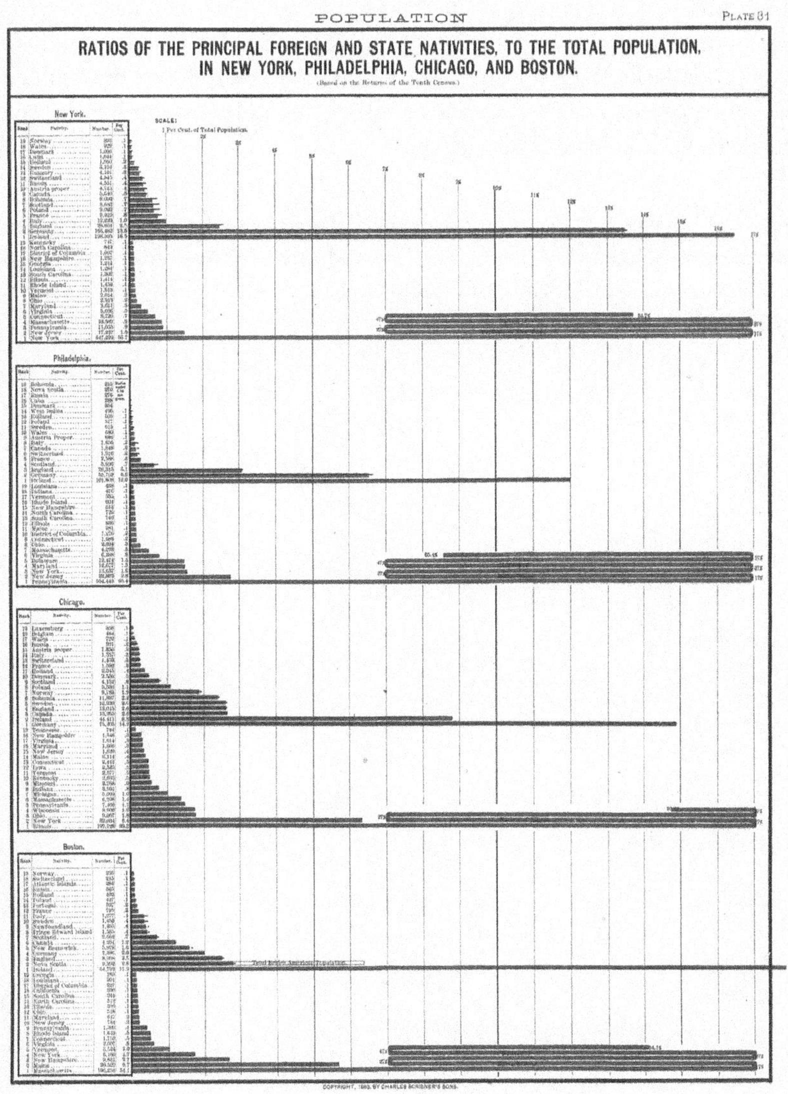

At the turn of the century other variations of the serpentine bar appeared in charts by W. E. B. Du Bois (1900, "Number") and in collaboration with his Atlanta University students (1900, "Religion"). Additionally,

Du Bois created another method for visualizing disproportionate numbers, shown in figure 1.11, an innovative bar chart that displays the location of African Americans in Georgia. People living in various sized "cities" appear in the zigzagging bars at the top (green, blue, yellow), and the rural population appears in the swirling red bar below, long enough to accommodate the large numbers. Extending the creativity of the "golden age," this ingenious chart illustrates another method of designing uneven data that's both efficient and elegant.[9]

Figure 1.11. Ingenious bar chart showing the distribution of the African American population in Georgia (Du Bois 1900, "City and Rural Population"). Courtesy of the Library of Congress, Prints and Photographs Division, Daniel Murry Collection [LC-DIG-ppmsca-33873].

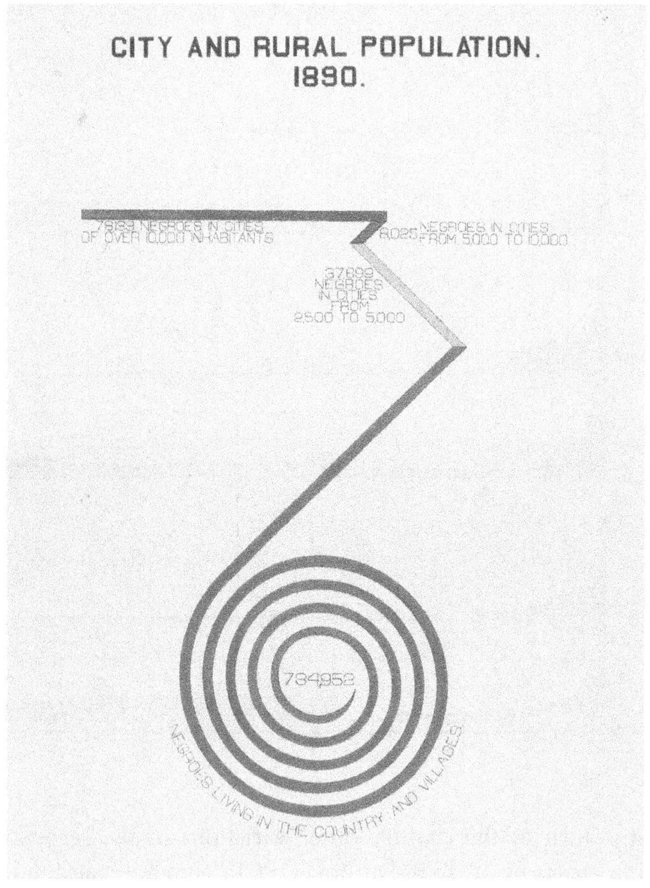

Vestiges of these charts with wandering bars—back and forth, in swirling circles—and with maverick lines poking up through the plot frame still appear occasionally today. However, any links to the charts in figures 1.9–1.11 have long been broken, their provenance long lost. In the "golden age" of data design, novel and experimental charts were springing up all over, and genre conventions were still congealing. Moreover, audiences at the time, who were new to the very idea of graphically visualizing data about human activities, were more malleable. They just went with the flow and savored the ingenuity. Today, some audiences might have a similar take, but designers must reckon with the rhetorical ballast of well-entrenched conventions if they want to foster that kind of response.

Chart designers are still faced with the problem of how to visualize highly variable data sets. To address this problem, they can place break marks in the Y-axis scale (and face the ethical dilemma this adaptation creates), they can use a logarithmic scale, or they can resort to a pie chart or a 100 percent chart, its cousin, forfeiting perceptual accuracy. Or, of course, data designers can cleverly snake bars around the plot frame or burst through it with maverick lines, claiming as their own these novel variations, the unwitting vestiges of their forebears.

Residual Picturing Conventions: Machines, Buildings, and Other Objects

Design conventions depend on the discourse communities that shape and sustain them and on certain rhetorical situations that recur, or emerge, along with the host of other factors that influence those practices, ranging from aesthetic taste and cultural values to economic conditions and available technology. Several traditional picturing conventions have contemporary residuals, most of which are hard to trace because they adapt or interweave past conventions to meet changing rhetorical needs as audiences continually bifurcate, realign themselves, and form more specialized enclaves. Many of these conventions become so normalized and so entirely taken for granted that their vestigial traces become invisible. But that doesn't make them any less rhetorically potent.

Picturing conventions have evolved over several centuries as drawing techniques, technology, and disciplines have emerged and the audiences that consume them have become more complex. The long genealogy of picturing conventions for practical purposes includes techniques like perspective, cross sections, cutaways, exploded views, and blowups (zooms), to name a few. In addition, the convention of envisioning human-made things in their spatial, environmental, and often social context—what I call *contextual picturing*—appeared early on in printed texts and still has rhetorical value in certain situations. Considering the evolution of contextual picturing, we might ask: To what extent are objects and buildings pictured as we see them, in three dimensions in the environment in which they function, or do they appear as flat, abstract entities detached from these elements? In other words, how much context does a picture visualize, and how have conventions for doing this evolved since the advent of print?

In the early days of print, mechanical devices began to appear in books that illustrated agricultural, military, mining, and hydraulic technology—that is, moving water to operate machines, pump garden fountains, or meet domestic needs. Some of this technology could be applied, and some of it was speculative and aspirational, like many of Leonardo's drawings (Zöllner and Nathan 2003, 570–671). Images of machines in early printed books were typically visualized in their contextual settings: in three dimensions and situated amid landscapes, castles, or towns and often included people operating the machines. Figure 1.12 shows a picture from an early engineering book, Agostino Ramelli's *Le Diverse et Artificiose Machine* (1588), that illustrates a mechanical device for opening doors and gates by removing the latch. Here the operator cranks the handle on the device to loosen the latch, presumably to break into a building where he's lost his key or, more likely, is not welcome. The three-dimensional rendering, derived from Renaissance painting techniques, envisions the technology in situ, in the context in which it would be used—in front of a castle, fortress, or large dwelling with a sturdy wood door, stone arch, and masonry walls. Penetrating this formidable barrier would apparently be short work with this mechanical device, and the contextual details in this picture bolster that argument.

This picture also includes another contextual convention of engineering drawings from this era: multiple versions of the mechanical device that supply detailed information. The picture of the fully assembled machine in the foreground, along with its parts lying on the ground and floating in space, illustrates those parts in the context of the whole apparatus as well as reveals how they all fit together.[10] These objects, which are lettered to correspond to explanatory notes in the text, provide technical details about how the apparatus is built and its mechanical method of operation: clamping the latch to the machine and pulling it away from the door by

Figure 1.12. Late sixteenth-century drawing showing a mechanical device for opening a door (Ramelli 1588, figure CLIIII). Courtesy of Iowa State University Library Special Collections and University Archives.

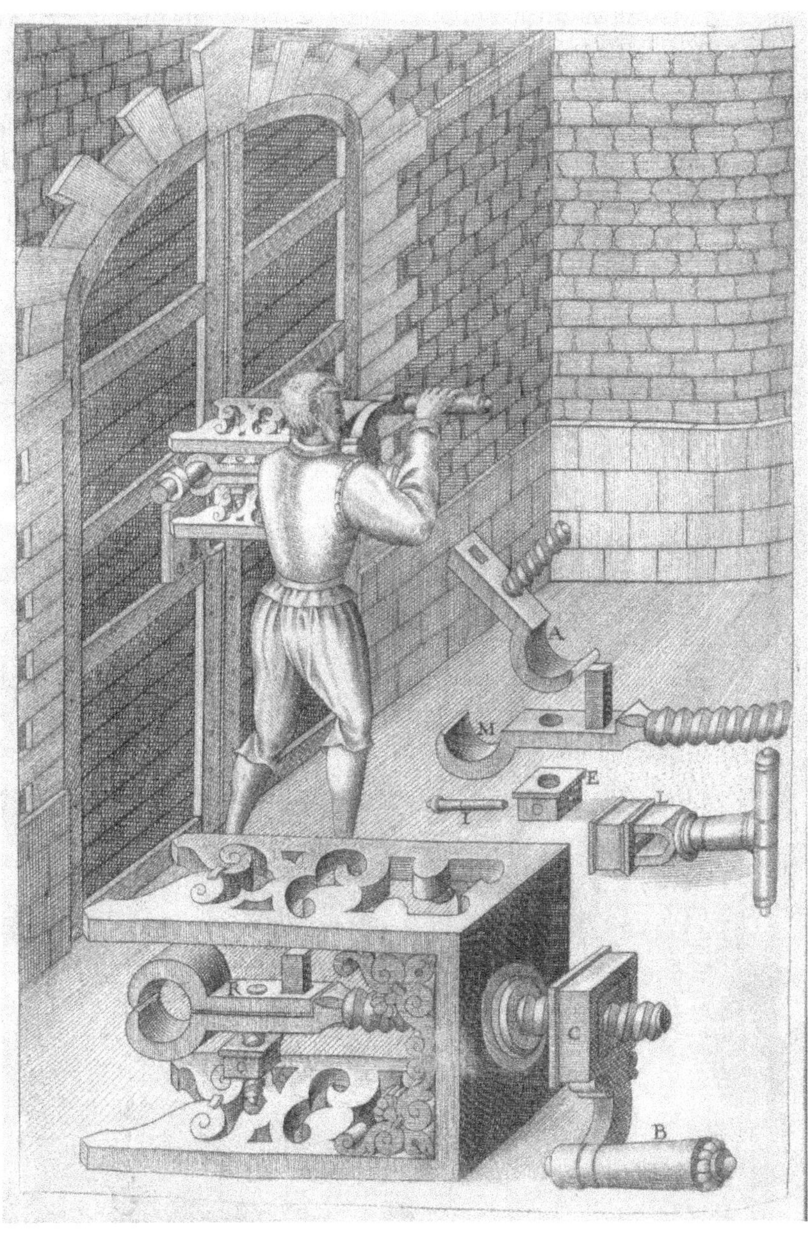

turning the giant screw. All of these contextual conventions coalesce to instruct and persuade the audience.

Contextual renderings of mechanical devices continued for some time and were often situated in social settings to illustrate their communal benefits. Figure 1.13 shows a water pumping system from Georg Böckler's

Figure 1.13. Seventeenth-century drawing showing hydraulic technology in its social context (Böckler 1673, fig. 87). Courtesy of Iowa State University Library Special Collections and University Archives.

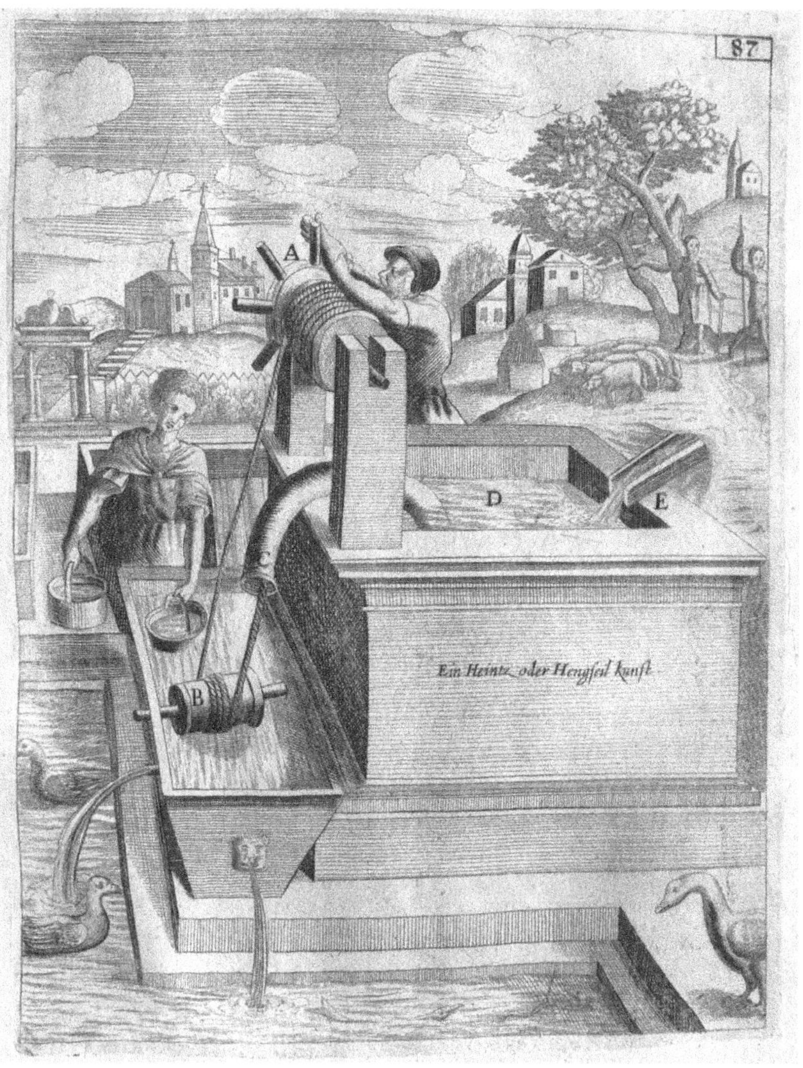

Theatrum Machinarum Novum (1673) nearly a century later. Like Ramelli's drawing, this one also includes plenty of contextual details, including the figures in the foreground operating the apparatus and collecting the water, the church and other buildings in the background that the hydraulic system is designed to serve, along with the livestock and farmers on the right and even the ducks on the pond below—indeed, all forms of life on the chain of being that benefit from the water pump. Together these elements compose a harmonic, socialized landscape, without which the technology would be arid and abstract and less powerful and persuasive.

These early engineering drawings follow the conventions of architectural rendering in three-dimensional space typical of this period (see Booker 1979, 37–47). By the early nineteenth century, engineering drawings had become flatter and more abstract, and contextual details had fallen by the wayside. Architectural drawings also evolved into leaner, less embellished forms. Figure 1.14, for example, shows an elevation of a classical façade for an English country house that appeared in Isaac Ware's 1756 treatise on architecture (which also featured designs by Inigo Jones, an influential seventeenth-century English architect). Although the façade in the drawing appears with full shades and shadows, a technique that continued until the late twentieth century, no contextual details appear: no foliage or human figures, no walks or driveways, no afternoon sky—no

Figure 1.14. Eighteenth-century elevation for an English country house (Ware and Jones 1756, plate 53). Courtesy of Iowa State University Library Special Collections and University Archives.

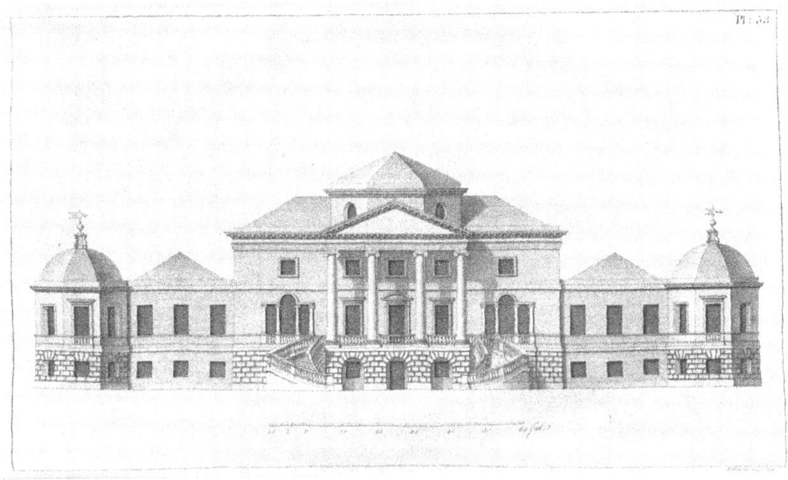

hint of the picturesque English landscape in which this building will likely be situated, nothing to distract its audience from its elegant forms and lines. Even though the design itself has copious details—columns, capitals, balustrades, pediments—the drawing is abstract and succinct and as rational and transparent as the emerging Enlightenment values.

What are the residual effects of these practices? What traces do we still have of them today? Engineering and architectural drawings have long been flattened to two-dimensional planes, stripped of shades and shadows, partly to ensure accurate dimensions—for engineers, manufacturers, builders, and other professional audiences. Nonetheless, a rhetorical exigency still exists for full three-dimensional rendering, especially for audiences that need to be persuaded to purchase, finance, or use a given design. Figure 1.15, for example, shows a drawing of a planing machine

Figure 1.15. Drawing of a planing machine from the *American Machinist* (Pryibil 1880, 1). Reprinted by permission of *American Machinist*, Endeavor Business Media. Courtesy of Iowa State University Library Special Collections and University Archives.

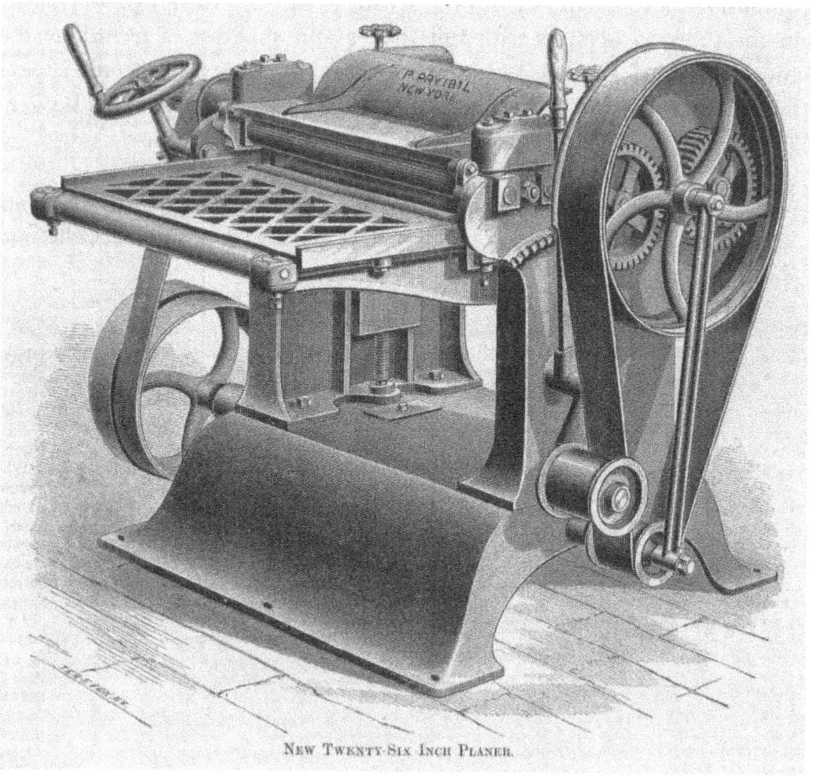

from an 1880 issue of the *American Machinist* (Pryibil 1880). Rendered in three dimensions with full shades and shadows, the picture reveals fine details about this device—the cranks, gears, lever, belts, and bolts—in a very realistic yet idealized form isolated from any contextual details—the operator, the wood being planed, the surrounding space in which the machine does it work. In these ways, this nearly photographic image becomes a beautiful, abstract object that fosters desire—to purchase it, operate it, and benefit from it financially.

Likewise, figure 1.16 shows an appealing early twentieth-century rendering of a house in full perspective, but with contextual details like landscaping, sidewalks, trees, a neighbor's house, and even someone sitting on the front porch. The picture is accompanied by floor plans showing the size of various spaces along with a textual description of how the house should be built, mostly by individuals seeking an affordable dwelling. The perspective view in figure 1.16 ties these drawings together into an affecting package—an ideal view of how the actual house might look once completed and occupied. This picture is a rhetorical precursor to similar renderings

Figure 1.16. Perspective view of an early twentieth-century house in its natural and communal context (Architects' Small House Service Bureau of Minnesota 1921, 77).

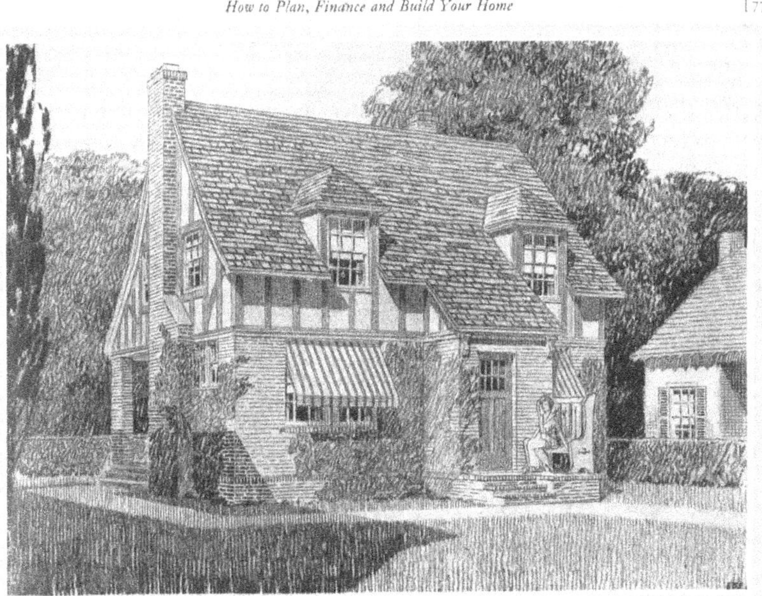

of 1950s suburban tract houses that sprang up across the U.S. after World War II. Drawings of ranch houses with low roofs, large picture windows, breezeways, and spacious yards with willowy trees conjured up visions of domestic bliss that young families yearned for in the Baby Boom era.

Another even more highly contextual picture appears in figure 1.17, which shows an architectural rendering of a parish hall designed by RDG Planning and Design for Sacred Heart Church in Boone, Iowa. The picture displays a wealth of contextual details so that the audience can envision the building as if it already existed as part of the church complex and neighborhood. The parish hall (in the center) is situated among existing buildings (offices on the left, the church on the right), with landscaping, pedestrians moving about, cars on the street, wispy clouds above, and even dappled tree shadows on the street pavement. And the planned building itself is envisioned in detail, with its sleek rooflines, large south-facing windows, inviting entryway, and stonework that matches the church façade. All of these contextual elements make the design seem a reality, and in doing so paint a very attractive picture of what might be—for building committee members, parishioners, potential donors, and city officials. By infusing this picture with the enargeia of elaborate description, contextual detailing brings an imaginary scene to life and enhances its persuasiveness.

Traces of the past can also be found in other picturing practices. For example, showing the disparate parts of an object, as we saw in Ramelli's

Figure 1.17. Rendering of a parish hall (center) for Sacred Heart Church in Boone, Iowa, in relation to other church buildings (RDG Planning & Design 2020). © RDG Planning & Design. Used with permission.

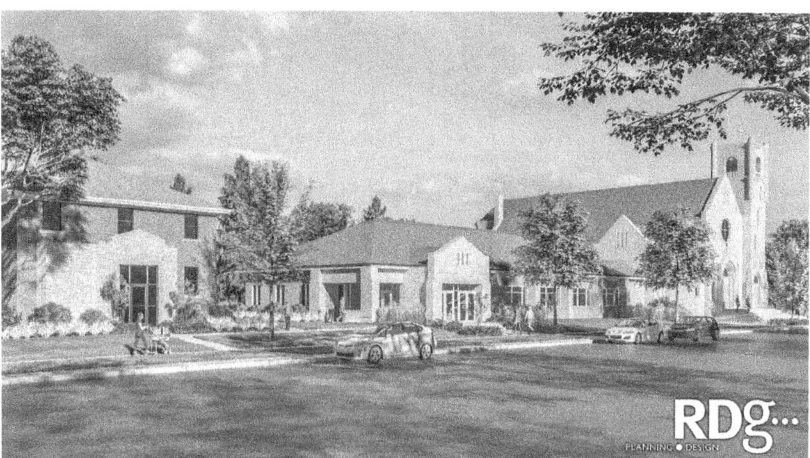

drawing in figure 1.12, still has a high degree of currency for nontechnical audiences that assemble personal computers, children's toys, and household furniture. These pictorial instructions typically show the parts of an assembled object, often numbered or lettered, and how they fit together. Figure 1.18, for example, shows a page from a set of IKEA instructions

Figure 1.18. Page from IKEA instructions for assembling a two-drawer dresser (IKEA 2010/2024, 11). © Inter IKEA Systems B.V. 2010. Used with the permission of Inter IKEA Systems B.V.

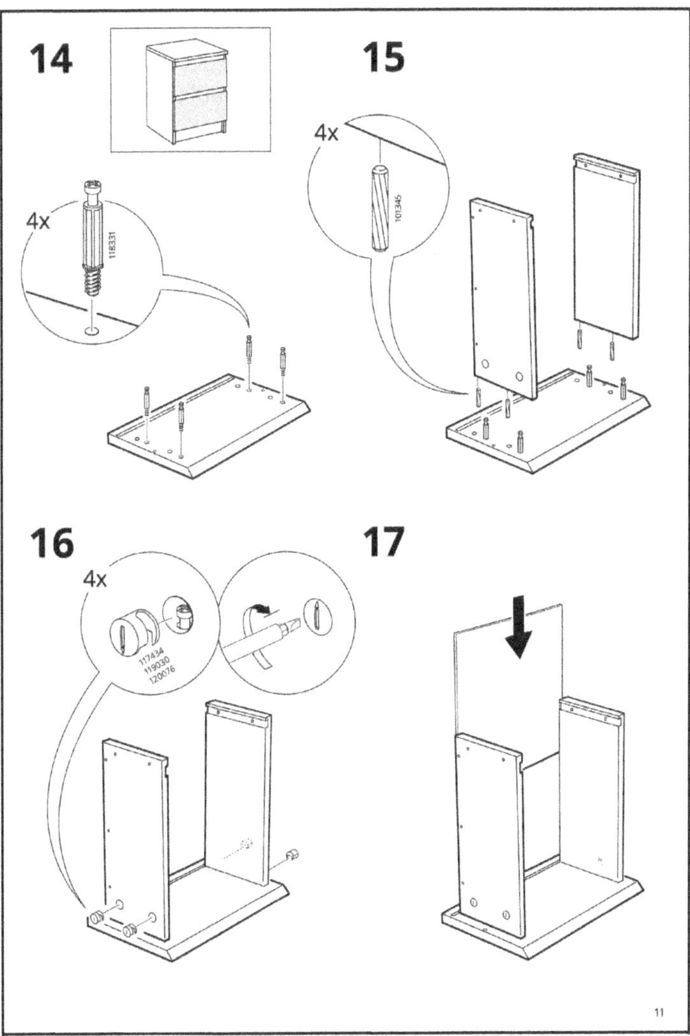

for assembling a small dresser. The pictures show how the pieces of the dresser fit together by using perspective to show multiple angles in the same drawing, blow-up images to zoom in on the details, exploded views to show how the fasteners hold the pieces together, and step numbers to guide the audience through the process. More than four centuries removed from Ramelli's drawing, the practice of picturing pieces of a disassembled object in relation to each other remains a residual convention, however remotely connected to its ancestors. On the other hand, unlike Ramelli's drawing, the IKEA instructions envision the objects devoid of physical or cultural context: no walls or windows in the room, no rugs on the floor, no other pieces of furniture, no human forms, just airy space. This minimalist approach responds to contemporary rhetorical situations: Busy global audiences won't be distracted as they assemble the furniture in their homes, and they won't be confused or put off by surroundings they can't identify with culturally.

Since the advent of print, picturing techniques for visualizing technology, buildings, and mechanical devices have morphed and adapted to a constantly shifting kaleidoscope of rhetorical exigencies. Technology has played a key role in setting and meeting audience expectations. Although contextual picturing for technical and professional audiences ceased long ago, envisioning objects in perspective has found its way back into engineering with 3D modeling and, more recently, augmented and virtual reality. Moreover, as we've seen in figures 1.16 and 1.17, contextual picturing has continued for lay audiences, especially for visualizing architectural spaces. Several years ago when my wife and I were searching for a house in a nearby state, a realtor snail mailed us photocopies and Polaroid photos of houses for sale. Those pictures were like images from a distant planet—miraculous but crude! Contrast those with contextual picturing today afforded by video tours and "virtual staging," which instantaneously give home buyers vivid insights about architectural space and character.

Picturing Styles: Ongoing Rhetorical Ballast

Picturing practices, then, wax and wane rhetorically as audiences and their needs and expectations shift. Those practices also include the style of drawing, which constantly evolves based on disciplinary standards, current technology, cultural values, and aesthetic taste. Ramelli's drawing of buildings and objects (fig. 1.12) reflects the functional and aesthetic values of the late Renaissance, Böckler's drawing (fig. 1.13) reveals the fine details and embellishments promulgated by Baroque aesthetics as well as a penchant for social context, and Ware's classical façade in figure 1.14 expresses the rational, stoic ethos of the Enlightenment. The *American Machinist* drawing (fig. 1.15) illustrates the Victorian and Realist emphasis on detail, along with the influence of burgeoning photography. The contemporary IKEA drawing (fig. 1.18) embodies the aesthetics of modernist minimalism, as does IKEA's lean, functional style of furniture design. In each of these cases, the pictorial style reflects certain values of a given time and place.

When we glance back at images from long ago, however, we primarily see only visual vestiges, the ghosts of picturing styles generated by cultural, disciplinary, and material conditions that no longer exist. Of course, these stylistic phantoms can be momentarily resurrected for rhetorical effect through imitation or reproduction, or the artifacts that embodied them can be purchased online and hung on walls as mementos, collectables, or conversation pieces. However, the residual aspects of style lie much deeper because practical images have *functional* and *rhetorical* purposes rather than expressive or aesthetic ones (though these often work in concert). In their own visual languages, these pictures inform, persuade, and instruct their audiences and motivate them to act, though the audiences and exigencies constantly change. At any given moment, a picturing style provides rhetorical ballast to meet these goals. In retrospect, each picture also serves as a precursor to ones that follow, with the vestiges of one style anticipating the emergence of a new one and each stylistic variant a means to achieving rhetorical goals rather than an end in itself.

Chronos and Technology: Affordances and Mutability

Visual conventions have to be efficiently deployed and reproduced in order to gain currency with their audiences over the long term. So when we survey the development of visual language through the lens of chronos, we need to consider how technology has shaped, intervened, and redirected those conventions—or simply cemented them in place, with the rhetorical weight of chronos having a crystallizing effect on them. In fact, sometimes new technology initially plays an adaptive role in promulgating existing conventions rather than an insurgent role in begetting new ones. For example, early print looked like manuscript text, typewritten text followed handwritten formats, and laser printers propagated Courier fonts. As Donald Norman observes in *The Design of Everyday Things* (2013),

"One way of overcoming the fear of the new is to make it look like the old" (159). So, in these instances, by conforming to existing conventions, new technology prolonged their vestigial lives—rather than curtailed them—before it enabled new visual language.

Still, designs are time stamped with the technology used to generate them. Designers rely heavily on the tools they use—cutting edge or old school—each of which has its own constraints and affordances for deploying visual language. As the affordances of print and digital technology changed over the centuries, so too did the capabilities of designers for creating and reproducing images. In early printing, woodcuts produced only the most basic kinds of illustrations, which lacked fine details and limited their content. For example, simple woodcuts visualizing classical and medieval learning appeared in William Caxton's *Mirrour of the World* (1490), the first scientific (and illustrated) book printed in the English language. Woodcut technology was pushed much further by Georgius Agricola's *De Re Metallica* (1556), which contained hundreds of detailed plates illustrating methods of mining and processing metals that captured the depth and complexity of these activities.

As science and technology advanced, more economical and efficient methods of visualizing knowledge were needed to meet the needs of ever larger literate audiences. For the next few centuries, copperplate engraving met those needs. Early engineering books like Ramelli's *Diverse et Artificiose Machine* (1588, fig. 1.12) and Jacques Besson and François Béroald's *Theatre des Instrumens Mathematiques et Mechaniques* (1578) demonstrated the detail and flexibility of the new copperplate technology, opening the way for visualizing increasingly sophisticated machines more clearly and efficiently. Penmanship copybooks like Edward Cocker's in the second half of the seventeenth century also exploited the capabilities of copperplate engraving to visualize his elegant handwriting samples. The eighteenth-century French *Encyclopédie* used engraving technology to reach a mass audience, with eleven volumes of the *Recueil de Planches* (Diderot and d'Alembert 1762–1772) containing thousands of illustrations like the one in figure 1.1.

In the nineteenth century, chromolithography made color more widely accessible for large-scale printing, epitomized by color engraving of the first three *Statistical Atlases of the United States* (F. Walker and the U.S. Census Office 1874; Hewes and Gannett 1883; Gannett and the U.S. Census Office 1898), examples of which appear in figures 1.6 and 1.8. Chromolithography also made print publication less expensive to produce, further enlarging its audience. With the widespread use of chromolithography, hand tinting of images gradually became obsolete. Color technology was further enhanced by photographic and then digital plates, which added detail, speed, and economy to producing images and paved the way for the ubiquitous online color that we now routinely encounter every day.

Today digital technology gives designers the flexibility to invent, revise, and comingle texts and images, often collaboratively, as John Logie argues, and to do so in ways that were inconceivable with print technology. Digital technologies provide avenues for novel designs, often (if not primarily) aimed at engaging audiences interactively with various forms of media, ranging from data design (like the stock chart in fig. 1.7), 3D imaging, and augmented and virtual reality. How do these technology innovations change the trajectory of conventions and their vestigial practices and traces? Do they merely generate hyper versions of existing conventions, or do they inspire entirely fresh ones? Sometimes digital technology enables designers to revive dormant conventions, like the mosaic charts I discussed earlier. Digital technology also enables designers to explore novelty, though novel designs often lack long-term practical value. Imitation is safer and more efficient than disruption. In the end, practical designs mainly echo and remix (and marginally depart from) one another as rhetors, initially at least, deploy new technology to meet the existing needs and expectations of their audiences.

Chronos in Motion: Streams and Pools

When we look at historical developments in information design over long periods of time from the perspective of chronos, we can see the big picture of how visual language emerges and takes its course. Like a mountain stream, it gains momentum as it rushes along and gains popularity, evolving in its descent and often mutating into side streams, then flattening and broadening its way in the slow-moving currents of conventional forms and in the course of time, coalescing in deep, obscure pools. Those pools of vestigial forms beckon our inspection because they not only give us clues about their past lives but reveal how these forms continue to function in the present as well as serve as rhetorical touchstones for the emerging forms that eclipsed them.

For the most part, designers and their audiences resist (or are indifferent to) the notion that visual language—whether in fast-moving streams or quiet pools—provides a lens to the past, either because historical knowledge of practical communication remains hidden or elusive to them or because the present requires enough creative or interpretive energy to occupy their rhetorical lives. As much as either of these might be the case, none of it matters to the reality we experience: that chronos remains a potent and pervasive force in the visual language we deploy and interpret every day.

Visual Kairos: The Rhetorical Dynamics of Short-Term Timing

While macro-level temporal changes often lie beneath the surface, time at the micro level has a more immediate and explicit role in information design. This temporal aspect of the rhetorical process, or kairos, constantly mediates our interpretation of visual language. Short term and situationally specific, strong kairos draws audiences to visual language. Like the mythological figure holding his balance on a razor (Hawhee 2002, 19, 25), rhetorical acts with heightened kairos seize the moment, maximizing their impact. With optimal temporality, an information design achieves full rhetorical force in a particular situation, and when that moment passes, it's rendered feckless and irrelevant—at least until it responds to a new audience and exigency.

This rhetorical process can unfold locally and immediately with a flyer on social media announcing tomorrow's garage sale, or it can occur globally and leisurely with a logo touting an upcoming concert or sporting event. Leading up to the Olympic Games, for example, the logo with five rings is visible virtually everywhere, especially in popular print and digital media, constantly reminding us of the events about to play out before us, generating ethos and emotion. Kairos reaches its apex as the games progress, but once the Olympic flame is extinguished, the rings immediately lose their immediacy and relevance. The air in the balloon escapes and its rhetorical power abruptly dissolves.

Opportune Design: Varying Levels of Visual Kairos

Kairos doesn't impose exact temporal constraints for visual designs: sometimes it remains fresh for long stretches of time; sometimes it lasts for only minutes, especially in digital media and data visualizations. On the longer-term end of the spectrum, some prominent designs continuously remain fresh and relevant over decades or centuries, like the *Mona Lisa*, the Eiffel Tower, and the Statue of Liberty. All of these dwell in a public realm where interpretation is influenced by culture, economics, engineering, politics, and art and architectural history. In popular culture, TV shows are especially vulnerable to temporal corrosion, so they try not to time stamp themselves so their content won't look dated in reruns. Inevitably, however, both TV shows and other designed artifacts—clothing, hairstyles, furniture, cars—go out of style. The corrosive effects of time take their rhetorical toll, weakening kairos: Indeed, if audiences don't consider a design timely and au courant, they might not trust or engage with it in the first place.

Of course, typefaces, color palettes, pictures, and graphical elements become stale and musty as well, their diminished kairos reducing their relevance and ethos. However, most practical information designs have narrow audiences and temporal spans, with the process of visual corrosion tied to situational variables. Instructional materials, like those for assembling an IKEA dresser (fig. 1.18), have heightened kairos when audiences first encounter them—in a living room, garage, or basement, tools in hand—before discarding (or losing) them. For some owners, however, these instructions have kairotic elasticity by serving as a future reference if the furniture needs to be disassembled or parts need replacing, extending their useful life well beyond the first encounter.

Some visual designs have kairotic elasticity because they get noticed only intermittently despite their constant visibility. At the point of encounter, the efficient perceptual act of *seeing* visual language speeds things up cognitively and semiotically (Bertin 1981). For example, a warning sign on a piece of equipment or in some public space will have optimal kairos every time an audience encounters it by alerting the audience to some imminent or potential danger. Kairos is continuous, as one audience member after another interprets the warning, though audience members encountering it multiple times might also become immune to the message. Consequently, the opportune moment of the initial encounter might be critical, especially where physical risk looms.

For example, for the bear crossing sign in figure 1.19, drivers and bicyclists can either take heed of the sign or flout it. From the looks of the bear pictured, heeding the sign—by slowing down a car, keeping the windows rolled up, or removing bear spray from a backpack—would serve the audience's self-interest. Similar decision-making is prompted by trail signs urging hikers not to climb on steep, slippery, or fragile rocks. The moment an audience encounters them, they can either obey the signs or disregard

Figure 1.19. Road sign warning of bear crossings (U.S. Department of Transportation, Federal Highway Administration 2023, 522). Courtesy of the Federal Highway Administration.

them, depending on how seriously they perceive the danger, though liability issues or environmental preservation may also be factors. Whatever the motivation of those who designed and posted these signs, their kairotic elasticity and resilience are evident every time someone encounters them.

This kind of perceptual and rhetorical immediacy also occurs with the New York Department of Health poster in figure 1.20, designed to enable laypersons to perform a short-term procedure for a choking victim. Some audience members might encounter the poster at a first aid or EMT training session while other audiences will have a more immediate encounter—in a restaurant, cafeteria, community center, or school lunchroom—places where people congregate and where a choking incident might suddenly occur. Most people who consult the poster in these spaces have probably never done this procedure before, so they need to be quick learners. They might have looked at the poster before the trauma occurs, but the immediacy of consulting it while the emergency unfolds speeds up the process, as they scramble to respond. Kairos suddenly goes into overdrive.

Other design genres offer additional options for optimizing kairos. Take for example organizational charts, which most large, complex organizations (corporations, universities, government agencies) use to visualize their leadership positions. In order to keep their charts relevant,

Figure 1.20. Poster showing how to perform emergency procedures for someone choking (New York State Department of Health 2023). Reprinted with permission from the New York State Department of Health, Public Affairs Group.

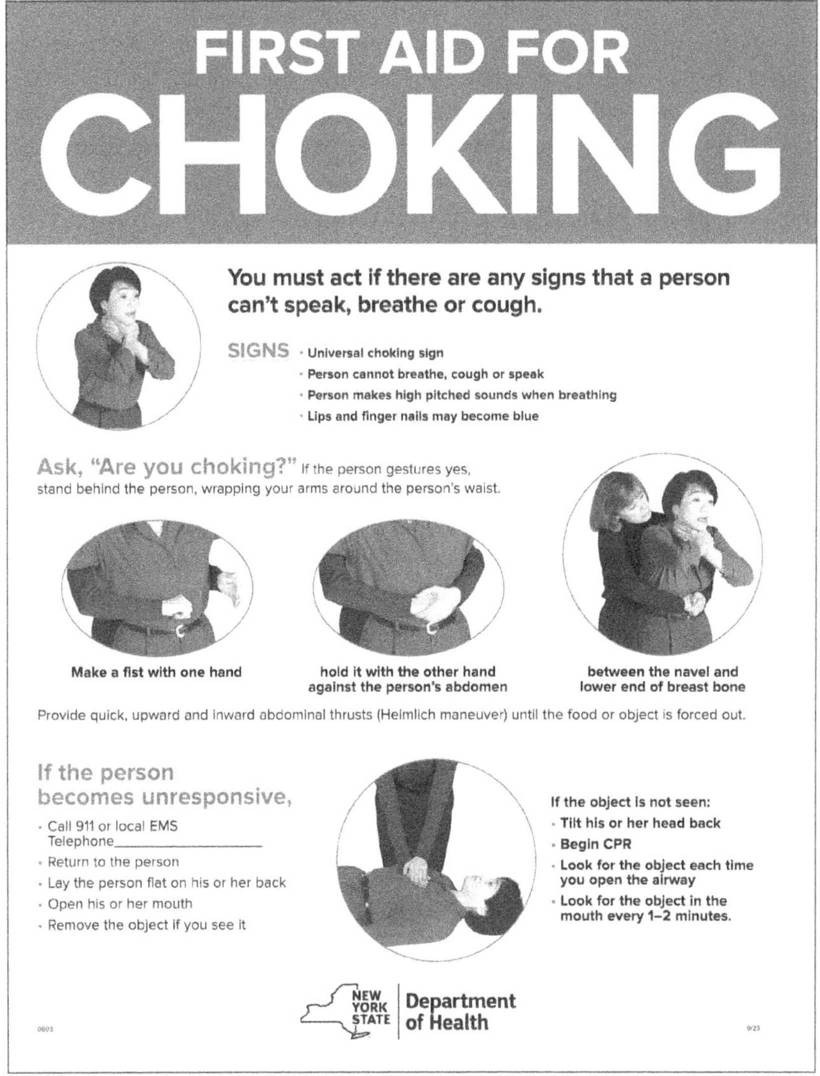

truthful, and timely, organizations periodically update them as leadership positions change hands or new ones are created. If specific names appear on their organizational charts, as they often do, the charts age quickly. As soon as someone steps down from a position, and someone else steps in, even as an "interim," the chart loses its temporal edge. Unless an organization is remarkably stable, its personnel chart has only a short window of immediacy and relevance and then is superseded by the next version—or languishes in confusing (and embarrassing) obsolescence.

In an age of digital design, however, organizational charts don't need to remain static, whereby they achieve optimal kairos for a few months and then suffer temporal degradation soon thereafter. Designers can sustain kairotic resilience by including interactive features that allow audiences to mouse over a given position to reveal a person's identity and other details. For example, the online organizational chart in figure 1.21 shows key positions in the U.S. National Weather Service. By including an interactive feature, the chart enables the audience to see the overall structure of the organization as well as to access more specific near-term elements—names, pictures, and contact information of key personnel and their array of responsibilities—giving the chart kairotic flexibility (and depth) unavailable in its static counterparts.

Figure 1.21. National Weather Service organizational chart with interactive information about key leaders and their responsibilities (U.S. Department of Commerce, National Weather Service, NOAA 2025). Courtesy of the National Weather Service.

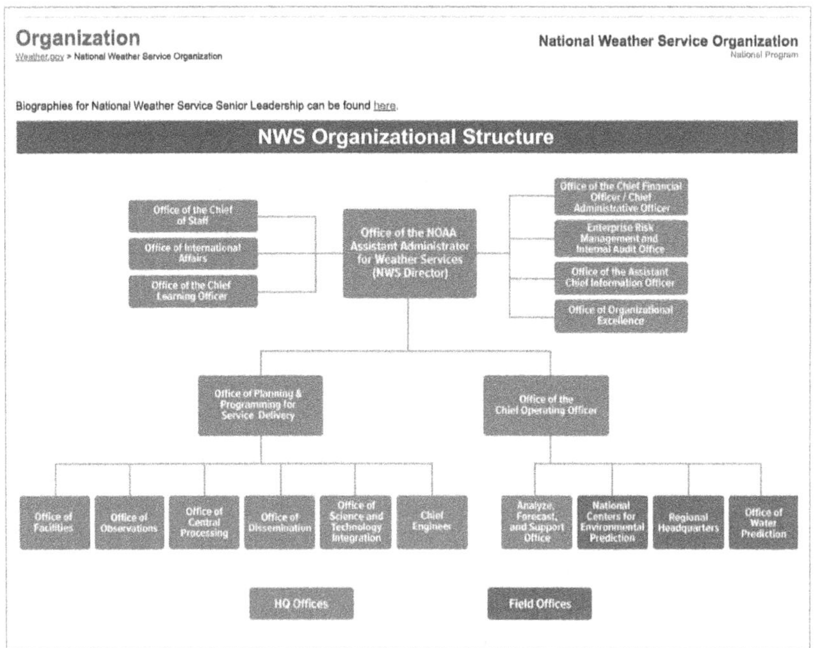

Hyperkairos: The Increasing Velocity of Digital Design

The temporal velocity of digital design often occurs at lightning speed, with audiences interacting with visual language in narrow time zones. This form of kairos coincides well with the mythical figure from ancient Greece who held a scale on a razor, always ready to act at a moment's notice (Hawhee 2002, 19, 25). Experiencing this kairotic immediacy, however, demands vigilant audience engagement while the communication still retains its temporal viability. Memes, YouTube videos, and Instagram or Twitter/X images can go viral only if potential audiences are paying attention. One trip to the refrigerator and the design's kairotic force might plummet! This kind of hyperkairos develops with visualizations of dynamic data. In today's digitally connected world, charts visualizing the prices of stocks, currencies, and commodities (corn, oil, precious metals) refresh the data in near real time, day and night, 24/7. For audiences around the world, the charts that greet them at sunrise have little value by noon, the noon charts less value by sundown.

On the day the stock chart in figure 1.7 appeared, its value to its audience skyrocketed because it revealed gains or losses in near real time. The next day, and in the successive days after that, however, this iteration of the chart wanes in value, as its moment quickly fades. It remains useful historically for those who track the market, but for most audience members, only its reincarnation the next trading day with fresh data restores its kairotic power. Digital displays and other ephemeral designs like them stand on the razor's edge, supercharged with short, rapid bursts of rhetorical energy, their relevancy fading once their moment passes. Hyperkairos also occurs in many professional and personal settings: Executives monitor dashboard charts visualizing sales, manufacturing, financials, and CO_2 emissions; runners check their steps, heart rate, and distance on their wearables; and homeowners consult their cell phone apps to monitor the electrical output of their solar panels. These hyperkairos visualizations, which have become part of many people's daily lives, will only continue to proliferate.

Digital Interactivity: Kairos in-the-Moment

Unlike spoken discourse, static documents (either print or digital) usually don't allow dynamic in-the-moment interactions between rhetor and audience. However, interactive elements invite (even impel) decisions and adjustments, similar to what Debra Hawhee (2002) calls "invention-in-the-middle," which lies "between" rhetor and audience (Hawhee, 17, 25) and which is neither premediated nor predictable. This shared agency between a rhetor designer and an audience creates the interactive in-the-moment space for decision-making (or invention) that Hawhee both traces to the mythological figure of kyros and ascribes to contemporary formulations of kairos that distribute and balance its power rather than concentrate it in the hands of an individual, autonomous rhetor. In this way, digital data design connects kairos to its traditional foundations in all three modes: spoken, written, and visual.

In this kairotic mode of interactive data design the rhetor designer supplies the affordances for the interactions, and the audience possesses the agency to manipulate and control them (Rawlins and Wilson 2014). The designer, of course, creates the framework and technical parameters for these affordances but doesn't know how or if a given audience-user will engage with these. However, users have the opportunity to customize the display at a given moment, in this way sharing agency with the designer. The stock market chart in figure 1.7, for example, enables users to choose from several display options—individual stocks (with line charts), time range (in two-day segments), and technical criteria, among others, and to mix and match combinations in the moment, creating visualizations that lie in "between" data designer and audience.

In our daily personal and professional lives, most of us engage every day with interactive displays visualizing money, health, travel, sports, and weather. For example, online weather maps constantly update conditions—temperature, wind, precipitation—that affect our daily routines or those of people we know. Because we can access weather maps on most any digital device, they are especially portable and potent, providing real-time data about current and impending events, including warnings about storms and other dangerous conditions. Figure 1.22 shows an online weather map from the National Weather Service with hazards around the country, county by county, color coded for strong winds, blizzards, gales, flooding, and avalanches. Depending on where audiences live, they can zoom in on their geographical areas in-the-moment and interact with dynamic data. Although the geographical size of a given county can disproportionately draw attention to it, this perceptual liability of choropleth maps using shading or coloring (like this one) doesn't really matter to audiences wanting, at a given moment, to see warnings about where *they* live.

Weather maps are only one of many kairotic interactions we experience digitally with heightened vigilance. During the COVID-19 pandemic,

Figure 1.22. National Weather Service map with county weather warnings that are color coded by type (U.S. Department of Commerce, National Weather Service, NOAA 2025). Courtesy of the National Weather Service.

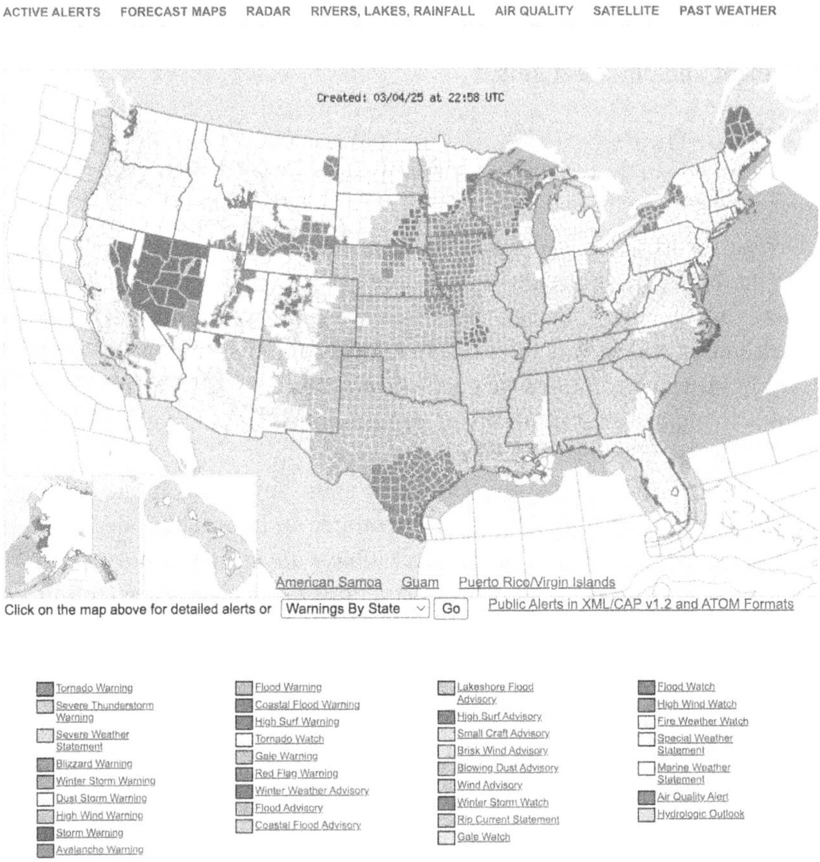

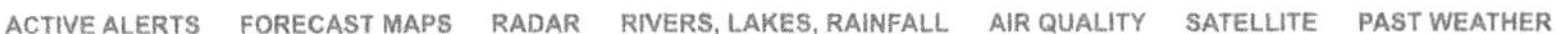

people around the globe were glued to data maps and charts visualizing the weekly number of new cases, with the mantra to "flatten the curve." As COVID began peaking in late 2020 and early 2021, hot spots blotted data maps everywhere, inviting explorations of our own neighborhoods and impelling us to zoom in and drill down into the displays for details. Depending on where we lived—or our close friends and relatives lived—the abstract swatches of data had the capacity to soothe or alarm us. We have similarly intense, albeit less consequential, interactive experiences with highway maps showing traffic conditions, diagrams tracking sporting events, and charts plotting restaurant activity—with each iteration of these fast-moving displays momentarily grabbing our attention, as we navigate from one to the next, never to look back. Today we take these kairotic moments for granted as we point, click, and scroll, active agents of designed interactions—both free and constrained digitally, in-the-moment and "in-the-middle."

Conclusion and Implications for Design

Design forms age and deteriorate, and they do so at varying velocities. Unlike high-profile designs in the fine arts, most practical designs fade away quietly with little fanfare. Often we don't even know that they've perished or realize that their residual forms still live among us—embedded in visual conventions we deploy every day, as obscure practices eking out a living, as reincarnations of designs long dormant, and sometimes explicitly as retro forms.

In the short term, the effects of aging are more immediate and rhetorically transparent. When we encounter an information design that's situationally relevant and readily accessible—a warning sign, stock chart, bus timetable, a set of wordless instructions—it fixates our attention for that kairotic moment so long as it remains useful, then quickly dissipates. We encounter these designs daily, however, with little awareness of time or aging, their temporal fragility escaping our purview. In this way, the short-term kairotic effects of aging parallel the long-term effects of chronos: To audiences that interact with these designs, the rhetorical *effects* of these temporal fluctuations are readily apparent, but the *causes* remain largely imperceptible. For information designers, however, understanding and externalizing the links between cause and effect can go a long way toward informing practice.

Staying the Chronos Course: Understanding and Deploying Conventions

As information designers we naturally pay most attention to design forms in close temporal proximity that are in our immediate purview—for example, noticing how a given genre behaves visually. If I'm designing a newsletter, a poster, or a set of instructions, I'm most likely to focus my attention on the well-worn paths leading right up to my door, which will tell me what my audience expects. I can then decide whether to follow or flout those practices. What about the more distant past—designs that thrived a decade ago, half a century ago, or longer? What role do they, or *can* they, play in practical design?

Earlier in this chapter I explored the long-term effects of time on information design, especially related to visual conventions. Time nurtures conventions like soil and water, but like most living entities, the time to maturation varies. Initial letters have evolved for centuries, web conventions for a mere few decades. However much time conventions need in order to develop and thrive, they are time capsules that embody sustained practices, and when we deploy them, the residual effects of time cling to them, along with the deep-seated interpretive habits of their audiences. In any given rhetorical situation, then, deploying conventions has two primary benefits:

- They meet audience expectations. Conventions simplify an audience's interpretive work and make them feel confident and comfortable. However, designers need to pay attention to details because audiences can easily spot design elements that breach conventional practices—hairlines that are too thick, spacing between text that's too tight or too loose, brand colors off by a shade, headlines and nameplates too tall or too short.
- They streamline the design process. Information designs don't have to be created from scratch, which can be time consuming, exhausting, and risky. Conventions provide pathways and guardrails to jump-start a design project, keep it on track, and maximize efficiency. So carefully studying existing conventions can be well worth the investment. Digital templates and AI-generated forms based on conventional practices can also streamline the design process; however, to be effective, they should only be a launchpad for invention (rather than a substitute) and carefully tailored to the rhetorical situation.

Introducing Novelty: Creating Tension Between the Old and the New

The flip side of conventions is novelty—the urge to flout precedent and stray from the well-worn path, something most information designers already do by bending and adapting conventions and improvising visually. However, forays into novelty that resist the past more aggressively, even subversively, create rhetorical tension between the old and the new. Despite the weighty ballast of visual conventions, novelty can infuse new visual language into design practice or even revolutionize it, laying the groundwork for new conventions. Novelty, however, also bears some risk. In order to engage their audiences, designers must minimize its interpretive demands by shortening the audience's learning curve—for example, by providing explanatory notes, legends, digital aids, and other interpretive scaffolding. Otherwise, novelty can fall flat as a fleeting "one-off."[11]

Seizing the Visual Moment: Designing for Kairos

Short-term factors also shape practice, typically in response to external events—seasons of the year, holidays, newsworthy events—and to situations that affect an audience's openness to the message. Kairotic design entails visual thinking that's responsive to both, and like the mythological figure with his balance on a razor, it constantly keeps designers and their audiences on their toes. For example, a print invitation to an autumn open house for college alumni might feature colorful leaves (an external factor) and a script font personalizing the message (a situational factor). Designers can also keep their communications fresh and their audiences engaged by doing the following:

- Extend the design's timeliness beyond the initial reception. Although the print invitation will reach all alums at the same time, elements of the design can enhance its kairotic elasticity: for example, by including a campus parking map alums can use on the day of the event, along with a QR code for donating to a scholarship fund. In these ways, the audience can have multiple timely interactions with the invitation.
- Keep audiences immediately engaged with digital hyperkairos. Posting the invitation on social media with interactive links to an archive of college pictures, along with a space for alums to post their own pictures, would give alums more agency—placing them in the moment and "in-the-middle." Creating an online chart with dynamic data about the number of open house RSVPs and scholarship donations would give the message both hyperkairos and kairotic elasticity.

These are just some of the ways that chronos and kairos can be applied to information designs to enhance their rhetorical effectiveness. We apply chronos in slow motion over long stretches of time through visual conventions, and we apply kairos to the here and now, nimbly pushing temporality into high gear. In these ways, considering the rhetorical effects of time can make us more thoughtful and deliberate designers, more conscious of the visual choices we make and how our audiences will interpret them.

Chapter 2

Epideictic Design

The Visual Rhetorics of Praise and Blame

Images appear all around us that celebrate achievements, memorialize milestones, and mark special occasions. These forms of visual language enrich our collective lives by calling our attention to exemplary people, places, and objects and by certifying the values that underpin their praiseworthy qualities. Other images occasionally do the opposite by warning us of the dangers or ills of certain actions and their displeasing or painful consequences. All of these kinds of rhetorical acts have traditionally been classified as epideictic: that is, they confer praise or blame on people and things.

In oral communication, epideictic rhetoric might take the form of a commencement address, a eulogy, or an awards convocation; in written communication, an obituary, employee evaluation, or letter of recommendation; in visual communication, the design of a plaque, diploma, or invitation, and a host of other ways that epideictic acts are conveyed in text, pictures, data displays, and material objects, including monuments and other forms of public display (Prelli 2006). Although we know a good deal about oral and written forms of epideictic expression, analysis of the visual epideictic in technical, professional, and other practical forms of communication has been sparse. In this chapter I explore this relatively uncharted visual mode.

To do so, I will trace concepts of epideictic rhetoric from classical to modern and describe the visual strategies that have been deployed to express it, for both praise and blame, in both the past and the present. I will also examine the socializing effects of the visual epideictic in promulgating the communal values of a given era. Finally, I will show how designers interweave the epideictic with other rhetorical strategies to achieve their goals, further illustrating its pervasive presence in our professional and personal lives.

Epideictic Displays: Amplifying Praise in Words and Images

The concept of the epideictic originated in the classical world where rhetoricians regarded it as a requisite and prevalent form of oratory. Gorgias practiced epideictic oration to display his rhetorical dexterity (Consigny 1992), and later Aristotle (2007) classified the epideictic as one of the three foundational modes: demonstrative (or epideictic), forensic, and deliberative rhetoric. Demonstrative rhetoric traffics in praise or blame; forensic rhetoric is based on logical argument and factual evidence and useful in courts of law (hence it is also called "judicial" rhetoric); and deliberative rhetoric entails political arguments about public policy in the future (Aristotle, 47–50; bk. 1, ch. 3, secs. 1–4). The Roman rhetorician Quintilian (1959–1963) also identified epideictic rhetoric as one of the three seminal modes, but he didn't separate it completely from the other two, claiming that praise (or blame) can also play a role in forensic or deliberative rhetoric (1:395–97; bk. 3, ch. 4, secs. 12–16). As a form of oratory, epideictic rhetoric could be practiced on many different occasions—at funerals, festivals, weddings, civic and religious events, and private gatherings. According to Aristotle, the praise of individuals should concentrate primarily on their "deeds" (encomium): what individuals accomplished, exemplary attributes of their actions, and perhaps how those actions compared to those of other praiseworthy people (80–82; bk. 1, ch. 9, secs. 33–39).

Early on, epideictic rhetoric became associated with "display" and "performance," whereby speakers could embellish their orations, though later this became less of a distinguishing characteristic (Chase 1961, 298). The key technique for enacting the epideictic that Aristotle describes entails the rhetor's capacity to achieve "amplification" (81–83; bk. 1, ch. 9, secs. 38–40). This ability to magnify praise (or blame) heightens the accomplishments (or deficiencies) of the subject and creates a powerful picture for the audience. Amplification can be achieved in a variety of ways: through comparatives and superlatives, through vivid description (enargeia), and through repetition and narration (Aristotle, 81–82; bk. 1, ch. 9, secs. 38–39; Jasinski 2001, "Amplification," 12–13). Many of these techniques appeal to the audience's emotions. For example, with stylistic effects like description—a "bringing-before-the-eyes" as Aristotle put it (219; bk. 3, ch. 10, sec. 6)—rhetors paint a vivid scene, experience, or person. According to Enlightenment rhetorician George Campbell (1776), the vividness of description heightens its emotional effect, for "whatever tends to subject the thing spoken of to the notice of our senses, especially of our eyes, greatly enlivens the expression" (Campbell 2:172; bk. 3, ch. 1, sec. 1). In the modern world, these and other stylistic strategies continue to amplify praise, and they do so not only orally but also visually, as we'll see with examples of text, picture, and data design from past and present.

Amplifying Text: From Pen to Print to Screen

Rhetoricians have traditionally emphasized the performative aspects of epideictic, the ability of an orator to dazzle an audience, to put on a good show. The agility and skill of the rhetor mattered a great deal to the success of these rhetorical acts, which were intended to heighten the glory and reputation of a person, place, or thing (see Crowley and Hawhee 1999, 338–44). To project this forceful level of praise, rhetors had to display a high degree of skill at amplification; audiences expected to be regaled with oratorical virtuosity, especially given that they probably already knew some of the content of the speech if they had any prior knowledge of the person or thing being praised.

How are these skillful celebrations displayed visually? In the classical world, the visible epideictic was embodied in the canon of delivery, in the kinesics that accompanied all forms of oratory through the language of gestures, facial expressions, and physical movements. However, the visual rhetoric of the epideictic also appeared in the ancient world in much more tangible and permanent forms: stone temples that praised gods and goddesses, statues and reliefs of powerful leaders, and monuments with inscriptions celebrating battlefield victories and other notable achievements. In these and other ways, image and text were literally etched in stone in public demonstrations of praise. In the modern world epideictic displays have also materialized in highly visible structures—the Arc de Triomphe, Nelson's Monument in Trafalgar Square, the Washington Monument—as well as in statues, markers, and memorials in public parks, along highways, and on college campuses, with many of these artifacts having long and sometimes complex and contentious histories.

The visual epideictic also has a ubiquitous, but more quotidian, presence in practical and professional communication in the form of handwritten and printed text. Before the invention of modern technology (typewriters, laser printers, digital interfaces), most practical documents were written by hand with pen and ink. Handwritten documents that served various practical purposes—business, personal, civil, and legal—typically displayed the visual dexterity of the writer, especially through "command of hand," the ability to write in a fluent and graceful style consistent with the conventions of the time (see Heal 1962; Fairbank 1968; Kostelnick 1994). Proficient writers engaged in a practice called "flourishing" by creating decorative elements on letters or ornamental designs alongside the text. Several signers of the Declaration of Independence (most notably John Hancock) used flourishing to embellish their signatures, expressing their distinct visual identities as they celebrated this historic proclamation.

Instruction in the act of flourishing began at least as early as the late sixteenth century and continued for hundreds of years, right up to the early twentieth century when typewriting largely supplanted handwritten text. The penwork of seventeenth-century "writing master" Edward Cocker exemplified flourishing at the highest level, which he displayed in numerous copybooks (Heal 1962, 34), including *The Pens Transcendency* (1660), *The Pen's Triumph* (1659), and *Arts Glory* (1685). Figure 2.1 from *Arts Glory*

Figure 2.1. Seventeenth-century illustration of writing and flourishing by Edward Cocker. *Arts Glory, or, the Pen-Mans Treasury*, The Rare Book & Manuscript Library, University of Illinois at Urbana-Champaign. Courtesy of HathiTrust.

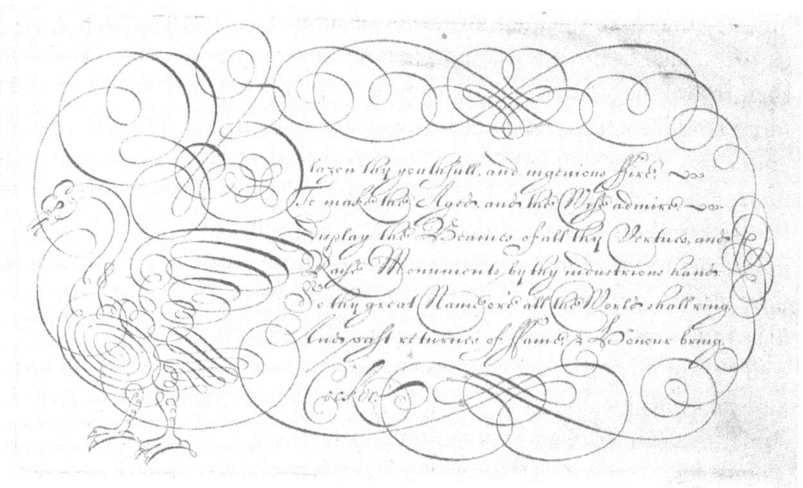

illustrates the "secretary" hand, which at the time was the conventional handwriting style for business and everyday correspondence. Cocker's dexterity with a pen, his "command of hand," undoubtedly exceeded that of his audience of pupils: the swirling strokes of his pen create complex knots and flourishes that frame the text, along with a playful bird above Cocker's signature—all engraved with copperplate technology that enabled fine strokes to be reproduced in print. Through his visual artistry and ingenuity, Cocker greatly amplifies the text, a moral commentary on the fulfillment and happiness that virtue brings.

The practice of flourishing proliferated throughout Europe in the eighteenth century, exemplified by the delicate, swirling penwork in figure 1.1 from the French *Encyclopédie.* This technique for expressing the visual epideictic also applied to professional communications on an individual level, shown in the document in figure 2.2 announcing a meeting with a German "Capitain" from Hamburg named Ludewig Martin Westermann (1768). The penwork here is typical for the later eighteenth century, with

Figure 2.2. Eighteenth-century meeting announcement for a German military officer from Hamburg (Westermann 1768). Courtesy of the Library of Congress, Prints and Photographs Division [LC-DIG-pga-11903].

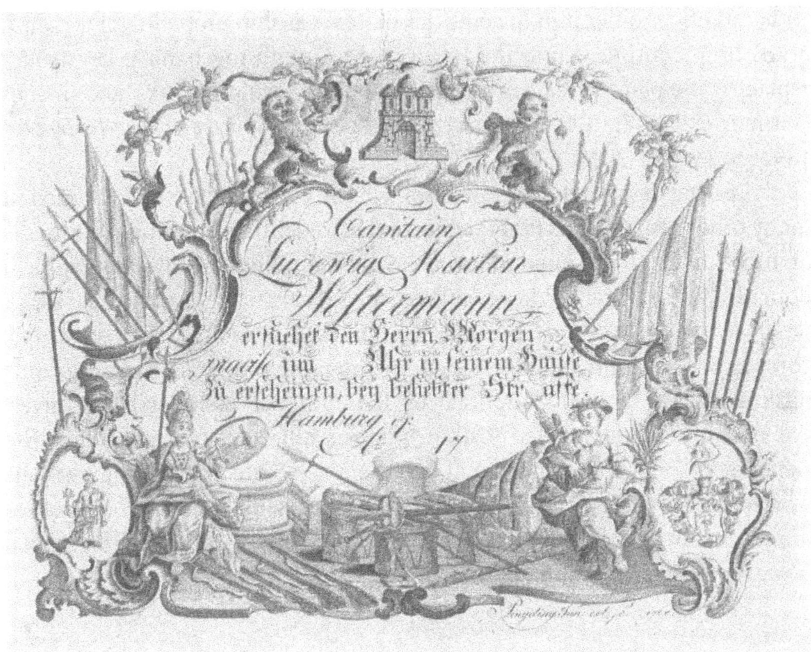

the capital letters of his name executed in a round, flourished style, with thick and thin strokes throughout his name, and an even more elaborately flourished German Gothic text immediately below. The text is further amplified with lavish graphical framing that lauds the credentials of the captain: images of lions with human heads and a castle keep with a grated entry; seated goddesses bearing military arms; an array of swords, spears, and other weapons; and profuse natural and architectural elements garnishing the cornucopia of graphical and pictorial forms. This highly ornamented framing, combined with the textual flourishing, amplifies the captain's stature and experience: He's not someone to be trifled with, and if you do business with him, you better come prepared.

The textual and graphic epideictic continued to thrive through the end of the nineteenth century in both Europe and North America as a means of celebrating and promoting the self. Figure 2.3 (Peale 1886), for example, shows handwriting techniques for creating professional and personal cards and albums, heralding each individual represented in the designs. The vegetation, birds, and abstract linework in this writing sample reflect the Victorian propensity for natural forms, presaging the aesthetics of Art Nouveau in the decades ahead. Like a chorus of trumpets, these visual amplifications heighten the rhetorical presence of each person they invoke along with the accompanying message. With all of the rhetorical and aesthetic momentum of centuries of penmanship propelling it forward, flourishing continued into the typewriter age with mechanical keystrokes replacing the pen to create ornate page borders, section breaks, and images (Pitman 1893, 37; O. R. Palmer 1893, 142–43, 196; *Typewriting Instructor* 1892; S. Walker 2018, 145).

Textual and graphical expressions of the epideictic have blossomed in many other visual forms that recognize the credentials or accomplishments of individuals (their encomium): résumés amplified with typographical flourishing (font size, boldface, bullets), award certificates with ornate script typefaces and decorative borders, and LinkedIn profiles with portraits, company logos, and digital badges. In interactive digital spaces, accomplishments are typically amplified by the audience members themselves. For example, social media posts in plain vanilla text that announce the rhetor's personal and professional deeds are celebrated by audience emojis (thumbs ups, hearts, smiley faces), animations with fireworks, and other digital forms of amplification that collectively endorse the achievements, however meager or momentous.

Figure 2.3. Examples of Victorian era flourishing from *The Home Library of Useful Knowledge* (Peale 1886, 145).

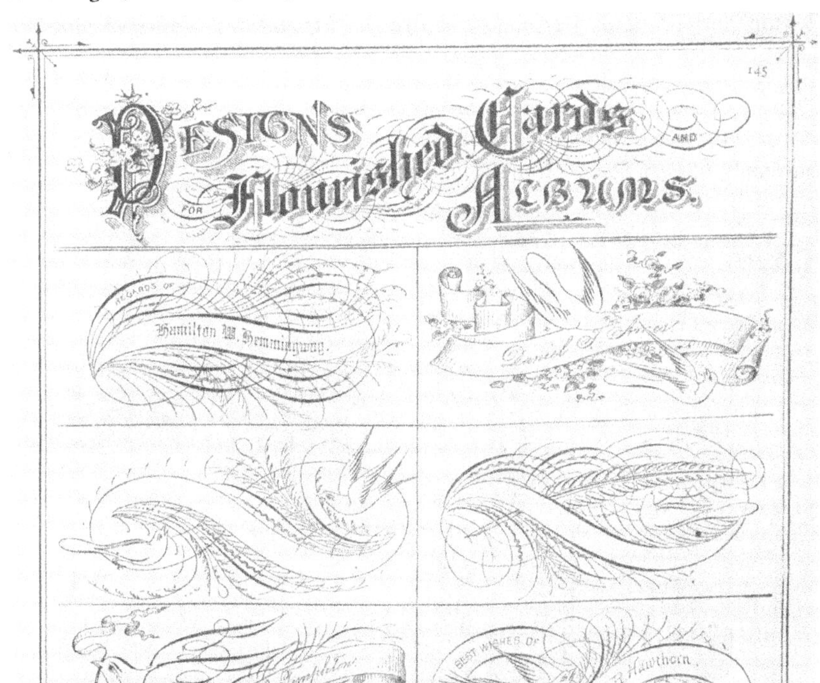

Visualizing Grand Entries: Frontispieces and Covers

The textual epideictic extends to other highly visual forms of self-promotion, most notably title pages and frontispieces, a genre that has flourished since the earliest days of printing—not only for belletristic books but also ones on engineering, architecture, science (Gigante 2018, 15–31), and military fortification (Pollak 1991). Figure 2.4 shows the frontispiece for Jacques Besson and François Béroald's *Theatre des Instrumens Mathematiques et*

Figure 2.4. Elaborate frontispiece for Jacques Besson and François Béroald's *Theatre des Instrumens Mathematiques et Mechaniques* (1578). Courtesy of Iowa State University Library Special Collections and University Archives.

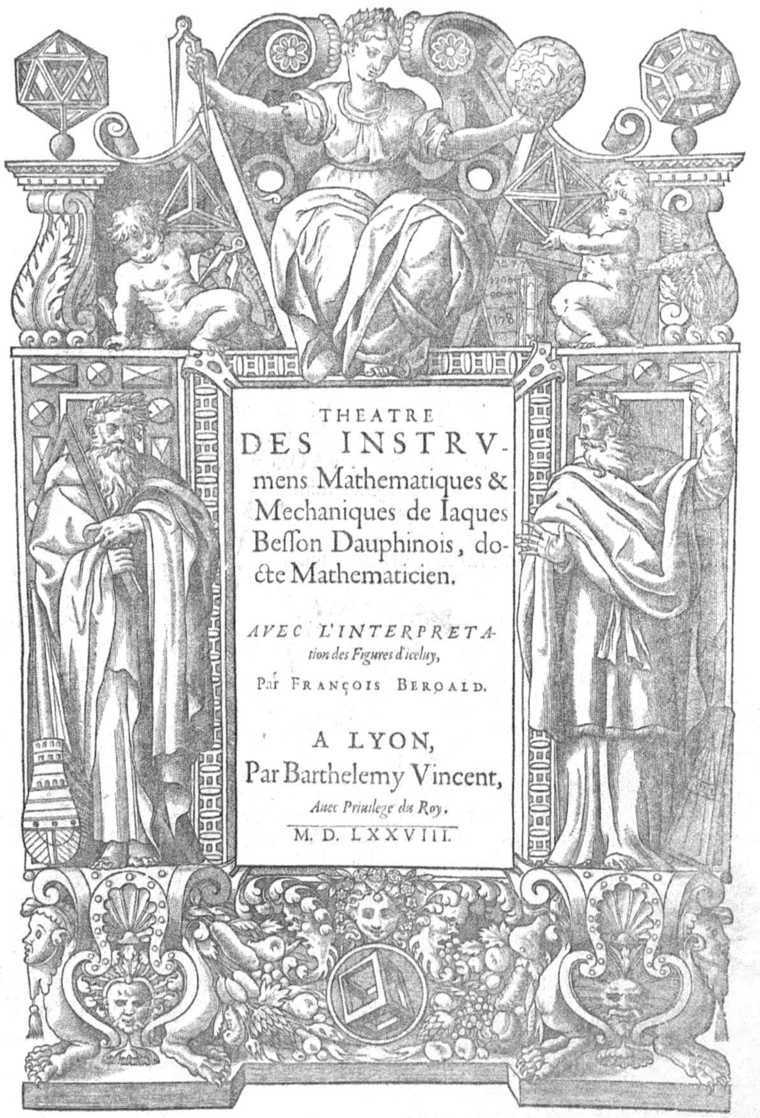

THEATRE
DES INSTRV-
mens Mathematiques &
Mechaniques de Iaques
Beſſon Dauphinois, do-
cte Mathematicien.

AVEC L'INTERPRETA-
tion des Figures d'iceluy,

Par François Beroald.

A LYON,
Par Barthelemy Vincent,
Auec Priuilege du Roy.
M. D. LXXVIII.

Mechaniques (1578) that explains in text and images numerous mechanical devices for moving objects, cutting wood, grinding grain, pumping water, and doing a variety of other practical tasks. Although the text in the center of the frontispiece announcing the title, author, illustrator, and publisher is the focal point, the graphical and pictorial elements surrounding it supply the epideictic energy. The goddess seated at the top, flanked by cherubs on either side and ancient figures below, some holding mathematical and geometrical tools, with a cornucopia of fruit, vegetation, and mythical creatures underneath—all of these extol and celebrate the book's extensive gallery of mechanical devices. This glut of celebratory images aimed to convince readers that they were entering a technological domain worthy of high praise (which it was). For centuries, practical books introduced themselves to their audiences with these kinds of epideictic displays that still have the power to enthrall and captivate us but whose visual amplifications now seem rather hyperbolic and self-serving.

Today practical books, pamphlets, and annual reports continue to announce themselves with amplified visual language—photos, icons, logos, mixed with textual and graphical elements—but do so more modestly, partly because aesthetic values changed radically with the emergence of modernist design principles, which engendered a new set of audience expectations. As a result, front matter for today's practical communications—covers, title pages, home pages—derive their amplification from lean, dynamic minimalism, a trademark of the modernist program. As practical communications open their front doors, then, their visual language still exudes epideictic energy, but they make their first visual impressions with more modulated flourishes beckoning us to enter.

Hyping the Hierarchy: Royals and Bosses

Textual amplification also materializes in tree diagrams that socialize the visual epideictic on a larger collective scale by displaying networks of praiseworthy people, typically in hierarchical patterns. Figure 2.5, for example, shows the French royal lineage as an elaborate family tree, with the heraldry of each family meticulously drawn for each branch and sprig of the royal line, right up to King Louis XV's son, the Dauphin de France and royal heir apparent (who never took the throne). This form of extreme amplification is created textually by arranging people in a hierarchy according to their ranks and graphically by visualizing each individual's heraldry (sized by importance)—all of which collectively celebrate the prestige and glory of the Bourbon line. Lineage played an important role

Figure 2.5. Family tree of the French monarchy illustrated in the *Recueil de Planches* (Diderot and d'Alembert 1762–1772, vol. 2, "Blason ou Art Héraldique," plate XXI) that accompanied the *Encyclopédie*. Courtesy of Iowa State University Library Special Collections and University Archives.

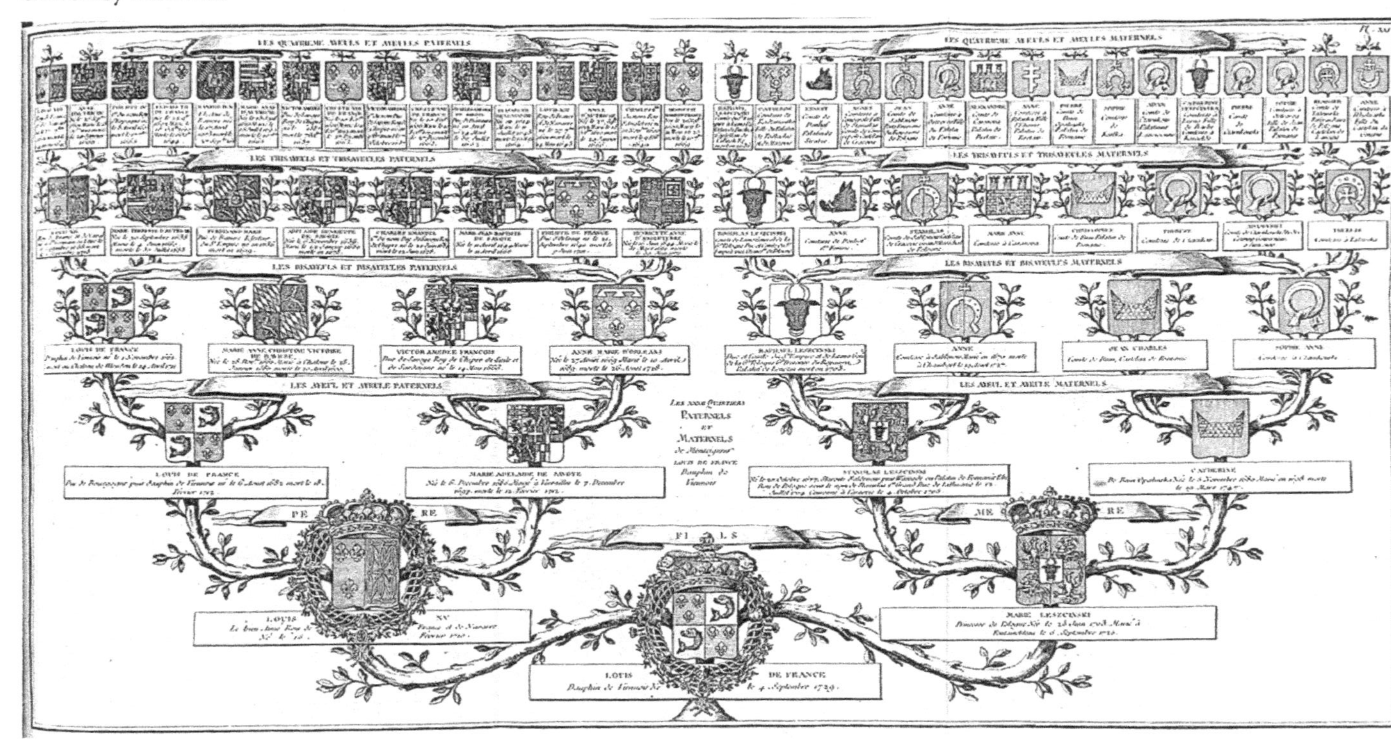

in epideictic rhetoric (Quintilian 1959–1963, 1:469; bk. 3, ch. 7, sec. 10; Pernot 2015, 95; Crowley and Hawhee 1999, 340) as a way to bolster the credentials of the person praised, and this display is a genealogical tour de force.[1] Although with less grandiose amplification, most family trees also serve this epideictic purpose, celebrating (or at least memorializing) matriarchs and patriarchs across generations, interspersed with a few notable individuals and a random rascal languishing in perpetual infamy.

Another highly socialized tree diagram, organizational charts embody the textual epideictic by amplifying leadership positions through visual arrangement. By convention, the most powerful positions appear in the top center of the chart and are often emphasized with large boldface type. These amplifications give the president, chief executive officer, or director a commanding visual presence, which implies certain abilities and achievements, while subordinates stationed in their hierarchical positions below receive lesser gradations of praise. In organizational charts with interactive features, as we saw in the National Weather Service chart in figure 1.21 (U.S. Department of Commerce 2025), that amplification becomes more explicit with mouse-overs and pop-ups that tout the duties of individual leaders as well as their past accomplishments (see also the 2008 interactive organizational chart for the U.S. Department of the Interior). However, unlike the lethargic flow of family trees, the temporal spans of organizational charts are relatively short, as people and positions continually rotate, and the epideictic energy is regenerated from one leadership team to another.

Picturing the Epideictic: In Praise of Places

In the modern world, then, epideictic rhetoric manifests itself in several forms of professional and personal communications through the visual amplification of textual and graphical elements. Amplification takes many other forms, including maps and pictures that represent places, buildings, and people.[2] All of these appear in figure 2.6, which shows a drawing of the city of Maastricht (in southern Netherlands) from the *Civitates Orbis Terrarum* (Braun, Hogenberg, et al. 1612–1618), an early seventeenth-century collection of drawings of cities across Europe and the Middle East. In this two-page spread, Maastricht's key buildings appear on a skyline in minute detail, with notations for each of them in the lower right and an inscription in the lower left touting the city's bridge, its fortifications, and distinguished people. These textual elements are framed with elaborate architectural elements, further amplifying the city's prestige and standing.

Figures appearing in the foreground celebrate the dress of the locals, reinforcing the city's identity as well as rendering the map feckless for potential Turkish invaders, for whom reading images with human forms was prohibited by their religion (Braun, Hogenberg, et al.; cited in Alpers 1983, 133, and in Barton and Barton 1993, "Ideology," 54–55). Showcasing the prosperity and status of prominent cities, the elaborate drawings from the *Civitates Orbis Terrarum* exemplified the pictorial epideictic. Descendants of these drawings can be found in modern bird's-eye views of towns and cities, college campuses, manufacturing plants, and amusement parks, all of which use aerial perspective to amplify their grandeur and status.

In the nineteenth century, the panoramic view became a staple of the visual epideictic, especially for commercial purposes. Businesses of all kinds heightened their reputations by picturing the spaces they inhabited: buildings, campuses, factories, hotels, and the like. Figure 2.7, for example, shows an epideictic bird's-eye view of a match factory in Portland,

Figure 2.6. City view of Maastricht from the *Civitates Orbis Terrarum* (Braun, Hogenberg et al. 1612–1618, vol. 2, fig. 21). Courtesy of the Library of Congress, Geography and Map Division [2008627031].

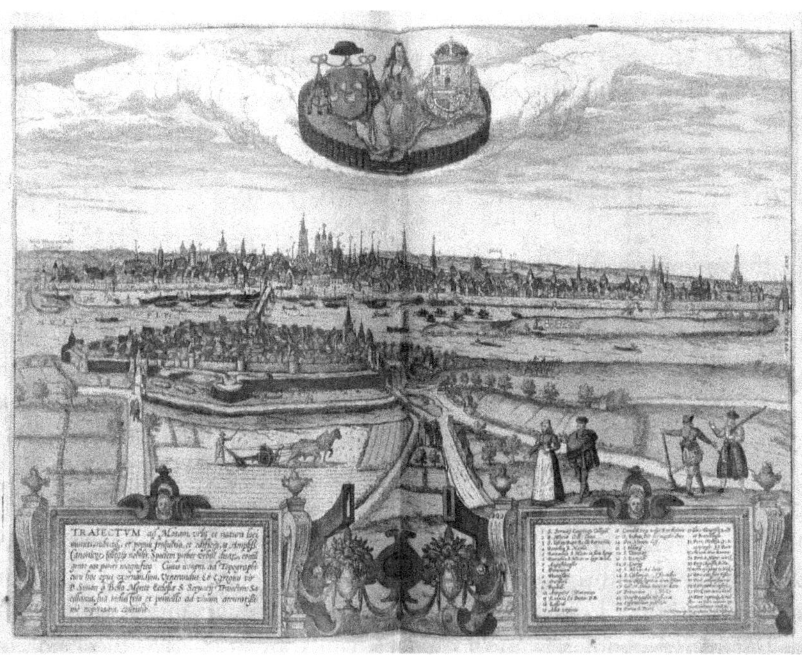

Figure 2.7. Bird's-eye view of the Portland Star Match Factory in Portland, Maine (G. Walker ca. 1860–80). Courtesy of the Library of Congress, Prints and Photographs Division, Popular Graphic Arts Collection [LC-DIG-pga-04222].

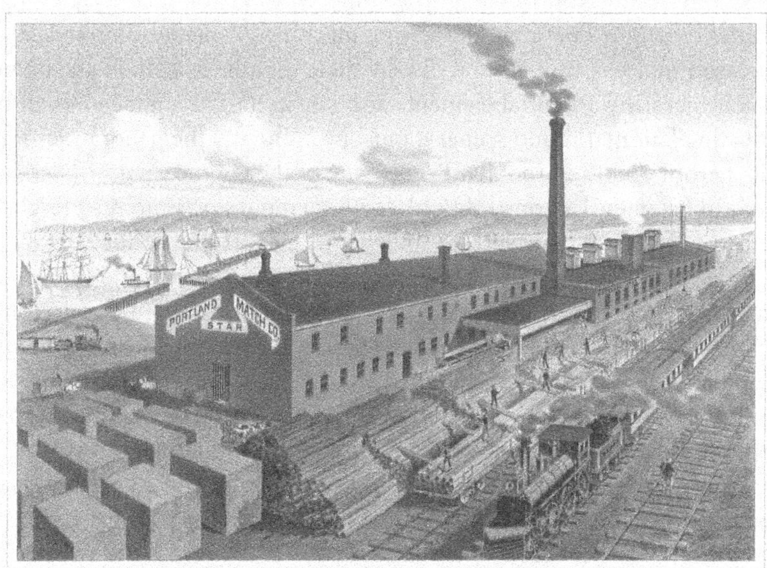

Maine, in the late nineteenth century. The aerial perspective emphasizes the factory's size and its proximity to the ocean and the railroad, a strategic location for shipping goods and procuring logs, which workers unload from railroad cars and saw up alongside the building. The bird's-eye view shows the entire expanse of the enterprise, with the towering chimney creating a distant focal point. Altogether, the picture projects an idealized view devoid of the clutter and messiness inherent in such operations, creating an epideictic statement about the company's efficiency, prosperity, and bright future. Many manufacturing companies during this era used this kind of pictorial epideictic to establish their identities, tout their success, and foster confidence in customers and investors—as well as send a message to their competitors.

Businesses also promoted themselves as images of success by associating themselves with other praiseworthy things or events. This kind of epideictic amplification is projected typographically and pictorially in figure 2.8, a promotional piece for the Empire Sewing Machine Company (1870), whose main office was located in New York City. An ornate title appears at the top in curved headline type, surrounded by pictorial vignettes of people operating the sewing machines, visual testimonials that showcase Empire's products and display their usefulness both in the home and commercially. Pictorial elements also situate Empire's machines amid captivating landmarks and scenery: most prominently the panoramic view of the harbor and the Brooklyn Bridge, one of the engineering wonders of the world (then under construction), further amplifying Empire's presence on nearby Bowery Street and its ability to transport its products by land and water. Immediately below the harbor scene, a stagecoach and train race across the landscape, showing the nation's westward growth and touting Empire's ability to deliver its machines across vast expanses. The confident, triumphant tone is further amplified by the lush, leafy embellishments framing all of these scenes. Like a rousing John Philip Sousa march, this elaborate design fills its space with self-assured verve and energy.

Some epideictic images that create strong business identities coalesce into conventional practices that recur for long stretches of time. For well

Figure 2.8. Promotional piece for the Empire Sewing Machine Company (1870). Courtesy of the Library of Congress, Prints and Photographs Division, Popular Graphic Arts Collection [LC-DIG-pga-02736].

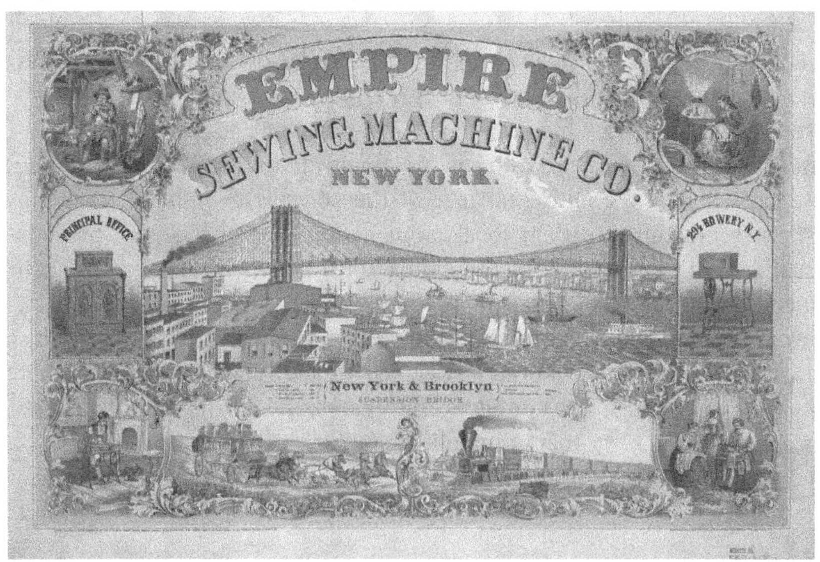

over a century, hotel pictures have appeared on various promotional and quotidian documents, primarily as letterheads on stationery, receipts, and invoices, but also on postcards, coasters, cups, napkins, and other artifacts that projected a cohesive brand centered around an impressive façade. In the late nineteenth and early twentieth centuries, these images created a public presence for hotel operations across North America and Europe, providing a unique visual identity for a given hotel and a rhetorical credential in the travel industry. Each hotel had its own signature façade and design, which attracted guests and differentiated it from its competitors. Guest rooms were stocked with stationery bearing the hotel image and creating a convenient means for visitors to correspond with family, friends, and business associates. In these ways, the hotel façade served a double epideictic function: for a hotel business to glorify and promote its image and for visitors to boast about their lodgings.

Hotel stationery with its prominent letterhead image also provided a medium for recognizing the lofty status of its visitors: presidents, athletes, entertainers, and other noteworthy guests.[3] Figure 2.9, for example, shows a letter from Alexander Graham Bell to his wife, Mabel, while he was staying at the Athearn Hotel in Oshkosh, Wisconsin, in 1896. The elevated view of the hotel's masonry façade, with a grand tower anchoring its corner, amplifies its stature, strength, and elegance. The detailing on the façade, a form of enargeia, further amplifies this massive object, and along with the letterhead's nameplate and graphic embellishments, magnifies and celebrates lodgings within which luminaries like Alexander Graham Bell wrote home to describe his travels in Wisconsin, closing his letter with a promise to write again that night, presumably on the same epideictic stationery.

Hotel images appeared on stationery (and other artifacts) throughout the twentieth century, especially during the first half—the golden age of independent hotels. Today hotels are often embedded in large chains that eschew their local identity in favor of corporate branding. As a result, the visualization of *exterior* façades, which had long supplied a resource for epideictic amplification, has lost some of its appeal, partly because many of those façades now mimic homogenous corporate designs and forego a unique architectural identity. Nonetheless, envisioning hotels as epideictic space continues today, with online images focusing on the *interior*: guest rooms, breakfast areas, pools, exercise facilities, and other amenities. In persuading online audiences, it's the *inside* that matters, that's worth boasting about, however much it lacks the grandiose amplification of the panoramic façade.

Figure 2.9. Letter written from the Athearn Hotel in Oshkosh, Wisconsin, by Alexander Graham Bell (1896) to his wife, Mabel. Courtesy of the Library of Congress, Manuscript Division, Alexander Graham Bell Family Papers [03910104].

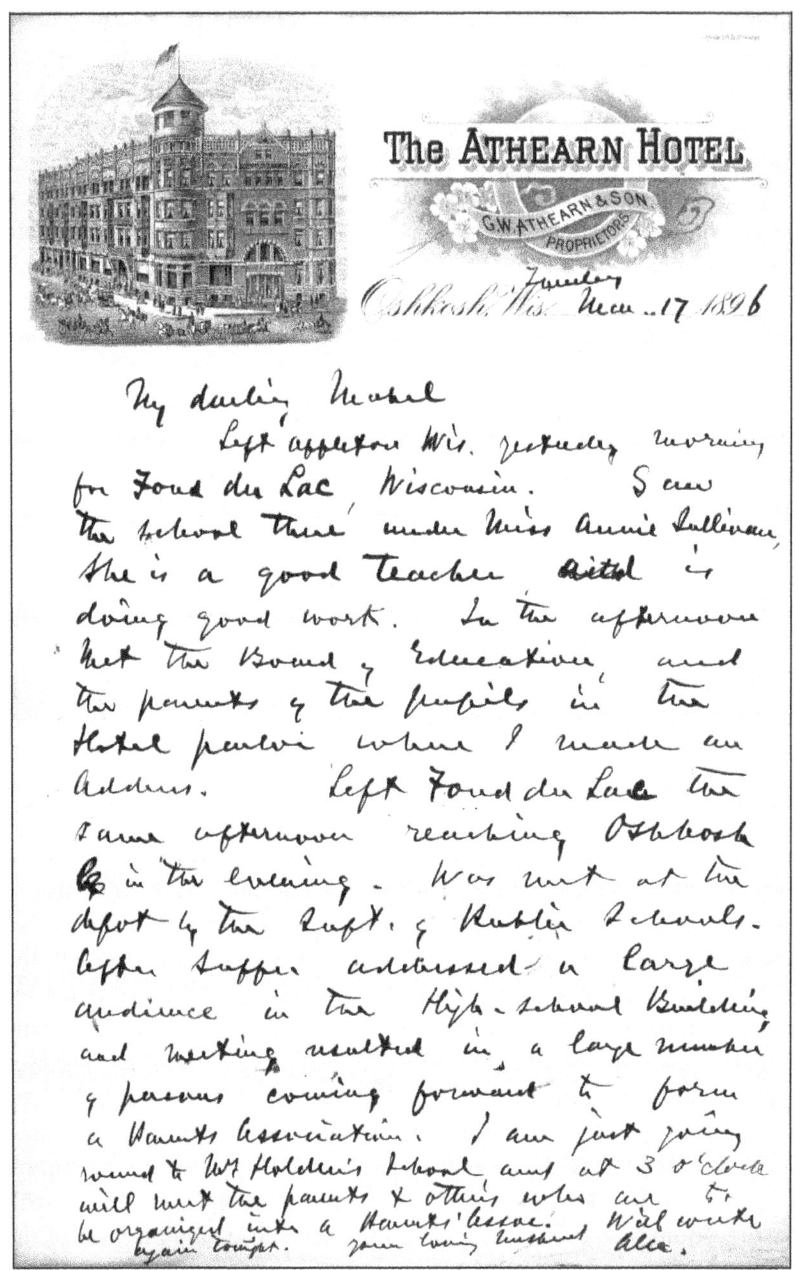

The ATHEARN HOTEL

G.W. ATHEARN & SON. PROPRIETORS.

Oshkosh, Wis. Sunday Mar. 17 1896

My darling Mabel

Left Appleton Wis. yesterday morning for Fond du Lac, Wisconsin. Saw the school there under Miss Annie Sullivan, She is a good teacher and is doing good work. In the afternoon met the Board of Education and the parents of the pupils in the Hotel parlor where I made an Address. Left Fond du Lac the same afternoon reaching Oshkosh in the evening. Was met at the depot by the Supt. of Public Schools. After supper addressed a large audience in the High-school Building and meeting resulted in a large number of persons coming forward to form a Parents Association. I am just going round to Mr Holden's school and at 3 o'clock will meet the parents & others who are to be organized into a Parents' Assoc. Will write again tonight.

Your loving husband Alec.

Drawing the Epideictic: In Praise of Things

Epideictic amplification is also projected pictorially to glorify functional objects like machines and consumer goods. Virtually every day we encounter pictures of cars, kitchen appliances, and exercise equipment extolled visually by their sales agents and manufacturers. Even though entirely functional, workplace machines can also be amplified pictorially for epideictic ends. As we saw earlier, drawings of existing, cutting-edge, and often aspirational machines lavishly graced early engineering books (figs. 1.12 and 1.13), including Jacques Besson and François Béroald's 1578 *Theatre des Instrumens Mathematiques et Mechaniques* (fig. 2.4). These extensive pictorial galleries showed machines in elaborate detail, in action performing work, and in doing so celebrated their efficiency and ingenious designs. Similarly, in the new industrial age in the nineteenth century, machines were glorified pictorially for their functionality, efficiency, and beauty. The picture of the planing machine in figure 1.15 amplifies the beauty and utilitarian virtues of this industrial machine through meticulous description. The designer delineates its many parts (gears, belts, bolts) using fine linework to create shades and shadows with near photographic realism—vivid description (enargeia) that amplifies the object as well as gives it emotional appeal. Visualized as a solitary object, the machine stands all on its own, a beautiful, useful assemblage worthy of celebration, both for the company that makes it and the owner that operates it.

This elegant machine drawing—and legions of others of like it from this era—stood on the cusp of a revolution in photography, paralleling the transition from handwritten to typewritten text, where fluid and flexible design gave way to mechanical methods. These changes in technology initially reduced the amplification tools of designers, whereby the pen and brush were supplanted by the lens of the camera, which mechanically transcribed what people saw in real life, though this process was hardly as transparent as it seemed. Today digital technology—with its capacity for detailing, color, and animation—affords the means to restore the design flexibility of drawings and their power to amplify their subjects.

Amplifying and Reifying the Moment: Ceremonies and Rituals

As a formal speech act, epideictic rhetoric traditionally marked events, ceremonies, or rituals to celebrate or mourn individuals based on their achievements or memories of their past lives. Visual language plays an important role in these events, including eulogies at funerals, which are a particularly poignant forum for amplifying praise. After Abraham Lincoln was assassinated, many memorial documents marked the occasion, including schedules announcing the railway route as his body was transported from Washington to his home in Springfield, Illinois. On its journey, the train made numerous stops, their times and places posted on a special schedule like the one in figure 2.10, which enabled local residents to gather along the route to pay their respects. The schedule in figure 2.10, which shows the last leg of the journey from Chicago, has several design elements that amplify its visual language: the centered text in all caps identifying the towns along the route, the lists on the left displaying the precise mileage, the notes explaining the protocols, and the dark mourning border framing the page, which collectively create a serious and somber tone. This is not just *any* schedule: It's a visual eulogy, a dirge, and a commemoration—and like the person it memorializes, one for the ages. Today, eulogies are made visible with a variety of artifacts: programs for funerals and memorials, photo collages at visitations, religious cards with inspirational inscriptions, and other printed tokens that accompany these rituals. The visual eulogy also extends to digital spaces, with testimonials, pictures, and emojis posted on social media.

Today most rituals that mark personal and professional milestones rely on visual language to communicate their proceedings: certificates, plaques, and photos for award ceremonies; invitations to weddings, groundbreakings, and open houses; posters and programs for convocations, concerts, and other performances; and ribbons, medals, and trophies for competitions. The visual language invoked at recurring social activities like graduations coalesces in several genres: a diploma, an invitation, a program listing speakers and graduates, an entry ticket, a photo with the university president—all of which also provide mementos to recollect the event. The designs associated with ceremonies and rituals use various forms of visual amplification: elaborate borders, centered text, traditional typefaces, plush colors (gold and silver), and the size and shape of the document. For example, figure 2.11 shows a document I created that contains several conventional elements of a university diploma: centered text in an ornate script typeface, signatures of high-ranking administrators, and the university's official seal—all on a landscape page with a thick border. Students receive documents like this one (or a cover as a stand-in) while actively participating in the graduation ritual: walking across a stage wearing gowns and betasseled mortarboards, shaking the

Figure 2.10. Railroad schedule of Abraham Lincoln's funeral procession from Chicago to Springfield, Illinois (Hale 1865). Courtesy of the Library of Congress, Rare Book and Special Collections Division, The Alfred Whital Stern Collection of Lincolniana [scsm000258].

Chicago and Alton Railroad Company.

TIME TABLE

FOR THE SPECIAL TRAIN, CONVEYING THE FUNERAL CORTEGE WITH THE REMAINS OF THE LATE

PRESIDENT

FROM

CHICAGO TO SPRINGFIELD,

Tuesday, May 2, 1865.

Total Distance.	Dist. betw'n Stations.				
		CHICAGO	Leave	9:30	P. M.
1.7	1.7	FORT WAYNE JUNCTION	"	9:45	"
3.5	1.8	BRIDGEPORT	"	9:55	"
12.0	8.5	SUMMIT	"	10:21	"
17.6	5.0	JOY'S	"	10:34	"
25.5	8.0	LEMONT	"	10:58	"
32.5	7.0	LOCKPORT	"	11:18	"
37.7	5.2	JOLIET	"	11:33	"
46.4	8.7	ELWOOD	"	11:58	"
48.6	2.8	HAMPTON	"	12:04	A. M.
53.0	4.5	WILMINGTON	"	12:16	"
58.0	4.8	STEWART'S GROVE	"	12:30	"
61.4	3.5	BRACEVILLE	"	12:40	"
65.0	3.8	GARDNER	"	12:51	"
74.0	9.0	DWIGHT	"	1:16	"
82.0	8.0	ODELL	"	1:38	"
87.4	5.2	CAYUGA	"	1:53	"
92.3	5.0	PONTIAC	"	2:07	"
97.8	5.6	OCOYA	"	2:22	"
102.6	4.7	CHENOA	"	2:35	"
110.6	8.0	LEXINGTON	"	2:58	"
118.5	7.9	TOWANDA	"	3:20	"
124.0	5.7	ILL. CENTRAL R. R. JUNCTION	"	3:36	"
126.0	2.0	BLOOMINGTON	"	3:42	"
133.0	6.8	SHIRLEY	"	4:05	"
136.5	3.6	FUNK'S GROVE	"	4:15	"
141.4	4.8	McLEAN	"	4:28	"
146.0	4.8	ATLANTA	"	4:42	"
150.0	4.0	LAWN DALE	"	4:58	"
156.8	6.7	LINCOLN	"	5:12	"
164.0	7.1	BROADWELL	"	5:32	"
167.6	3.7	ELKHART	"	5:48	"
173.5	5.9	WILLIAMSVILLE	"	5:58	"
178.3	4.8	SHERMAN	"	6:12	"
180.0	2.1	SANGAMON	"	6:18	"
185.0	5.0	SPRINGFIELD	Arrive	6:30	"

The following instructions are to be observed for the above train:

1. All other Trains on this Road must be kept thirty minutes out of the way of the time of this Train.
2. All Telegraph Stations must be kept open during the passage of this Train.
3. A Guard with one red and one white light will be stationed at all road crossings by night; and with a white flag draped by day, or after day-light, on Wednesday morning.
4. A Pilot Engine will run upon this time, which is to be followed by the Funeral Train, ten minutes behind.
5. Pilot Engine must not pass any Telegraph Station, unless a white flag by day, or one red and one white light by night, shall be exhibited, which will signify that the Funeral Train has passed the nearest Telegraph Station. In the absence of said signals, the Pilot Engine will stop until definite information is received in regard to the Funeral Train.
6. The Funeral Train will pass all Stations slowly, at which time the bell of the Locomotive must be tolled.

By order of Brevet Brigadier General D. C. McCullum, 2d Div., in charge of Military Railroads.

ROBERT HALE,
General Superintendent.

Figure 2.11. Design conventions of a college diploma that enhance its epideictic purpose.

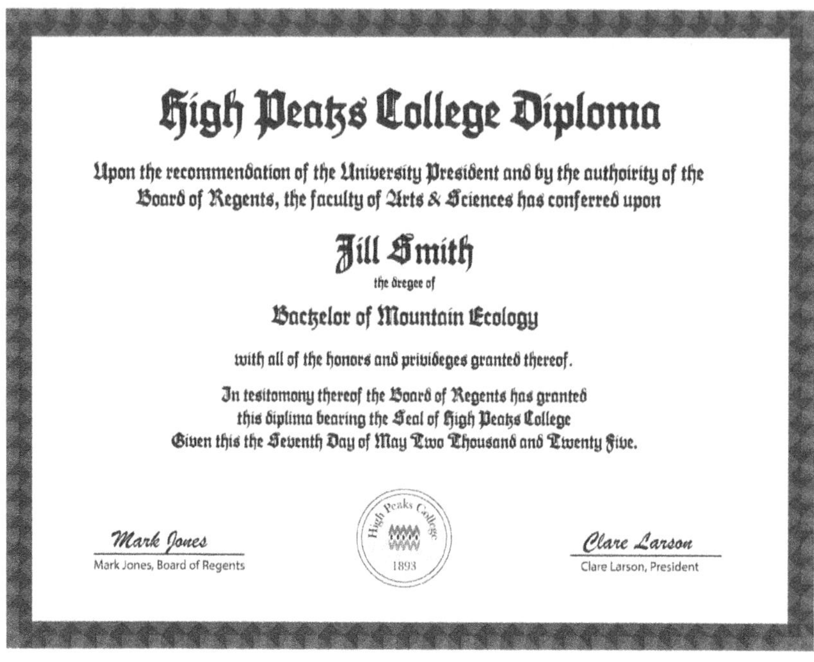

president's hand, and receiving their diplomas in full public view. Rituals like graduation ceremonies are repeated over and over, and the visual rhetoric of the diploma reifies these events, heightening their epideictic potency. For individual graduates, the diploma embodies a storehouse of rhetorical energy, retaining its epideictic glow long after the ceremony when it's framed and hung on the wall.

Some epideictic designs praising individuals have a global online presence, like those for the World Food Prize, which was initiated by Norman Borlaug to celebrate agricultural luminaries from around the world who have made major contributions to achieving food security. Figure 2.12 from the World Food Prize website pictures the 2023 winner, Heidi Kühn, in various scenes—in the fields and among farmers from around the world—which give context to her influential work in restoring farmland in war zones. These pictures also create repetition, a key strategy of amplification, as well as an upbeat tone that reinforces the festive occasion. The large picture of Kühn towering over these images and the headline text with her name further amplify the praise. The cool blue and green colors of the website add a serious, professional tone to her accomplishments, and the World Food Prize logo in the upper left further certifies their global value to humanity.

Figure 2.12. Page from the World Food Prize website celebrating the 2023 winner, Heidi Kühn (World Food Prize Foundation 2023). Credit to World Food Prize Foundation and Heidi Kühn.

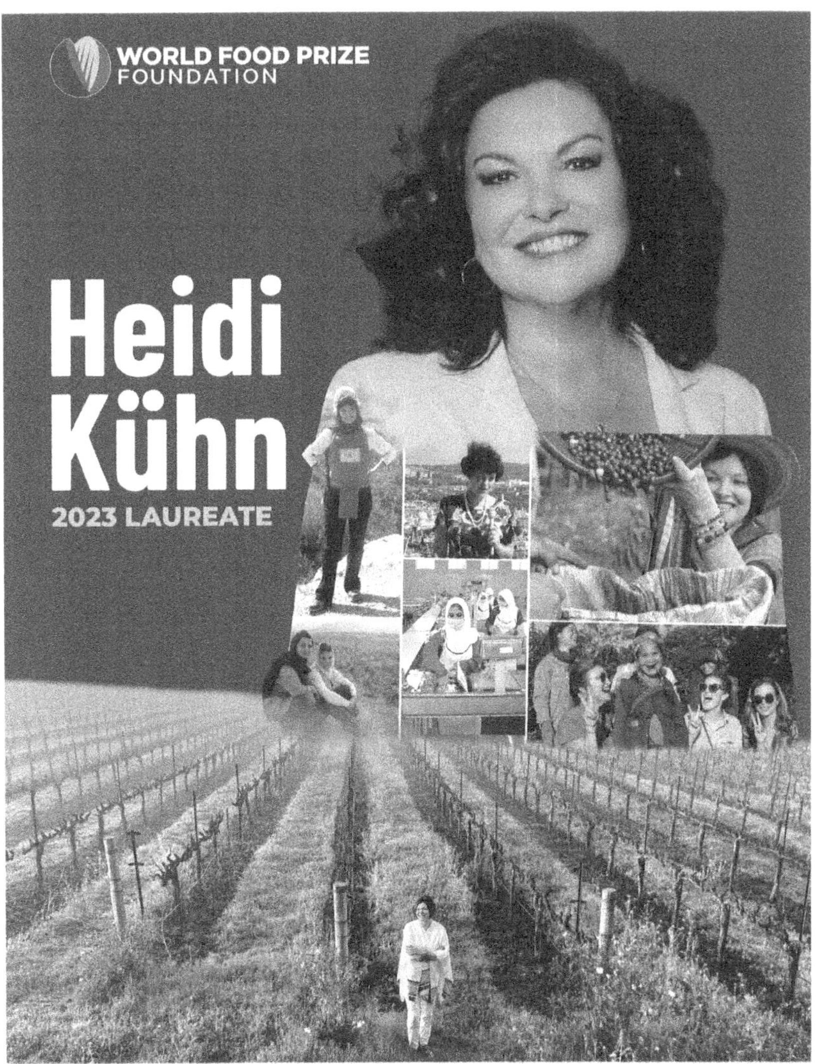

This award is enshrined not only digitally on the website but also physically in the World Food Prize's Hall of Laureates, a prominent building in Des Moines, Iowa, that houses memorials to food pioneers and plaques for each annual prize winner. The online image in figure 2.12, therefore, initiates a chain of epideictic visuals—from the website image and story announcing the winner, to the award event itself, to the physical space memorializing the accolade for posterity.[4]

Sometimes epideictic events praise *future* accomplishments rather than those that have been achieved in the past—for example, activities, buildings, or spaces that have been planned or designed but have not yet materialized. This form of epideictic rhetoric permeates architectural renderings, which amplify designs of future office buildings, churches, shopping malls, and homes by including perspective views with contextual enhancements like landscaping and people, as we saw in the drawing of the parish hall in figure 1.17. Oftentimes these visionary pictures are associated with a particular event like the one in figure 2.13, an invitation to a groundbreaking for the Biorenewables Research Laboratory at Iowa

Figure 2.13. Invitation to the groundbreaking ceremony for the Biorenewables Research Laboratory on the Iowa State University campus (Iowa State University 2008). Image credit: Iowa State University.

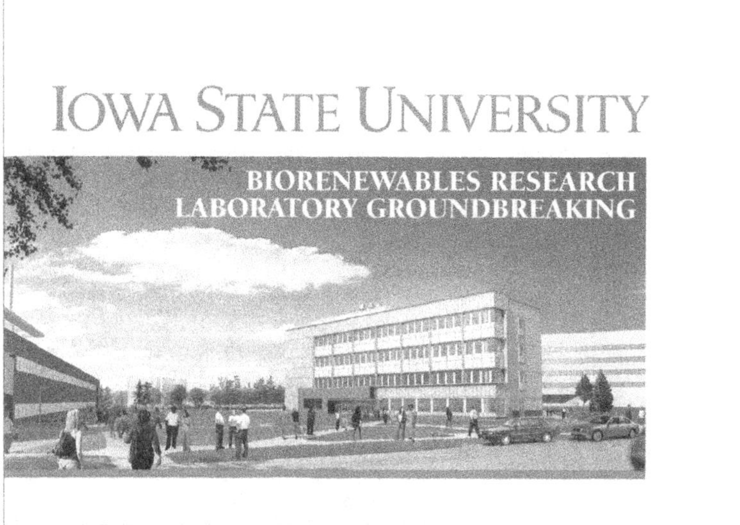

State University. This vivid color drawing envisions a bright, festive image for the audience—mostly administrators across the campus along with external donors and other stakeholders. Even though the building had not yet been constructed, the picture shows it in full operation, replete with people milling about on a sunny day and cars parked nearby—a beehive of activity. The visual enhancements in this picture illustrate that this project (and the purpose it serves) warrants celebration and praise, motivating the audience to attend this event and lend their support to this project as well as to the economic, scientific, and environmental values that inspired it.

In all of these ways, visible forms of the epideictic serve to mark time by pausing to recognize an achievement, milestone, or anniversary. In the contemporary world, it's hard to imagine celebratory events without visual language to amplify them, if not to become the focal points for engaging their audiences. Graduations without diplomas, award ceremonies without certificates or plaques, weddings or groundbreakings without invitations, concerts or memorials without programs, tournaments without medals and trophies—all seem culturally, socially, and rhetorically inconceivable. Visual language not only crystallizes these epideictic moments, but also extends them far into the future. After spoken words of praise are rendered mute, visual artifacts sustain—and even further amplify—their epideictic force.

Taking Temporal Stock of Amplification: Then and Now

Visual amplification, then, materializes in many different forms—in handwritten and typographic text, in graphical embellishments, in pictures of people and things, in awards and certificates. Like traditional written and spoken modes, over the past several centuries these visual techniques have been deployed to achieve epideictic effects: from praising the accomplishments of an individual or a collective group (a royal family, a business) to praising the qualities of a place or thing. Techniques for visual amplification—flourishes, lavish embellishment, detailed description, repetition, and other means—are intended to dazzle and delight audiences, stimulate the emotions, and produce an edifying, uplifting, and memorable experience.

So how does visual amplification today differ from displays of it in the past? As technology has changed, the strategies and affordances for amplifying praise have evolved as well—from handwritten to digital flourishes, from bird's-eye views to web galleries, from static charts to interactive data visualizations. Digital tools for amplification (color, typography, graphics, stock drawings, and image editing) have become readily available, enabling even novice designers to engage in the practice—for example, by amplifying an invitation to a local tea garden with script typefaces, pastel colors, and floral icons and images. Of course, designers constantly bear the risk of turning amplification into hyperbole and excess, but that's a risk worth taking to engage and persuade audiences by giving praiseworthy subjects their due.

These rhetorical acts of visualizing praise, of course, are never performed in a vacuum. The exigencies for visualizing them are bound to both short-term and long-term factors, and moreover, the collective values that underpin them are embedded in certain social, cultural, and historical circumstances. These diachronic forces, which are intrinsic to epideictic acts, are constantly shifting, a topic I'll turn to in the next section.

The Social Epideictic: Visualizing Community Values

In the contemporary world, the epideictic does more than just to recognize a praiseworthy achievement or quality of an individual, place, or thing—as satisfying, important, and necessary as those rhetorical acts are to marking milestones, achieving justice, and enhancing understanding. The epideictic also has a collective function that celebrates entire communities (Jasinski 2001, "Epideictic," 209–15). As Chaïm Perelman and Lucie Olbrechts-Tyteca (1969) theorize, epideictic rhetoric also has a vital argumentative function by creating "adherence" (50) to the beliefs and standards that underpin communities. Perelman and Olbrechts-Tyteca claim that "epidictic oratory forms a central part of the art of persuasion" (49), that it "has significance and importance for argumentation, because it strengthens the disposition toward action by increasing adherence to the values it lauds" (50), and that it does so by upholding "traditional and accepted values" (51), which the audience likely shares and respects. According to Laurent Pernot (2015), then, epideictic rhetoric aims at "solidifying the social order" (99); moreover, he claims, "It is a question not only of aesthetic quality but also of moral and religious worth" (99). In supplying the rhetorical cohesion that binds individuals into communities, epideictic rhetoric fosters the "identification" Kenneth Burke describes in his *Rhetoric of Motives* (1950) that makes those individuals "consubstantial" with one another (20–22). In other words, by praising the behavior or status of someone (or some place or thing), we make a judgment that such qualities have value and, moreover, that they do so collectively in the wider social realm in which audiences dwell.

So how does this aspect of epideictic rhetoric play out visually? Whatever the genre or mode, epideictic design almost invariably bolsters certain underlying values. For example, the invitation to the groundbreaking in figure 2.13 underscores the belief that biorenewable research, and renewables themselves, is worth pursuing—a conviction audience members who attended the event must have collectively shared. Similarly, the online epideictic for the World Food Prize (fig. 2.12) advances the argument that food security should be a global priority for virtually everyone on the planet, largely as a quest for social justice. This kind of epideictic energy infuses other figures we've seen as well, with visual amplification fostering collective "adherence" to the values they embody.

The visual epideictic also embodies cultural and aesthetic values. For example, the classical architecture, decoration, and mythology in Besson and Béroald's frontispiece in figure 2.4 reinforce the aesthetic values of the late Renaissance, where classical mythology and elaborate ornamentation coalesced, generating elaborate designs in all the arts and fostering audience expectations within these cultural paradigms. Aesthetic values are also embedded in Cocker's handwriting flourishes (fig. 2.1), which both reflect and bolster the Baroque style of the period that featured ornate curvilinear forms. Furthermore, the detailed drawing of the planing machine (fig. 1.15) emulates the aesthetics of Realism that pervaded the nineteenth century, and the promotional piece for the Empire Sewing Machine Company (fig. 2.8) exemplifies the Victorian appetite for profuse details. In each of these cases, epideictic design promotes "adherence" to aesthetic values that were widely accepted at the time but that have long since been supplanted by modernism.

The socializing effect of epideictic visuals can also foster strong local or regional identity. The map of Maastricht in figure 2.6 includes symbols of the city floating in air (at the top), suggesting a divine or mythological presence, and the Latin inscription in the lower left includes a laudatory description of the city's edifices. The figures in the foreground that picture a farmer plowing a field, a noble-looking couple, and two well-armed fellows indicate the city's self-sufficiency, prestige, and ability to defend itself. Collectively, these images project the soul of the city, spiritually and materially, and inspire "adherence" to its values among members of its collective audience, primarily its own residents but also like-minded people beyond its walls. In the *Civitates Orbis Terrarum* (Braun, Hogenberg, et al. 1612–18)—from London and Vienna to Nuremberg to Granada—have similar forms of visual language that reinforce the values and identities of their respective cities.

The "adherence" to values socialized by the visual epideictic can also be rooted in personal achievement. A college diploma (simulated in fig. 2.11) constitutes a visual argument on several levels: that higher education is worth the time, money, and effort and that it has economic, social, and intellectual value. By making these values visible, the diploma amplifies and argues for them and binds together the community that shares them—mainly, those who collectively acquire a college diploma and those who support them in doing so, including family, friends, teachers, advisors, and administrators. Displayed on a wall or stashed away in a drawer, the diploma reaffirms those values whenever an audience encounters it. Similarly, an award certificate or plaque might also engender certain communal values within an institution or organization—for example, recognizing outstanding teaching at an awards ceremony attended by others that highly regard that kind of achievement. On the other hand, a ribbon bestowed by a local garden club for the largest tomato of the season might bolster its own local values, but members of the surrounding community (and beyond) might shrug their shoulders. Thus, the values promoted in a given epideictic act might have a widespread socializing effect or be confined to a narrow audience.

This socializing effect outlined by Perelman and Olbrechts-Tyteca should apply to any epideictic expression, though the values to which it fosters "adherence" might be more or less substantial. For example, a newsletter for a business might feature a section showcasing significant milestones (birthdays or years of employment) that includes pictures of employees or clip art that conveys the organization's gratitude for long-term service. Or the newsletter might picture employees recently promoted within the company, emphasizing its belief in upward mobility within its own ranks. Along similar lines, managers in an organization might give employees framed certificates for having completed a series of training sessions, showing that they strongly endorse professional development. In these cases, the "adherence" to community values might be more modest and pedestrian, but the visual epideictic has the same goal: to refresh, enliven, and buttress those values.

The "adherence" to values promulgated by epideictic rhetoric, however, can be complicated and nuanced if members of a community don't share the same values—for example, about education, food security, or political power. For loyal monarchists in the eighteenth century, the family tree in figure 2.5 that envisions the royal French lineage served to reinforce their belief in the social, political, and religious system that undergirded royal power. On the other hand, to those favoring an egalitarian system, this elaborate family heraldry demonstrated extreme excess and privilege, an egregious and unjust allocation of resources to those empowered by the monarchy. An epideictic design can promote common values but also antagonize audience members who reject them—and who in this case contemplated a revolution against them.

Epideictic visualizations like the royal family tree adhere to wide-ranging social and cultural values. Two such strands of epideictic "adherence" emerged in the nineteenth century in the U.S.: images that capture the wonders of burgeoning technology and visualizations that make large-scale comparisons. These two forms of epideictic expression are steeped in the values of their times, and in the next few sections I'll examine both of them, which are often intertwined.

Burgeoning Technology: Celebrating the Future

Advances in technology have long been the subject of the pictorial epideictic. Legions of detailed drawings of existing, experimental, and often aspirational machines graced early engineering books—Ramelli (1588), Besson and Béroald (1578), Böckler (1673), to name a few—heightening their credibility and persuasiveness, as did pictures of machinery in the new industrial age like figure 1.15. However, the scale and impact of technology and engineering vastly changed, especially in the U.S., with the emergence of what David Nye (1994) calls the "technological sublime." American railroads, steam engines, bridges, skyscrapers, and electrification became instilled in the public imagination as visible testimonials to progress. According to Nye, "Since the early nineteenth century the technological sublime has been one of America's central 'ideas about itself'—a defining ideal, helping to bind together a multicultural society" (xiii–xiv).

Sublimity is typically associated with the sheer scale of an object—"vastness," as Edmund Burke put it in his *Philosophical Enquiry into the Origin of Our Ideas of the Sublime and Beautiful* ([1759] 1967, 72)—which makes it appear powerful and impressive and even dangerous and terrifying. By any measure, one of the most celebrated engineering feats of the nineteenth century was the Brooklyn Bridge, which appeared prominently on the promotional piece for the Empire Sewing Machine Company (fig. 2.8). The Brooklyn Bridge had both practical and symbolic value for Empire: it would provide easy access to their Brooklyn customers from their New York office, and it wedded its sewing machines to the technological progress of the new industrial age.

The epideictic potency of the "technological sublime" was emphatically displayed at world's fairs, which began with the Great Exhibition in the Crystal Palace in Victorian England (1851) and which for over a century grew to immense heights in venues around the world. Figure 2.14 shows a picture from the Columbian Exposition in Chicago (1893) illustrating the Electricity Building. The Exposition introduced the world to lighting (and electricity) on a grand scale, heralding a new age of industrial and domestic electrification that would generate prosperity and enrich everyone's life. The Electricity Building showcasing this new technological wonder glows in a waterfront scene, its powerful lights bathing everything around it—people, plants, boats, and the skies above—celebrating this spectacular feat and the ingenious inventors (Thomas Edison and Nikola Tesla) who made it possible. This transformative technology was hugely

Figure 2.14. The Electric Building at the World's Columbian Exhibition in Chicago (Winter's Art Lithography Company ca. 1892). Courtesy of the Library of Congress, Prints and Photographs Division, Popular Graphics Arts Collection [LC-DIG-pga-03063].

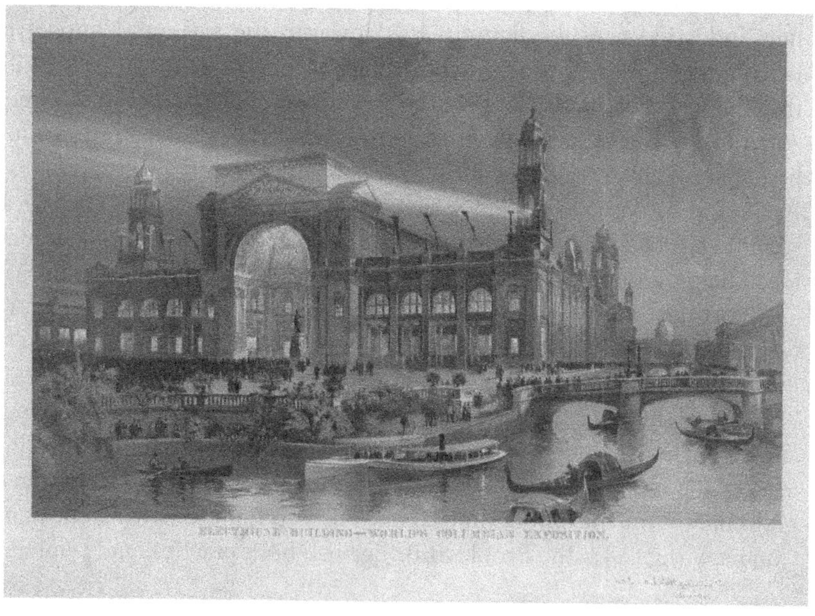

popular with fairgoers, especially the lighting on water fountains and the Ferris Wheel (Nye 1994, 145–49). The picture of the Electric Building, no doubt, became a memento for those who attended Chicago's Columbian Exposition, refreshing and gilding the experience as they shared memories with family and friends. Even today, its epideictic glow still resonates as a milestone in the history of technology, world's fairs, and Chicago—a glimpse of the future still worthy of praise.

Pictures celebrating the "technological sublime" appeared frequently in the U.S. and Europe in the late nineteenth and early twentieth centuries, often using familiar landmarks to amplify the magnitude of a given building, bridge, or machine. Modern industrialism was capable of creating immense objects: ever more powerful steam engines, ocean liners like the *Titanic*, and other mechanical wonders that vied with each other for the title of "biggest" or "longest." Figure 2.15, for example, from the *Scientific American Reference Book: Edition of 1913* (Hopkins and Bond 1913, viii) shows the longest ship (the *Imperator*) and the tallest building (the Woolworth Building) standing next to each other. Both of these immense structures were built almost simultaneously, so many audience members would have heard about them, and some would have witnessed firsthand the construction of the Woolworth Building. This imaginary and rather whimsical scenario juxtaposes these two titans, with the German-made ship getting the better of the match. Standing the ship upright in the middle of Manhattan next to a new landmark building enables the audience to experience the magnitude of this object in familiar surroundings, amplifying its stupendous length. The comparison amplifies and celebrates the technological feat of building such a huge seafaring craft, while also paying homage to the world's tallest building. In this case, the comparison actually creates epideictic reciprocity: two remarkable achievements standing triumphantly side by side.

Amplification through immense scale also occurs in figure 2.16 (also from the 1913 *Scientific American Reference Book*), which shows the amount of natural gas consumed in a week (presumably in the U.S.). This quantity—over two billion cubic feet—is visualized as a gigantic imaginary gas tank, the size of which is compared to familiar structures such as the Eiffel Tower, U.S. Capitol Building, Washington Monument, and Great Pyramid, all of which are dwarfed by the gas container towering above them. Although comparing the volumes of these objects presents perceptual challenges for the audience, whatever epideictic charts like this lacked in precision they made up for in familiarity. Landmark structures like these

Figure 2.15. Comparison of the world's longest ship, the *Imperator*, and the world's tallest building, the Woolworth Building in New York City (Hopkins and Bond 1913, viii). Illustration by F. Worms.

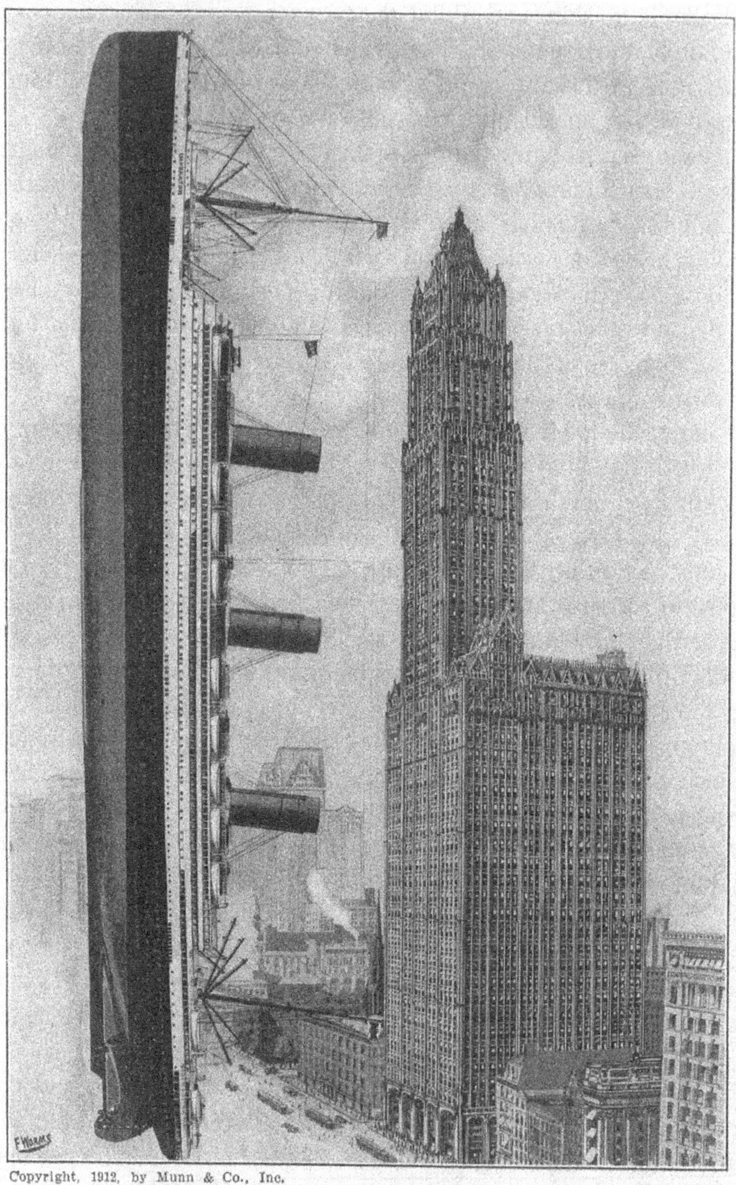

Copyright, 1912, by Munn & Co., Inc.

THE LONGEST SHIP AND THE TALLEST BUILDING.

The "Imperator," 900 feet; Woolworth Building, 750 feet.

Figure 2.16. Comparison of natural gas consumption and landmark structures of the world (Hopkins and Bond 1913, 131). Illustration by C. M. Knight-Smith.

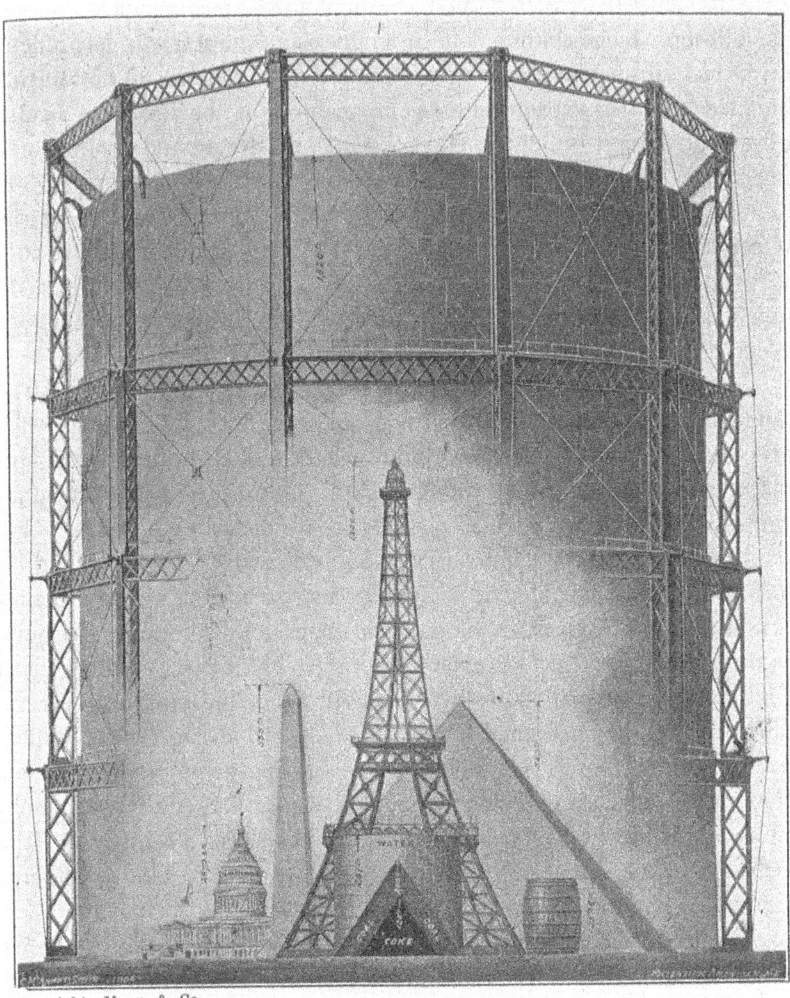

Copyright, Munn & Co.

THE MAGNITUDE OF THE GAS INDUSTRY.

A Week's Supply of Gas. Fuel for 23 Million Horse-Power Hours. The Gas Holder contains 2,163,207,368 Cubic Feet.

appeared in other pictorial comparatives, becoming epideictic benchmarks for the industrial wonders of the era. Below the gas container, additional comparisons are rendered in the form of a water tank and other energy commodities (coal, coke, and oil), which further amplify natural gas as the most voluminous energy resource on the planet, its size comprehensible only through comparisons relatable to the audience.

The values that fostered the celebration of these technological wonders were bolstered by an abiding faith in progress and the ability of technology and industrialism to create a more prosperous and powerful nation that benefited all of its citizens. Epideictic pictures of the Brooklyn Bridge and gigantic ships, buildings, and gas tanks illustrated and promulgated "adherence" to those values, as did the poetry of Carl Sandburg (1950), who sang the praises of Chicago's skyscrapers, railroads, and industrial culture. Enthusiasm for the wonders of technology also had its advocates in Europe, primarily under the aegis of Futurism, an Italian movement launched before World War I that extolled industrialism and machine-age aesthetics and that subsequently spread across the Continent.

Although we don't still celebrate these same kinds of industrial wonders (unless they serve primarily as entertainment, as Las Vegas–like spectacles), the "technological sublime" is still fertile ground for the epideictic. For example, space exploration still captures the public imagination, as we seek new, unexplored worlds beyond our own, wonders that (ironically) aren't always quite visible but still worth celebrating. To do so, NASA generates epideictic images for its space programs to visualize current and future missions, along with images of aspirational journeys. Figure 2.17 shows a NASA Kepler-16b poster (2020) that visualizes travel to a distant planet that circles two stars. An astronaut looks out over the colorful landscape, presumably in awe of the expansive view, a double shadow trailing behind. This sublime scene satisfies the public's appetite for otherworldly adventure, fueled by popular science fiction films. Although not as powerful or universally acknowledged as it was a century ago, the "technological sublime" still provides rhetorical glue to bind together a fragmented nation.

Values also change over time, and along with them the persuasive power of "adherence." Over a century ago the audience saw that humongous, hypothetical gas tank (fig. 2.16) as a laudable feat, comparable to other technological achievements like the Eiffel Tower, a perspective that today's audience would reject or feel ambivalent about. In addition, the values subject to "adherence" might shift to other epideictic forms: The milestone of electrification celebrated at the Chicago World's Fair seems irrelevant today, though the underlying values about progress remain, promulgated by new technologies realized both here on Earth and beyond.

Figure 2.17. Poster of a mysterious planet with two suns (NASA and the Jet Propulsion Laboratory 2020). Courtesy of NASA and the Jet Propulsion Laboratory.

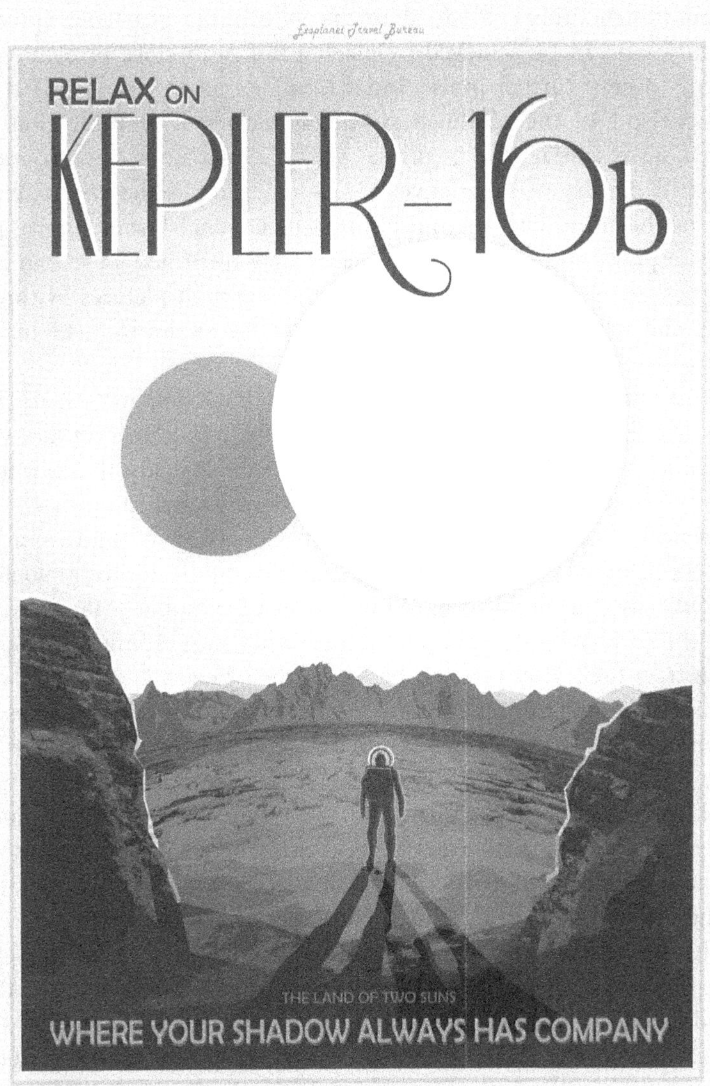

The Comparative Epideictic: Sizing Up the Competition

As we've seen with the pictures of the ship next to the building and the landmark structures embedded in the gargantuan gas tank, epideictic amplification can be achieved by placing things in close proximity as a strategy to heap further praise. Indeed, as Laurent Pernot (2015) claims, "Comparison is one of the best-known means of amplification, and it plays an essential role in praise" (88). Comparisons are also closely related to the use of superlatives (88)—for example, the wisest senator, the most beautiful city, the most spectacular arena. *Visual* comparatives are widespread and easy to spot because they often juxtapose one entity and another in the same visual field, either through pictures or through charts and graphs, where we can see a praiseworthy element in close proximity to other elements.

Like other strategies for visual amplification, comparatives articulate "adherence" to certain communal values. In the nineteenth century visual comparatives found their roots in the competitive zeitgeist that permeated the era. This was the age of empires, laissez-faire capitalism, industrialism, nationalism, and natural selection, all of which fostered and (to a large extent) celebrated competition: geographically, politically, technologically, economically, and biologically. Comparison on a global scale appears in figure 2.18, which juxtaposes many of the world's most prominent structures to illustrate their relative sizes, based on their height and girth, thereby extolling some and minimizing others. This epideictic mosaic includes both ancient and modern buildings: churches, towers, temples, pyramids, and other structures like the Statue of Liberty and U.S. Capitol (far left), the Brooklyn Bridge (far right), and the Washington Monument poking up its head in the center, its sheer height commanding the most praiseworthy place of all. So what's the rhetorical effect of this swarm of epideictic amplifications? Although each structure receives its rightful place, which enables the audience to celebrate them incrementally, inserting several American structures into the mix bestows enormous status on them, seated among the world's great monuments, ancient and modern. These American landmarks also give the drawing scale and relevance, though situating most of them on the margins (left and right) puts them in their place internationally. With such a feast of architectural wonders, such a glut of superlatives, the competition reaches an impasse. At this point, the epideictic mosaic invites audiences to appreciate the amplification in

Figure 2.18. Comparison of seventy-six of the world's largest buildings, structures, and monuments (Goodchild 1885). Courtesy of the Library of Congress, Prints and Photographs Division [LC-DIG-pga-12910].

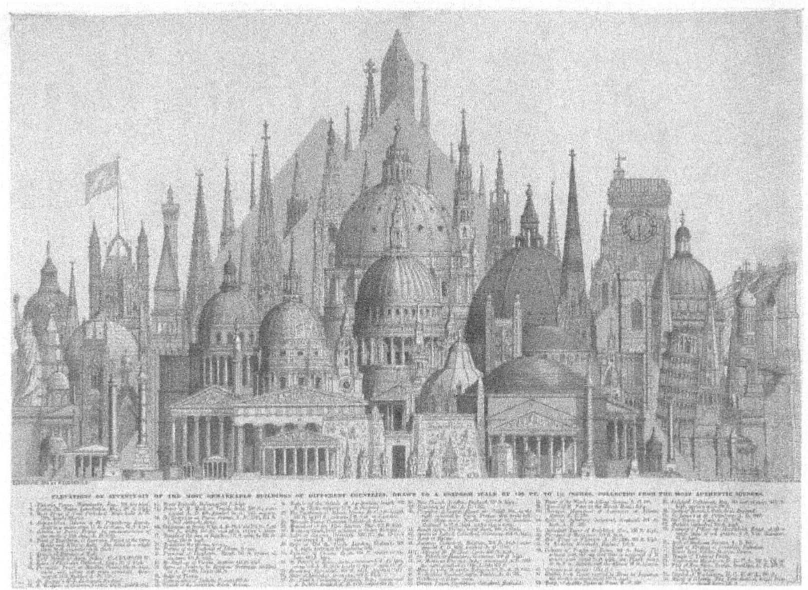

its totality, to celebrate humanity's great building achievements, however fierce the competition.

Nature has also been amplified through visual comparisons, a rhetorical practice that became widespread in the second half of the nineteenth century. As colonialism reached every corner of the globe, and exploration, cartography, and science opened up new avenues for observing and recording nature, a geographical mode of epideictic emerged in the form of maps, pictures, and charts.[5] Figure 2.19 illustrates this passion for chronicling the wonders of the earth: a chart by G. W. Colton (1874) that displays the largest rivers and mountains of the world based, respectively, on their length and height. The rivers begin in the upper left with the Mississippi, which holds first place when combined with the Missouri and Yellowstone, followed by the Amazon and other notables like the Yang-Tse and Nile in stair-step succession down to the shorter rivers on the far right (the Hudson, Thames, and Merrimack), with their mouths at the top and their sources at the bottom. The mountains follow a similar incremental

Figure 2.19. Chart comparing the longest rivers and highest mountains of the world (Colton 1874, nos. 4 and 5). Image credit: David Rumsey Map Collection, David Rumsey Map Center, Stanford Libraries.

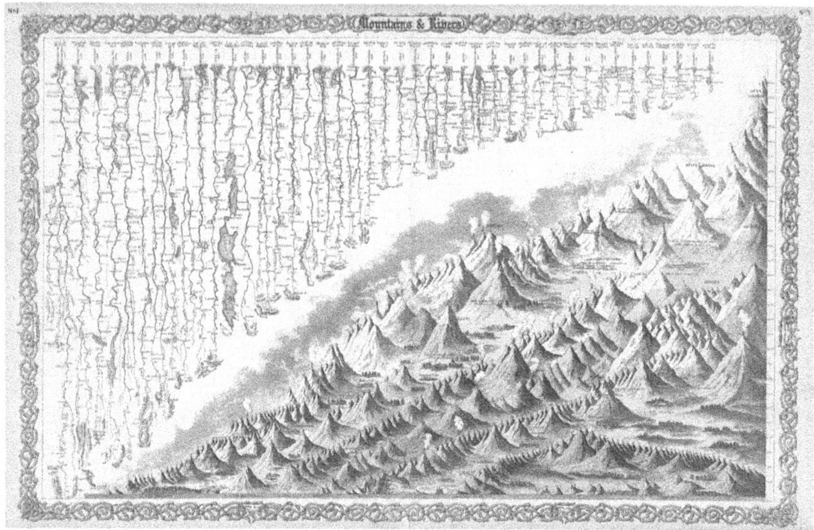

pattern from the bottom to the upper right as they get progressively taller, culminating with those in Asia and the Pacific. To accommodate the space constraints, Colton straightens out and compresses the rivers and tributaries, and he grossly elongates the mountains to differentiate their profiles. At the same time, he provides details for sorting and grouping: color coding mountains by continent as they move to the right, inserting fiery puffs of ash on volcanoes, and altering the vegetation as the mountains get progressively taller. The incremental changes in length forcefully compel the audience to compare and rank these entities, amplifying their grandeur or diminishing their stature, especially those in the British Isles, which rise meekly in the lower right. Across the planet, then, winners and losers abound, including nation-states as well, depending on whose empires the mountains and rivers currently reside in. However, unlike comparatives of human artifacts (buildings, monuments, gas tanks), this visual epideictic has temporal resilience: despite which empire once claimed them, the rivers and their tributaries still flow, and the mountains still stand.

The comparative epideictic is especially emphatic and perceptually transparent in data displays, the very purpose of which is to engage the eye in judging differences. An early example appears in figure 2.20, a line graph designed by William Playfair (for his 1801 *Commercial and Political Atlas*) that celebrates the success of the English economy in the eighteenth century. A pioneer in charting statistical data, Playfair visualizes England's trade advantage throughout the century, with exports shown with the red line above and imports with the yellow line below. Playfair amplifies the difference between the export and import lines by shading the area in green and labeling it the "Balance in Favour of England"—with only a tiny segment (in the early 1780s) appearing in red. The perceptual immediacy of displaying these data graphically provides additional amplification: As Playfair explains in his introduction to his *Commercial and Political Atlas* (1801) charts (unlike tables) reveal data quickly and efficiently, often in an instant (x). Even in the subdued confines of a plot frame, an entire nation-state basks in the epideictic glow of its financial and commercial success. Playfair's trade chart (fig. 2.20) strongly reinforces nationalist values: that a trade "Balance in Favour of England" benefits the country, that it's a source of national pride, and that it can and should be sustained.

Later in the nineteenth century, vivid and even more explicit amplification through comparison appeared in rank charts, which visualized changes with sloping lines. Figure 2.21 from *Scribner's Statistical Atlas of*

Figure 2.20. William Playfair's chart of England's exports and imports over a century (Playfair 1801, plate 1). Image credit: David Rumsey Map Collection, David Rumsey Map Center, Stanford Libraries.

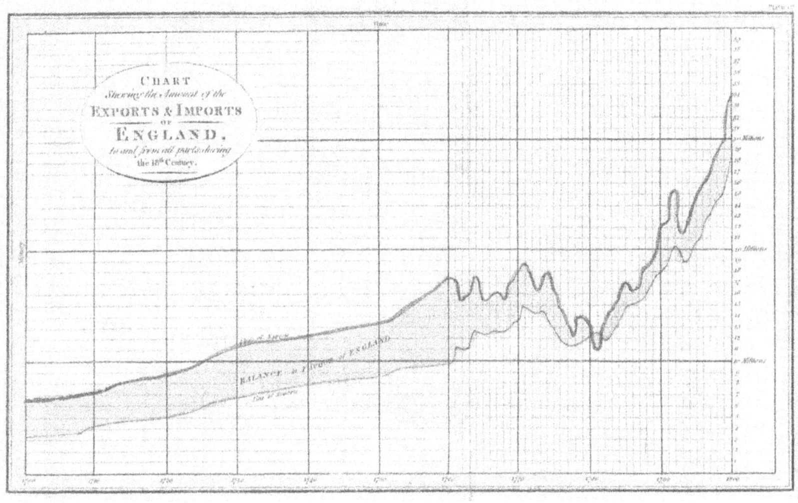

Figure 2.21. Chart from *Scribner's Statistical Atlas of the United States* showing the rankings of states by population from 1790 to 1880 (Hewes and Gannett 1883, plate 18). Courtesy of the Library of Congress, Geography and Map Division [a40001834].

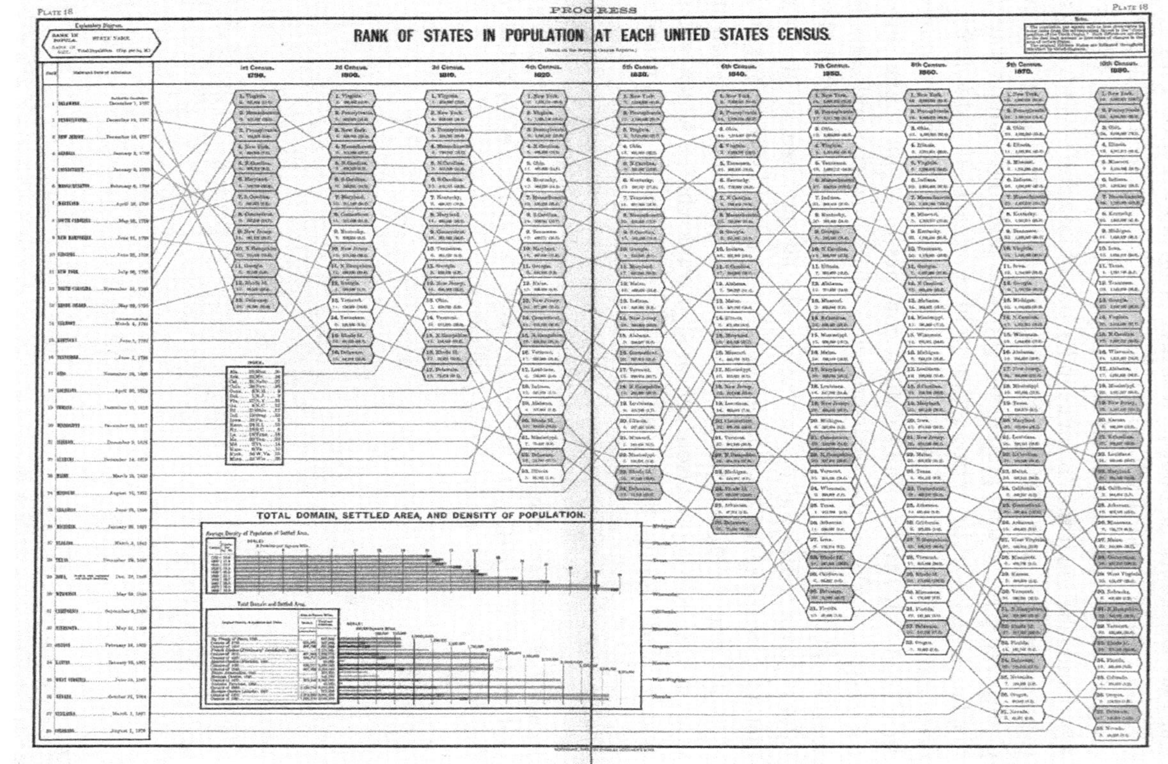

the United States (Hewes and Gannett 1883) plots the rankings of state populations from one census to the next. As state populations grow and as states are added, many of them supersede the populations of the original thirteen states, which are shaded light red to emphasize their changes in position. Lines connect each state from one census to the next, with their slopes enabling the audience to compare their rankings over time. In these ways, the chart shows a dynamic nation that's growing by leaps and bounds as new states assert their places in the rankings. Although large-population states like New York, Pennsylvania, and Ohio continue to dominate, states like Virginia and Maryland and some of the other thirteen original states gradually sink over time, giving way to newcomers farther west. Unlike Playfair's monolithic national narrative, this display is more complex and fluid, with states rising and falling over time as their populations grow and shift, creating epideictic dynamism from one census to the next. However, seen in its totality on the macro level, the chart celebrates the nation's westward expansion, statistically and graphically affirming the values of Manifest Destiny. That Maryland falls in the rankings with each census is less an acknowledgment of its diminished status than an affirmation of the nation's rapid growth.

Contemporary data displays have the same capacity for epideictic amplification through visual comparison and hierarchy. Figure 2.22, a bar chart from the U.S. Census Bureau (2019), shows the fifteen fastest growing U.S. cities in a single year, most of them in the South and West, with the growth rates topped by Buckeye, Arizona, at an astonishing 8.5 percent annually. The thick red bars on the chart celebrate the surging growth these cities have collectively achieved, while arranging the cities in descending order simplifies the comparisons and heightens the differences among them. While Buckeye, New Braunfels, and Apex are particularly worthy of praise, all these cities have achieved remarkable growth, even Round Rock with its relatively modest 4.3 percent. The abstract buildings on the bottom of the chart, whose arrow-like roofs all point up, further amplify the soaring growth. Like the collage of world buildings in figure 2.18, epideictic energy flows all through this chart, with comparative gradations modulating the superlatives.

So which collective values does this U.S. Census chart foster by visually amplifying growth? That population growth is desirable, that cities should constantly strive to achieve it, and that cities far from traditional strongholds are most likely to succeed. The chart promulgates "adherence" to values by implicitly arguing that these cities are highly desirable places to live because

Figure 2.22. Chart from the U.S. Census Bureau showing the percentage increase of rapidly growing cities in the U.S. (U.S. Department of Commerce, U.S. Census Bureau 2019). Courtesy of the U.S. Census Bureau.

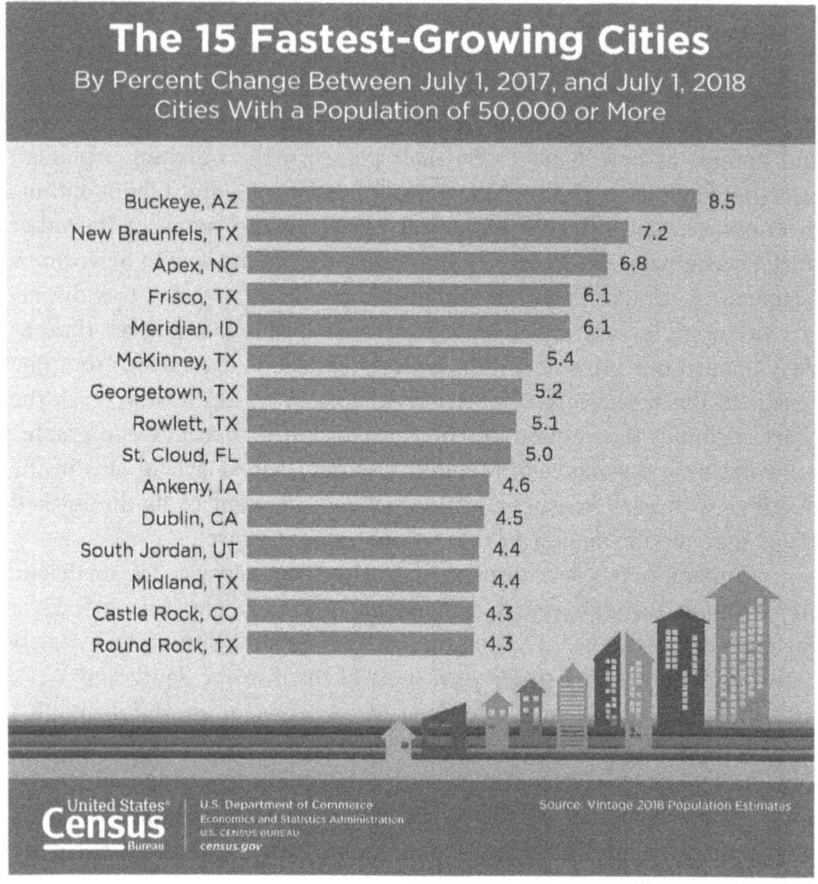

they have such a powerful attraction for new residents, amplifying the value of local growth and expansion, primarily in the Southwest. On the surface, these values seem to mirror those a century and a half earlier in the rank chart of state populations. However, "adherence" to those traditional values has been compromised by current values: less appetite for growth and increasing concern about environmental degradation and dwindling natural resources caused by burgeoning development.

Interactive bar charts are further infused with epideictic comparisons. Data on digital bar charts can be sorted from high to low or animated over time based on performance measures (DataRoyals 2025; VGraphs 2025), and interactive slope charts enable dynamic ranking over time.[6] The treemap in figure 1.7 ranks companies by area—from the largest in the upper left to the smallest in lower right, a convention of the genre—while affording users a host of interactive options. In a nutshell, digital data displays with interactive features enable users to compare data more efficiently and flexibly than static charts. However, because users can reshape displays according to their own needs and preferences by rearranging the data and zooming in on the details, visual hierarchies are destabilized, along with the values underpinning them. As a result, designers forfeit some control over "adherence" as the epideictic energy is distributed and diversified.

Blame and Shame: Flipping the Epideictic Coin

Epideictic rhetoric also includes the less prevalent but equally potent mode of assigning blame or finding fault, in which the rhetor casts aspersion on the subject. For Aristotle, just as the epideictic rhetoric of praise is associated with "virtue" and things "honorable," so blame is associated with the opposite—with "vice" and things "shameful" (Aristotle 2007, 75–79; bk. 1, ch. 9, secs. 1–27). Many recurring situations compel rhetors to assign blame: criticizing a rival in politics or business or a defendant or witness in a judicial proceeding, writing a negative job recommendation or performance appraisal, or warning an audience against certain dangers or risky behaviors and describing their consequences. In less formal settings, negative online product and service reviews have become a staple of contemporary consumer culture.

So how is blame *visualized*, both in the past and the present, and what are the impetuses for doing so? Visualizing blame in the past millennium began with a strong moral imperative—for example, with scenes from the Bible (Adam and Eve, Cain and Abel, the Passion of Christ) in medieval stained glass, religious murals, and illuminated manuscripts. In Renaissance painting, stories from the Bible and from classical mythology that visualized sin and corruption—and its consequences—were valued not only for their aesthetic qualities but also their didactic value of guiding the moral behavior of their audiences. Secular examples of epideictic blame in modern culture were exemplified by William Hogarth's pictures, which drew attention to moral evils of mid-eighteenth-century life in England.

In his drawing *Gin Lane* (1751) he visualizes the devastating effects of drinking: profligate behavior, depression, even suicide. In a series called *A Harlot's Progress* (1732) a young woman who arrives in London falls into a life of prostitution, her life thereafter spiraling downward; in *A Rake's Progress* (1735) the main character Tom Rakewell squanders his inheritance on a life of dissolution: drinking, gambling, prostitution, and extravagant spending. In one of these drawings, shown in figure 2.23, the hapless Rakewell languishes in debtor's prison surrounded by his distressed wife and former girlfriend, a man and a boy hounding him for money, and delusional people—poignant details that amplify his downfall. All of these pictures by Hogarth place blame on those who engage in, enable, and benefit from destructive behaviors, serving as strong cautionary warnings to their audiences.

Figure 2.23. A drawing from William Hogarth's *A Rake's Progress* showing Tom Rakewell in prison after a life of dissolution (Hogarth 1735, plate 7). Courtesy National Gallery of Art, Washington. Rosenwald Collection.

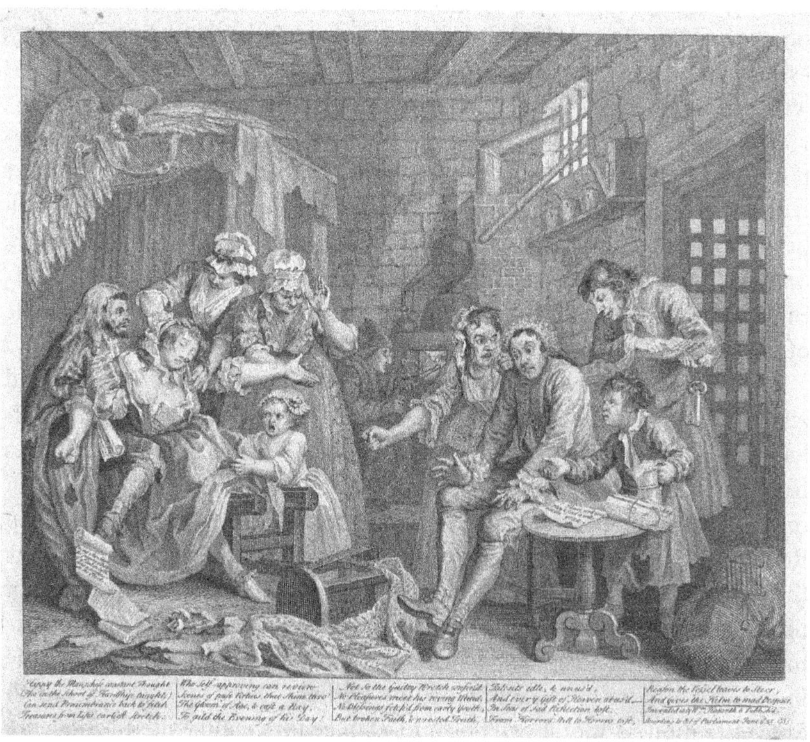

Warnings and the Fallout from Flouting Them

Today, warning signs and symbols are one of the most ubiquitous forms of epideictic blame, telling people what not to do and implying the consequences of failing to comply with the proscriptive message. For example, a circle and slash sign telling an audience to refrain from using their cell phones, as shown in the warning I created for figure 2.24, will direct the blame toward those who flout the command or who simply forget about it. When a phone rings in the middle of a lecture, committee meeting, church service, or airplane take-off, the guilty party becomes the brunt of shame and disgust among the distracted others who have already complied. Likewise, figure 1.19 warns the audience about the presence of bears and implies that something bad might happen if the warning isn't heeded. Similarly, if audiences flout a rock climbing warning and they fall or start a rock slide, they have no one to blame but themselves. Similar warnings using the circle and slash tell audiences to avoid all kinds of routine activities—smoking, parking, swimming, taking photos, drinking, walking dogs without a leash, and so on—as a means of protecting people or property from harm.

Figure 2.24. A warning sign forbidding the use of cell phones.

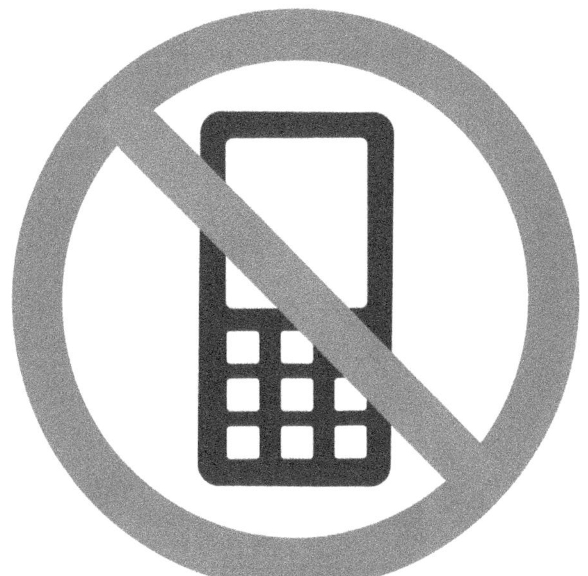

Warning signs use various kinds of visual amplification, both textually and graphically, to draw attention to their messages. Large black and red text, boldfacing, all caps, and underlining typically appear on warning signs to ensure that we pay attention, implying that those who don't comply must accept the blame because they've been duly warned. Pictorial warnings are often amplified by explicitly visualizing the consequences of noncompliance: slipping on wet floors, injuring ourselves by lifting heavy boxes, or the more gruesome effects of breathing poisonous gases, getting electrocuted, falling off a garden tractor, or getting fingers entangled in a machine, as shown in the Clarion Safety Systems sign in figure 2.25. Many warnings exaggerate the injury, using amplification to heighten the danger by envisioning elongated fingers wedged between gears (fig. 2.25) or bodies wrapped around drive shafts like snakes. At the same time, the abstract style of these signs depersonalizes the trauma and directs the blame toward a hypothetical individual, an "Everyperson" (Kostelnick 2019, *Humanizing*, 42–46, 153–56), with the expectation that such an agent will never actually materialize. To the extent that an agent does experience the trauma, the warning potentially immunizes the designer (or the organization posting it) from legal liability. Visualizing blame, then, ensures the safety of its audience in many of the same ways that Hogarth intended—amplifying the effects of behaviors the audience

Figure 2.25. A warning sign that amplifies the danger of working around machinery (Clarion Safety Systems 2025). Illustration courtesy of Clarion Safety Systems (www.clarionsafety.com).

should find offensive and harmful—though many contemporary warnings also act in the rhetor's self-interest.

Data Blame and Risk Assessment

Data design is particularly powerful mode of epideictic blame because, as we saw earlier, graphical visualization makes comparisons explicit, and the objects of blame often have very little place to hide, especially on a map. In Charles Joseph Minard's famous *Carte Figurative* (1878) of Napoleon's Russian campaign, shown in figure 2.26, he visualizes the devastating effects of the French retreat from Moscow. Using lines of diminishing width to show the increasing loss of life during the retreat, Minard tells a brutal tale about the horrendous costs of this foreign military adventure. By the time Napoleon's army returns to its starting point in the west, its black line of troops has diminished to barely a trickle. The chart's stark, somber design reflects the dire conditions faced by Napoleon's devastated army, which Minard documents by visualizing climate and geography. Although Minard doesn't identify the object of blame, the chart clearly warns his audience about the folly of conquering distant lands. The chart also serves as a visual eulogy, a grim memorial to Minard's contemporaries

Figure 2.26. Chart by Charles Joseph Minard showing the disastrous loss of troops during Napoleon's Russian campaign (Marey 1878, 72).

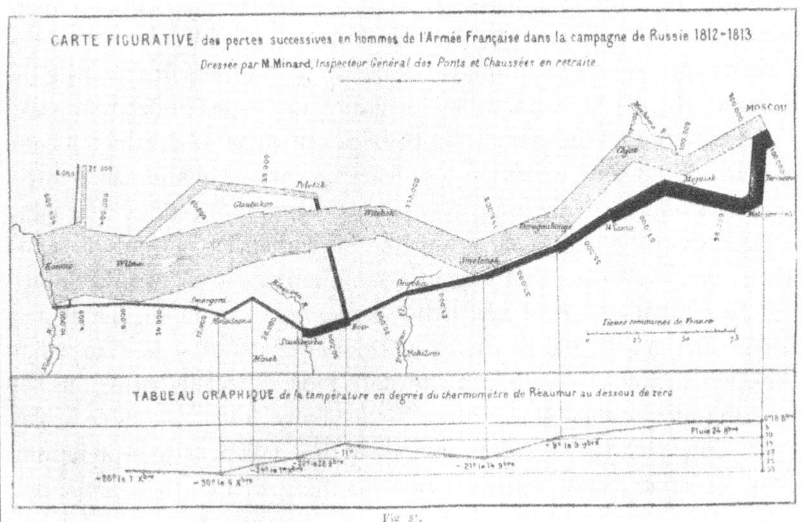

who served on this calamitous campaign, especially civil engineers (like himself) who provided logistical support. Minard's famous chart, then, has double epideictic value: as a blunt warning for France's political and military leaders and as a eulogy for the lost peers of his youth.

Today data maps that assign (or imply) blame can more immediately modify behavior as a form of risk communication. Interactive maps identifying outbreaks of diseases (COVID-19, flu, malaria) prompt users to wonder what went wrong in those areas, who's to blame, and what steps they can take to protect themselves and others. Users can also assess risk with crime maps, a particularly potent form of visualizing epideictic blame. From the earliest versions in nineteenth-century England and France, crime maps have a long and contentious history (Cook and Wainer 2016). Contemporary crime maps amplify blame with interactive features that enable users to zoom in on specific locations, see the frequency of crimes, sort them by type, and mouse over specific events to retrieve details. Crime maps have both spatial and temporal value: They enable users to assess the risks of visiting, traveling through, or living in a given area and to gauge these risks on specific days and times.

Figure 2.27 shows a Crimemapping.com (2024) visualization over a week's time with icons that display and amplify various criminal acts: a mask (burglary), a car (car theft), a tilted car (stealing from a car), a spray can (vandalism), a wine glass (drugs or alcohol), a dollar sign (larceny), and a fist (assault), among others. Blame is assigned to anonymous perpetrators, and users don't immediately know if they have been caught and prosecuted, nor do they know what caused them to commit these crimes—moral turpitude, immaturity, drug use, poverty, homelessness. Whatever the cause, something has gone awry, and users interpreting the map make their own assessments, including how to avoid becoming victims. A variation of the crime map is the sex offender map, which locates the dwellings of individuals on a registry so that neighbors can protect themselves and their children. As epideictic communications that vilify certain acts, crime maps bear a warning for audiences, not only to avoid the behaviors visualized but also to assess the risks in a given location.

As a potent form of epideictic blame, other forms of digital data design—line graphs, bar charts, scatterplots—enable designers to assign blame by making variations in the data look especially egregious: by stretching plot frames, starting the Y-axis above 0, using double scales, or including pictorial elements that skew the audience's interpretations. If data designers want someone or something to look bad and point

Figure 2.27. Interactive map by Crimemapping.com that shows various crimes committed in Fort Worth, Texas, over a week in December 2024.

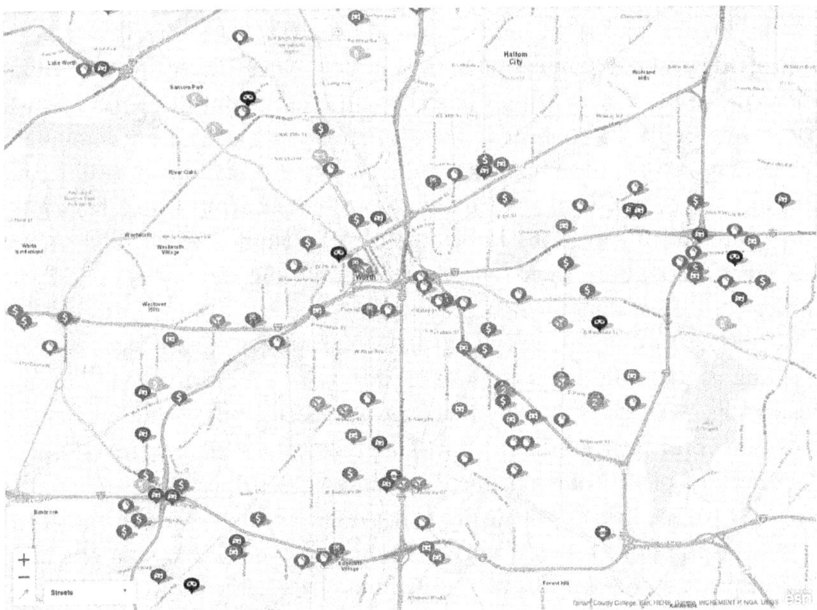

their digital fingers, they have the power to blame at will, data shaming that might motivate audiences to pay attention but that can also breach ethical boundaries. On the other hand, data blame can identify patterns and causes of certain problems that virtually everyone has a stake in solving—disease, poverty, crime, environmental degradation—and motivate users to confront the reasons for the blame and take action.

Epideictic Synergies: Interweaving Rhetorical Strategies

As Quintilian pointed out, epideictic rhetoric can't be completely separated from deliberative and forensic modes, "for all three kinds rely on the mutual assistance of the other" (1:397; bk. 3, ch. 4, sec. 16). So the epideictic aspects of the examples I've examined so far are often intertwined with other functions, which heighten the epideictic elements and vice versa. For example, the chart of tall buildings (fig. 2.18) serves the forensic function of visualizing the facts about their heights in close proximity, and the river and mountain chart (fig. 2.19) displays vital data about the planet's geography. The royal family tree (fig. 2.5) documents the lineage of the French monarchy, and the rank chart (fig. 2.21) precisely reveals census data about population trends across states. These forensic functions argue on a factual basis for a certain truth about these data, which of course might be contested. The distortions in the river and mountain chart (fig. 2.19) could impair the audience's interpretations of reality, and the chart showing the world's buildings and monuments (fig. 2.18) shows only their front elevations rather than their volume, which might tell a different story about their relative size, prompting the audience to draw other conclusions about their grandeur.

Several examples we've seen have deliberative functions as well by serving as planning tools or pointing to future activities. Playfair's trade chart (fig. 2.20), for instance, visualizes a trajectory for a vibrant English economy, creating a roadmap for making political decisions or investments. The detailed picture of the planing machine (fig. 1.15) serves as a marketing tool for persuading factory owners that it's a productive, well-engineered piece of equipment that's worth purchasing for their operations. The frontispiece to Besson and Béroald's book (fig. 2.4) tries to persuade its audience that it merits reading—and believing. The drawing of the huge natural gas tank (fig. 2.16) shows the dominance of this energy source, a cue for potential suppliers on where to focus their efforts. Hogarth's drawing of Tom Rakewell (fig. 2.23) warns against the vices of profligate behavior, urging the audience to assess the moral direction of their lives, and the crime map (fig. 2.27) enables travelers to plan trips, lodgings, and daily activities. And of course, with digital technology, images can be easily repurposed or remixed to praise or blame their subjects, often with ironic or paradoxical outcomes.

As Quintilian claimed, then, epideictic rhetoric doesn't necessarily act in isolation. Sometimes it functions with other rhetorical modes and strategies like a symphonic orchestra, amplifying certain praiseworthy or derogatory features of the subject, with other objectives driving the ensemble in one direction or another. Depending on the rhetorical situation, this intermingling of epideictic elements can be visually emphatic or subtle;

it can brighten the subject with sunlight or eclipse it with clouds, and in doing either, steer the audience's interpretations toward larger rhetorical goals.

Conclusion and Implications for Design

Epideictic rhetoric saturates the visual language we encounter every day that honors (or disparages) its subjects. Through amplification that brandishes the rhetor's design dexterity, visual language casts a positive (or negative) glow on the subject, making it more brilliant, attractive, and engaging (or odious and offensive). Comparisons and superlatives, along with vivid descriptions, further heighten these effects. All of these design elements aim to elevate their subjects—people, buildings, technology, natural phenomena—beyond the mundane world. For some audiences, such conspicuous attention to praise or blame might seem like justice well served in giving these subjects their rightful due; to other audiences, it might seem to devolve into embellishment and hyperbole. Either way, the techniques for deploying the visual epideictic have evolved with advances in technology, which have increasingly democratized these rhetorical acts, giving everyone with an interface the tools to create ornate certificates and awards, deploy congratulatory pictures and emojis, and design charts and graphs visualizing praise or blame. All of these digital affordances, along with interactive digital media, make the visual epideictic more accessible than ever, both to rhetors and their audiences.

On a broader level, the rhetorical processes for designing and interpreting the visual epideictic also respond to diachronic conditions. To praise or to blame is to judge—and judging requires criteria and standards. At a given cultural and historical moment the community values that underpin these judgments are constantly affirmed and promulgated by epideictic visuals, providing social and rhetorical cohesion that binds communities together. Visual language that accompanies ceremonies and rituals explicitly fulfills this purpose, while other visuals do so implicitly by glorifying (or disparaging) their subjects and the values they embody. Epideictic designs inspire us collectively to realize a higher standard by making visible the people we strive to emulate and the things and places we aspire to possess or experience—and conversely, those we try to eschew. Visually amplifying collective values both energizes and crystallizes them, creating a rhetorical magnet for audiences searching for paragons and ideals within—or just beyond—their reach.

Amplifying Praise with Visual Language

When we want to praise people for things they've achieved, visual amplification intensifies the epideictic acts. We practice these acts both formally and informally, with long-term and short-term temporality:

- In formal situations when we're looking retrospectively at something someone has achieved, visual language bridges past, present, and future. Respectful, substantive praise will usually be amplified by the well-worn path of visual conventions—a framed certificate, a plaque, or other tactile object that memorializes the event. Although these artifacts may eventually collect dust, their material presence extends the rhetorical act well into the future. Digital design lacks this material temporality—virtual Oscars won't cut it.
- On the other hand, when we're praising less momentous accomplishments, digital media offers a more effective and timely medium for visualizing praise. Thumbs-ups, hearts, smiley-face emojis, cross-posted pictures, and other mini-epideictic acts give us the instantaneous ability to praise personal or professional achievements that pop up on the screen.
- For any of these epideictic acts, designers also need to consider the collective values that underpin them, which are subject to diachronic forces and cultural variations. Designing an award that touts exemplary behavior—for example, green energy efforts by employees—reflects "adherence" to the current values of that organization. Similarly, a thumbs-up (or other visual affirmation) on social media adheres to the collective values of others making similar gestures. Aligning praise with audience values will certify the amplification.

Praising things and places through visual amplification also applies to persuasive situations for selling products or experiences—from cars, furniture, jewelry, and real estate to hotels, resorts, and concerts. Vivid description (energeia) arouses audience interest and engagement and can be amplified with selected views, color enhancements, favorable angles and lighting, and the presence of human forms. The picture of the parish hall plan in figure 1.17 uses some of these techniques to celebrate the building after several years of planning and deliberation.

Visualizing Blame Constructively

Well-designed blame can be constructive, especially if it aligns with communal values shared by our audiences. Warnings and exhortations not to smoke, overeat, do drugs, or otherwise put yourself in harm's way are a routine part of our daily lives, taking the form of posters, infographics, and icons. Here are some suggestions for amplifying blame:

- Impersonal and abstract images using modernist minimalism, like the Everyperson in figure 2.25, enable designers to amplify harm without personalizing it. Everyperson images also allow diverse audiences to identify with them.
- Digital data design provides a potent and nimble public medium, with maps and charts plotting harmful conditions—crime, pollution, flooding—to raise awareness and modify behavior. Audiences can customize the data by using interactive features that display micro-level details.
- In the tradition of Hogarth's pictures, comics and graphic novels with recurring characters provide powerful epideictic tools for amplifying risk. Designing these pictorial stories can be daunting for novices, but comics software and AI tools can simplify the work.

Because we are compelled to make judgments about the people, places, and things we tout or disparage, we must carefully consider how closely those assessments adhere to the communal values of our audiences, which guide their interpretations. In short, as designers we constantly need to remind ourselves that the visual epideictic is a highly social activity.

Chapter 3

The Pathos of Memory

Sentiment, Nostalgia, and Rhetro Design

The art of rhetoric, contends Richard Weaver (1970, 161–84), must rely not only on rational thinking but also on emotion and an abiding awareness of the past in order to fulfill its potential for enriching modern life. Every day in our personal and professional lives we encounter all kinds of images—popular, practical, and aesthetic—that arouse emotion by invoking the past. In these encounters with visual language time heightens the rhetorical power of emotion through sentiment, memory, and nostalgia. These emotions engender an attachment to a past that audiences may never have experienced firsthand but nevertheless perceive as more authentic and stable than the present. Fostered by a seemingly fixed and immutable past, rhetorical memory (memoria) aligns with the concept of chronos (Tirrell 2015, 174), though audiences often hold different versions of the past (Kurlinkus 2018), based on their values and lived experiences. The past itself is also subject to ongoing scholarly scrutiny and revision. However variable and negotiable our perceptions of the past, emotion triggered by time and memory serves some potent rhetorical purposes in information design—first and foremost to persuade, like other emotional appeals, but also to attract and engage audience members and to connect them to each other through a shared memory of a past.[1]

Viewed from the expansive vista of chronos, sentimental values provided the long-term cultural and rhetorical impetus for fusing time and emotion. With its origins in Romanticism, the sentimental permeated nineteenth-century literature and visual art; it also found practical applications in commercial designs promoting products ranging from farm equipment and flower seeds to clothing, domestic supplies, and other consumables. The sentimentality of Victorian aesthetics, however, was later supplanted by modernism and its emphasis on lean universal forms that maximized perceptual efficiency, subdued emotion, and eschewed the past. Today retro designs elicit emotion by invoking the sentimental with nostalgia, often (ironically) with the aid of digital technology.

In this chapter, then, I'll explore the connections between emotion and the past as they evolved in visual language over the past few centuries. I'll begin by tracing the place of pathos in the rhetorical tradition and identify some of its key elements. I'll show how the rise of the "sentimental" in the eighteenth century cultivated an aesthetic based on sensibility, memory, and moral empathy; how the sentimental intertwined with the cultural and aesthetic values of the picturesque, the Gothic Revival, and Romanticism; and how these values were visualized in painting, architecture, and landscape gardening and how they later saturated Victorian design. I'll then explain how arousing emotion through sentiment and memory was subverted by early twentieth-century modernism, which rejected sentimental values and replaced them with a functional aesthetic embodying the utilitarian rationality of the industrial age. Today, partly as a reaction to modernism, emotion is re-evoked with "rhetro" designs—the rhetoric of evoking the past with vestigial elements—in the form of handwritten text, traditional typefaces, pictures, data design, color, and simulations of older technology. Finally, in recognition of the deeply personal aspects of time and emotion, I'll examine how memory and sentiment mediate our interpretation of visual language in a single lifespan.

Rhetoric, Memory, Sentiment: Springboards for Visual Pathos

To explain how time heightens emotional appeals, I need to delineate the intellectual, rhetorical, and aesthetic factors that foster this affinity. To do so, I'll examine three interwoven strands. First, I'll trace the foundational role of emotion in the rhetorical tradition, as concepts about pathos appeals evolved from antiquity through the Enlightenment, revealing the diachronic capacity of visual language to express emotion. Second, I'll explore how memory, both collective and individual, serves as a linchpin connecting time, emotion, and visual rhetoric. Finally, I'll examine the concept of the "sentimental" and how it became a powerful cultural force from the Enlightenment through the Victorian age, and how its residual effects infuse information designs we encounter today in our everyday lives.

Pathos and the Rhetorical Tradition

Emotion has long played a central role in the rhetorical tradition, beginning with classical rhetoricians. Aristotle, for example, believed that rhetors must combine pathos with logos and ethos to create effective arguments (Aristotle 2007, 112–13; bk. 2, ch. 2, secs. 1–8), and he describes the disparate emotions of "pain and pleasure" (113; bk. 2, ch. 2, sec. 8) and its various dualities as a means for understanding and appealing to audiences (116–47; bk. 2, ch. 2–11). In his *Institutes of Oratory* Quintilian (1959–1963) also describes a range of emotions, both strong and mild (2:421–23; bk. 6, ch. 2, secs. 8–17), that rhetors need to understand and strategically deploy to persuade audiences when reason alone won't suffice (Katula 2003). According to Quintilian, in order for rhetors to argue cases in court, they must harbor a deep and genuine commitment to emotion: "Consequently, if we wish to give our words the appearance of sincerity, we must assimilate ourselves to the emotions of those who are genuinely so affected, and our eloquence must spring from the same feeling that we desire to produce in the mind of the judge" (2:433; bk. 6, ch. 2, sec. 27). If we intend to use emotion to persuade others, Quintilian claims that we must first internalize it: "The prime essential for stirring the emotions of others is, in my opinion, first to feel those emotions oneself" (2:431–33; bk. 6, ch. 2, sec. 26). This requires a rhetor's genuine empathy, a principle that dovetails with Quintilian's maxim that only a "good" person can create authentic ethos (1:315; bk. 2, ch. 15, sec. 33).

In the eighteenth-century Enlightenment, the emotional aspects of rhetoric were rediscovered and redefined by, among others, George Campbell in his *Philosophy of Rhetoric* (1776), where he argued for the primacy of the passions, claiming that "when persuasion is the end, passion also must be engaged" (1:199; bk. 1, ch. 7, sec. 4). Indeed, persuasion isn't feasible without passion: "To say, that it is possible to persuade without speaking to the passions, is but at best a kind of specious nonsense" (1:199; bk. 1, ch. 7, sec. 4). So how does passion persuade? Largely, claims Campbell, it's aroused by external elements: specifically, "passion must be awakened by communicating lively ideas of the object," as "passion is most strongly excited by sensation" (1:207–8; bk. 1, ch. 7, sec. 4). According to Campbell, several other factors can stimulate an audience's emotions, among them the "connexion" (or "proximity") of time, place, and people. For example, talking about more recent events (the "connexion" of time) will be more affecting than talking about those in the distant past or the distant future (1:221–22; bk. 1, ch. 7, sec. 5). The "Connexion of place" (1:224–26; bk. 1, ch. 7, sec. 5) has an even stronger effect on audiences: "Who is not more curious to know the notable transactions which have happened in his own country from the earliest antiquity, than to be acquainted with those which have happened in the remotest regions of the globe, during the century wherein he lives?" (1:225; bk. 1, ch. 7, sec. 5). Finally, relationships with people—family, friends, other citizens—are even more powerful than place (1:226–27; bk. 1, ch. 7, sec. 5). When we identify with others with whom we share common bonds, emotion can be aroused more deeply. In short, to appeal to an audience's emotions, a "connexion" must be established temporally, spatially, or personally.

How do pathos appeals play out visually relative to Campbell's concepts? If someone, for example, receives a meeting invitation that prominently displays the date and hour in the near future (proximity of time), includes a picture of a local community center where the meeting will take place (proximity of place), and is hand signed by a close friend who belongs to the sponsoring organization (proximity of people), those emotional appeals will likely succeed. These various pathos appeals can intermingle in different ways. For example, including a picture of a famous structure (the Brooklyn Bridge, the Golden Gate Bridge) in a tourist website would cultivate proximity of place, given that many Americans relate to them, even if they have never seen them in person, though people living in New York City will have a stronger connection to the Brooklyn Bridge, as would residents of San Francisco with the Golden Gate. Moreover, including in an election fundraising letter the picture and signature of a local representative from the audience's home area (proximity of place, people, and time) will induce a greater emotional connection than a letter from someone in another state. Making emotional appeals with visual language, however, can be subtle and unpredictable: Some New Yorkers might loathe crossing the busy Brooklyn Bridge, skewing their interpretation; or a recipient of the meeting invitation might have a conflict with the specified time, weakening temporal proximity.

Memory as a Catalyst for Visual Pathos

Memory and rhetoric have long been deeply connected, though that affinity has shifted dramatically over time. In the classical era, memory had oratorical value as one of the five canons of rhetoric by enabling speakers to deliver speeches without reading directly from a text. Memory was also closely associated with visualization in the form of "loci memoriae," practiced by the Roman rhetoricians Cicero and Quintilian (Den Boer 2008, 19–20). This mnemonic technique, which rhetoricians have employed for millennia, entailed cataloging ideas and information into architectural space, as "rooms" in the mind. Sharon Crowley (1993) argues, moreover, that memory provides a repository of shared understandings about the past that feed rhetorical invention (36–39); as Crowley observes, "Memory and rhetoric thrive in cultures that celebrate community and plenitude" (43). Our ability to remember is also aided by "prosthetic" tools that Plato regarded as damaging to the mind (Tirrell 2015, 165–66). These tools so dominate the contemporary world that the art of memory has largely been lost because our methods for storing ideas and delivering messages have radically changed, with PowerPoints, handheld devices, and teleprompters keeping us constantly on track.

Today the associations between rhetoric and memory have broader implications for designing and interpreting messages, especially regarding the power of memory to elicit emotion. Emotion and memory are kindred spirits, and they constantly filter the way we perceive the world, often without much effort or conscious reflection. In the perceptual realm, Rudolf Arnheim (1969) explains how the "past" constantly mediates the perceptual acts of the "present," with our prior experiences filtering and categorizing objects and images we encounter (80–96). However, those encounters often entail more than image recognition: Visual stimuli filtered by memory often trigger emotion. When we visit a place from our childhood, for example, or hear a song from long ago, those sensory experiences immediately "bring us back." Perceptually, they create such close temporal (and spatial) proximity to our lives that we can't suppress the emotion.

Oftentimes memory triggered by visual stimuli fosters a nostalgic yearning for the past, which can be experienced both publicly and privately. Svetlana Boym (2001) differentiates between two forms of nostalgia: "restorative" and "reflective" (xviii). Nostalgia that's "restorative" seeks to preserve the past by memorializing it in monuments, refurbishing iconic buildings and artifacts, and witnessing to the "truth and tradition" of the historical past (xviii, 41–53). In contrast, "reflective nostalgia" is more personal and contemplative and "lingers on ruins, the patina of time and history, in the dreams of another place and another time" (41). Although these two modes of nostalgia aren't completely separable—they often overlap and are invoked together—they give us a rhetorical compass by which to consider the purposes and effects of emotions that are stimulated by the visual language of the past.

Drawing on memory to arouse emotion with visual language has a long and abiding history—perhaps most ubiquitously with public monuments like statues, triumphal arches, parks, murals, and other memorials. Scholars in the field of memory studies have examined the rhetoric of these artifacts, which are constructed to celebrate, mourn, or reflect on the past. Anyone walking the streets of central London, Paris, or Rome can't travel a hundred meters in any direction without finding an artifact intended to activate memory. In the U.S., famous artifacts like the Washington Monument, Lincoln and Jefferson Memorials, World War II Memorial, and Vietnam Veterans Memorial have a highly visible national presence, both for people who visit them and for those who experience them vicariously through films, photos, and drawings. Moreover, in squares and parks in virtually any town center on the planet, less celebrated monuments tell local stories about notable people and events.

The rhetorical power wielded by these artifacts derives heavily from their ability to stir the emotions of their audiences. Greg Dickinson, Carole Blair, and Brian Ott (2010) claim that "public memory is typically understood as animated by affect" in that it "embraces events, people, objects, and places that it deems worthy of preservation, based on some kind of emotional attachment" (7). And based on George Campbell's analysis of emotion, it's easy to see why: People in the U.S. have a "connexion" with influential Americans based on place (e.g., Lincoln, Edison, Clara Barton, Martin Luther King Jr.), just as they do with local figures in their own hometowns. And the same phenomenon occurs globally with people and events celebrated visually in close proximity to their audiences.

Time also figures substantially in connecting audiences with artifacts. Virtually all memorials, by definition, call attention to people and events from the past and in doing so derive their rhetorical energy by fusing time and emotion.[2] We respond to these artifacts because they recall an event or a person that often has residual effects on us today: a decisive battle, a life-changing discovery, a pioneer who settled nearby, the founder of a town, company, or university—all "connexions" that foster emotion. Memory triggered by visual language is also fed by larger forces, what Astrid Erll (2008) calls "cultural memory"—the shared history, values, traditions, and aesthetic tastes of a nation-state or region that its respective social or cultural formations externalize with memorials, spaces, texts, and rituals (1–5). Memory envisioned on this scale creates the rhetorical glue that binds audiences together through a variety of modes and media: sculptures, plaques, street names, buildings, parks, their accompanying textual inscriptions, and other visual artifacts that evoke the past.

Of course, many of these artifacts don't last forever, as values change and memories fade. Statues and buildings are torn down, plaques are removed, and park and street names are changed—and with their demise, the emotions they invoked are also erased and redirected to new artifacts recalling the past. The past, then, is not a stable, monolithic entity. As William Kurlinkus (2018) points out, we inhabit a world with a plurality of pasts (4), some of which conflict or compete with each other or barely attract attention.[3] In walking across my campus, I sometimes spot worn, algae-covered benches and stones that memorialize alumni long deceased, and I wonder whose past they represent and who among the living still connect with and celebrate it. Herein lies the fine line between "restorative" and "reflective" nostalgia, with the former urging me to preserve these forlorn little monuments for posterity and the latter moving me to speculate wistfully on their hidden past.

The Emergence of Sentimental Values

Visual designs that elicit emotion are often associated with the "sentimental," an aesthetic approach that can be traced to eighteenth-century philosophy, literature, and culture. Viewed from the expansive vista of chronos, sentimental values provided a long-term cultural and rhetorical impetus for fusing time and emotion. Seminal moral and philosophical elements of this pivot toward emotion, especially the sentimental, appeared in Adam Smith's *Theory of Moral Sentiments* ([1759] 1774), where he formulates the concept of "fellow-feeling" (3–5, pt. 1, sec. 1, ch. 1) whereby an individual can, through the imagination, put one's self in another's shoes—for example, someone experiencing pain or deprivation. That kind of emotional empathy, which everyone shares to some extent, provides a social foundation for moral behavior that unites individuals and makes them accountable to one another. The arousal of emotion through visual stimuli was theorized by Edmund Burke in *A Philosophical Enquiry into the Origin of Our Ideas of the Sublime and Beautiful* (1759), in which he describes the varieties of pain and pleasure induced by sublime and beautiful objects. Burke's explanations and analyses of the sensory experiences that stimulate these emotions provided both an intellectual rationale and a psychological playbook for painters and poets to foster emotion through visual images, especially through the sublime.

During this same period new modes of intense feeling figured prominently in literary works like Jean-Jacques Rousseau's *Julie, ou La Nouvelle Hélöise* (1761) and Laurence Sterne's *A Sentimental Journey Through France and Italy* (1768). The sentimental also had a close affinity with "the man of feeling," popularized by Henry Mackenzie (1771) and widely adopted as one of the aesthetic and cultural trademarks of the period. As the Enlightenment transitioned to Romanticism, sentimental values infused literature and art. Early Gothic novels, especially those by Ann Radcliffe like *The Mysteries of Udolpho* (1794), blended the sentimental and the sublime, as did poems by William Wordsworth that featured rural characters and settings.[4]. In his treatise *On the Naive and Sentimental in Literature* (1795–1796) Friedrich Schiller defines sentimental aesthetics as a distinctively modern phenomenon. He describes two differing modes of aesthetic expression: the "naive" mode that directly engages with nature, is instinctive (and even childlike), and characterizes much of classical poetry; and the "sentimental" mode that comes from within, is reflective and emotive, and characterizes much of modern poetry.

Throughout the nineteenth century, sentimental values continued to infuse art and literature, both in Europe and America (De Jong and Bennett 2013; Howard 1999; Kaplan 1987), where they also animated social movements, most notably through abolitionist writings like Harriet Beecher Stowe's *Uncle Tom's Cabin* (1852) in the form of sympathy for the oppressed (De Jong 2013, 1–3). As a broad cultural, social, and aesthetic force closely allied with Romanticism, sentimentalism was also appropriated for commercial purposes, especially in the later nineteenth century in designs promoting agricultural and domestic products. As we'll see later, practical communications could hardly escape the reach of such a widespread and deeply rooted cultural phenomenon that embodied potent emotional appeals.

Discord and Symbiosis Between the Sentimental and Sublime

Although sentimentalism inundated nineteenth-century culture, later it was disparaged for its hyperbolic expression of emotion, especially in literature (Wilkie 1967), a perspective that has been forcefully countered more recently (Howard 1999) but still lingers among critics of literature and visual art. The denigration of sentimental values stems partly from the perception that it resisted progress. While the sentimental looked nostalgically to the past, fostering traditional communal values and domesticity, the "technological sublime," as we saw in chapter 2, pushed in the opposite direction by celebrating industrial miracles, unleashing powerful forces that elicited wonder and hope mixed with anxiety and trepidation.

The coexistence of these two conflicting views shouldn't be surprising. As Svetlana Boym (2001) observes, "Nostalgia inevitably reappears as a defense mechanism in a time of accelerated rhythms of life and historical upheavals" (xiv). Instead of pitting the sentimental and the "technological sublime" against each other, then, we might regard them as two sides of the same subjective coin: The sentimental idealized the past to escape from the dehumanizing effects of industrialism, while the "technological sublime" idealized the future through the spectacle of human ingenuity. Both injected emotion into the visual culture they temporarily shared in the late nineteenth century. Today we can look back in wonder at images of giant steam locomotives, masonry-clad skyscrapers, and *Titanic*-sized ships while at the same time experiencing the sentimental nostalgia of retro design.

Romantic Emotion: From Landscape Garden to Victorian Page

In addition to the sentimental, the penchant for expressing emotion was strongly tied to an emerging visual culture that had its modern roots in Romanticism and the cultural and aesthetic factors that conceived it, beginning in the mid-eighteenth century. These seminal developments included the presence of ruins in landscape gardening, picturesque aesthetics, and the revival of the Gothic—all of which were intertwined and often looked backward, to the past, for inspiration.

Ruins and the Picturesque: Heightened Emotion in the Eighteenth Century

The synergy between time and emotion developed rapidly as the incipient values of Romanticism began to coalesce in the eighteenth century. A growing fascination with ruins—real, artificial, or imagined—became a focal point for experiencing and expressing emotion, with time past as the rhetorical fulcrum. By the middle of the eighteenth century, visiting ruins or installing them in private gardens became highly fashionable, eliciting powerful emotions, ranging from delight and reverie to melancholy, dread, and terror. Thomas Wharton's poem "The Pleasures of Melancholy" (1747) captures this mindset:

> Beneath yon' ruin'd Abbey's moss-grown piles
> Oft let me sit, at twilight hour of Eve,
> Where thro' some western window the pale moon
> Pours her long-levell'd rule of streaming light; (5)

By visualizing time in stone, ruins evoked dissolution and abandonment—whether the remnants were from classical antiquity or medieval abbeys, castles, and churches, which were abundant in England and northern Europe, or they took the form of natural ruins in America. Picturing ruins of course had precedents, with the landscapes of Nicolas Poussin, Jacob Van Ruisdael, Salvador Rosa, and Claude Lorraine paving the way the century before, often by situating ruins in classical or mythological settings. The practice of visualizing ruins, however, exploded in the eighteenth century with the paintings and drawings of Hubert Robert, Claude Vernet, Giovanni Battista Piranesi, and Paul Sandby, along with legions of other artists during this era that depicted classical, medieval, or natural ruins, real or imagined.

The visual language of ruins perfectly embodied features of the emerging picturesque aesthetic. Ruins typically had rough, irregular surfaces that were worn by the forces of time, nature, and neglect, leaving them weather-beaten and disheveled. These physical attributes, which conflicted with neoclassical principles of order and symmetry, were strongly endorsed by advocates of picturesque landscape beginning in the early eighteenth century. In his *New Principles of Gardening* (1728) Batty Langley, for example, denounced the "stiff regular" garden plans imported from the Dutch (iv), preferring instead gardens with a "variety" of features "that will present new and delightful Scenes to our View at every Step we take" (iii).[5] The new method of gardening often included mock ruins created from painted canvas or from brick and plaster (to mimic stone) and placed in the landscape (xv).

Because of their physical attributes, their ability to stimulate memory, and therefore their capacity to arouse emotion, ruins became a staple of both art and landscape design as the eighteenth century unfolded.[6] In his *Observations on Modern Gardening* (1770), Thomas Whately declares that ruins "are a class by themselves, beautiful as objects, expressive as characters, and peculiarly calculated to connect with their appendages into elegant groups" (130). Physically and psychologically, they were well suited to picturesque gardening, given that "imperfection and obscurity are their properties" (131) and that "certain sensations of regret, of veneration, or compassion, attend their recollection" (132).

Ruins fostered these kinds of emotional responses because they often were pictured with shrubs, trees, and other forms of vegetation sprouting from their remains, temporally fusing them with nature through a process of reclamation. Figure 3.1 shows an etching by artist Paul Sandby (1758)

Figure 3.1. Drawing by Paul Sandby of a ruin with figures (1758). Courtesy of the National Gallery of Art, Washington. Ailsa Mellon Bruce Fund.

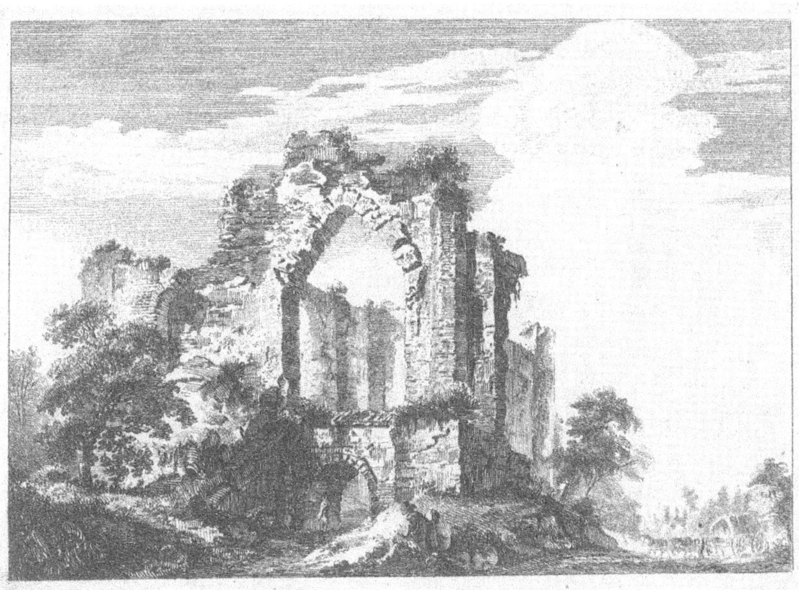

that illustrates this assimilation, here on a massive stone church or monastery. Vegetation shoots out of the crevices all around and atop the Gothic arches, while a figure in the foreground contemplates the forlorn scene, and in the background (on the right) a wagon crawls up the road with two figures on foot. Similar affecting scenes of ruins undergoing natural reclamation appeared in the plates (*Recueil de Planches*, 1762–1772) that accompanied the French *Encyclopédie*. A series of drawings titled "Antiquities" envision classical Roman ruins where tourists, probably embarked on the Grand Tour, mill about the massive, slowly deteriorating edifices embellished with vegetation. One of these drawings, slightly cropped, appears in figure 3.2.

Collectively, these images of ruins represent a tiny sampling of what quickly became a flood of such images—in paintings and drawings of ruins, like those of Piranesi (1748); in mock ruins and "follies" on country estates; and in Gothic novels, romances, and "Graveyard" poetry. Ruins also became stock features of books on landscape gardening and picturesque sketching and travel.[7] In England the fascination with picturing ruins continued into the Romantic era and beyond—for example, with J. M. W. Turner's pictures of Tintern Abby (1794) and the River Wye (1812). Images of ruins also became popular on the European continent as picturesque aesthetics spread beyond the British Isles (Hunt 2002; Wiebenson 1978).

Figure 3.2. Ruins in the Roman forum of Nerva pictured in the *Recueil de Planches* (Diderot and d'Alembert 1762–1772, vol. 1, "Antiquités," plate IV, fig. 3) that accompanied the *Encyclopédie*. Courtesy of Iowa State University Library Special Collections and University Archives.

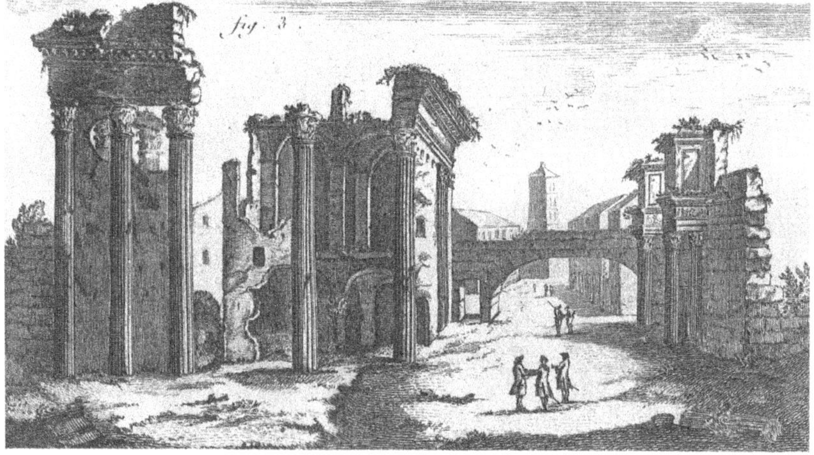

In all these incarnations, ruins made time and memory visible, provoking emotional responses that ranged from the sentimental to the sublime.

Gothic Revival: The Cultural Nostalgia of Handmade Objects

Gradually, the immersion in ruins—especially medieval ruins—evolved into a revival of Gothic architecture and decoration, a resurgence that acquired both aesthetic currency and social and moral capital. Evoking the memory of the Christian medieval world, from which it found its inspiration, the Gothic Revival inundated the design of domestic, civil, and religious buildings during the Romantic and Victorian eras, including the Parliament Building in London and innumerable churches and buildings throughout England, as well as hospitals, libraries, courthouses, and schools across the United States. A. Welby Pugin's *Apology for the Revival of Christian Architecture in England* (1843) articulated his vision for a national revival of the Gothic, epitomized in his drawing (fig. 3.3)

Figure 3.3. A. Welby Pugin's vision for a Gothic city in his frontispiece for *An Apology for the Revival of Christian Architecture in England* (1843, plate I). Courtesy of the City College of New York Cohen Library.

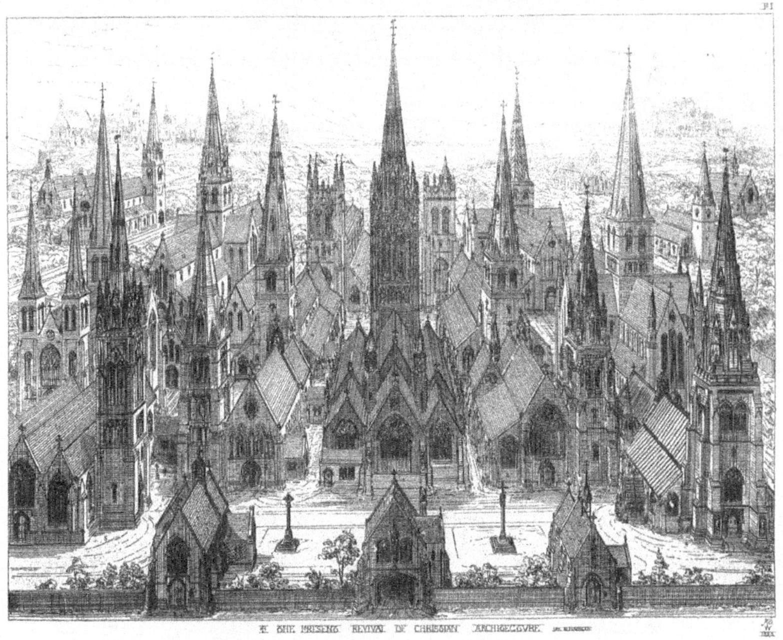

of a massive assemblage of towers, pinnacles, pointed arches, and steeply gabled roofs—a veritable Gothic Jerusalem. Like Gothic ruins and romances in the eighteenth century, the resurgent forms of the Gothic continued to stir its audience's emotions, with its towering forms and indigenous style mingling the sentimental and the sublime with feelings of nationalism.

Gothic aesthetics influenced virtually all aspects of visual culture, including ordinary Victorian dwellings replete with steep roofs, corner towers, multiple gables, and intricately decorated porches. Its influence also materialized in interior spaces with ornate, hand-carved furniture, wallpaper with profuse floral designs, and elaborately decorated utensils. The influence of the Gothic, and medievalism more generally, also extended to information design. Figure 3.4 from Hermann Esser's *Draughtman's Alphabets* (1877) shows "north" symbols for architectural drawings with their elaborate finials and fletching that mimic the tracery of Gothic stone and that required labor and physical dexterity to execute. Handmade designs, inspired by the Arts and Crafts movement, emphasized the moral and social value of hand labor and a resistance to mass-produced goods. Many of these practices were inspired by John Ruskin (1867), who advocated for Gothic design methods that allowed for individual creativity, which was thwarted by the dehumanizing effects of industrialism.[8] In this way, as cultural nostalgia and a means of social reform, contemporary design was allied to past practices and their presumed ideals. The elaborate north symbols in figure 3.4 illustrate one small example of this deep commitment to these design principles.

Figure 3.4. Highly ornamented north arrows from a book for engineers, architects, and draftsmen (Esser 1877, 27).

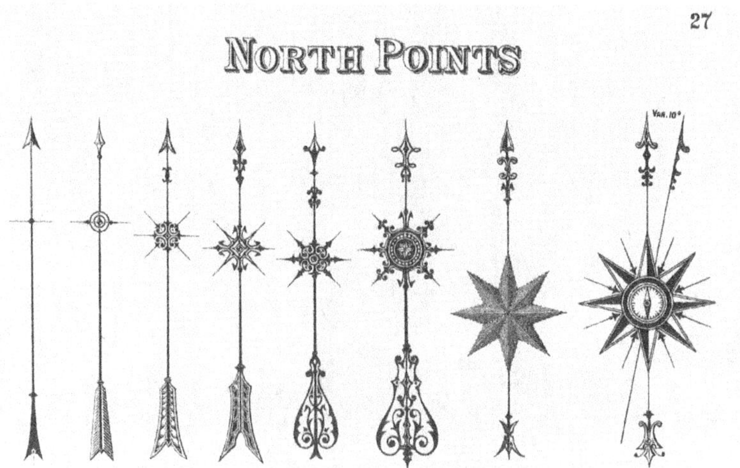

The Rhetoric of Victorian Emotion: Romantic Values in Action

All of these design concepts and forms—the sentimental, the picturesque, the Gothic—found their origins in the nexus of memory and emotion that began in the eighteenth century and that propelled design right up to the doorstep of twentieth-century modernism, both in Europe and America. These ideas thoroughly infused the visual culture of the nineteenth century when they were reinvented and repurposed for commercial applications. Saturated with sentiment, typography and images during the Victorian era trafficked heavily in emotion.

Romanticizing Nature: Sentimental Scenes of Agrarian Life

This cultural and aesthetic diffusion in the United States materialized most conspicuously in pictures that romanticized nature and agrarian life.[9] Despite increasing industrialization, the U.S. remained largely an agricultural society throughout the nineteenth century, and the popular hand-colored lithographs created by Nathaniel Currier and James Merritt Ives captured this lifestyle. Figure 3.5, for example, shows a Currier & Ives picturesque

Figure 3.5. Picturesque harvest scene by Currier & Ives (n.d.). Courtesy of the Library of Congress, Prints and Photographs Division [LC-DIG-pga-09065]. 1

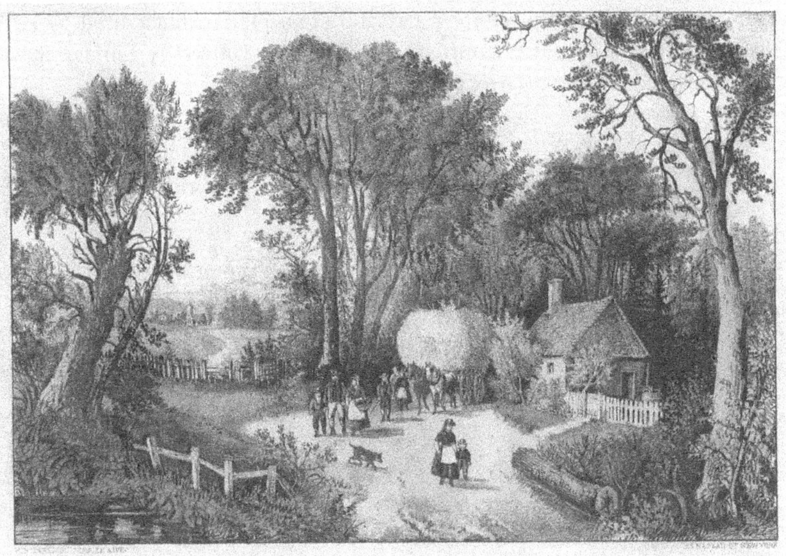

harvest scene with elements that are typically "rough" and sentimental: a rustic cottage, a blasted tree on the left, irregular fences, a small body of water in the foreground, rural people on the road with a hay wain, and a distant church with mountains beyond. This composition aimed to delight and stir the emotions of its viewers—to recollect some such scene from their own lives or to imagine themselves in this ideal setting, ensconced in nature amid the company of their relations. Today Currier & Ives has become synonymous with these kinds of sentimental picturesque scenes that have widespread appeal in the U.S. and can still be seen hanging on the walls of homes, offices, and public buildings. Indeed, today in the U.S. when we utter the word "Victorian" we usually mean something akin to a Currier & Ives image infused with sentiment and emotion.

Sentimental pictures of rural landscapes were widely used by U.S. manufacturers of agricultural equipment to promote their products. Harmonizing human figures, technology, and nature, the picture in figure 3.6 shows a promotional piece for an Ohio implement company, A. W. Coates, that manufactured mechanical rakes for harvesting hay and grain. The operator of the horse-driven machine labors alone in nature in a very Romantic and Emersonian setting—a private preserve, wild and remote—that contains several picturesque elements: a body of water on the left, fir trees lining its banks, and a distant barn and fences. A wagon

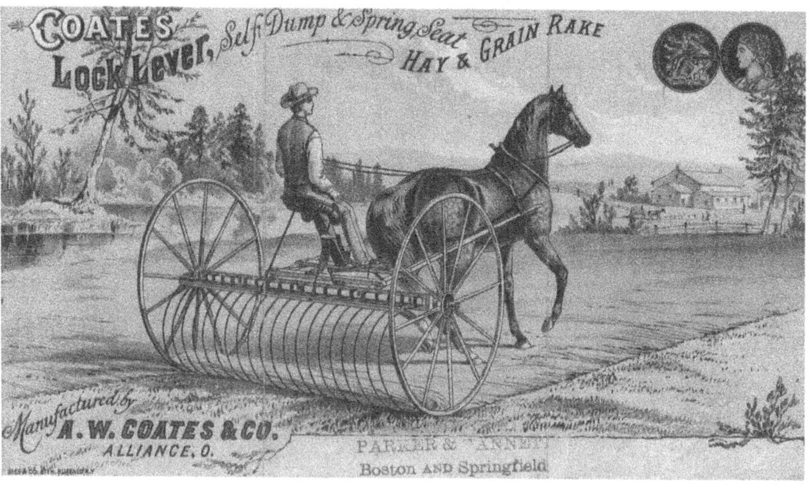

Figure 3.6. Promotional piece for a hay and grain rake manufactured by A. W. Coates & Company (n.d.). Courtesy of Iowa State University Library Special Collections and University Archives.

and a few small figures also appear in the distance, socializing the scene. These kinds of compositions appeared frequently in promotional materials for farm implements, projecting a sentimental emotional appeal that envisioned the potential buyer blending harmoniously with nature while easefully operating the new technology that enabled this lifestyle.

This Romantic ideal also appears in figure 3.7, another promotional piece for farm machinery, here grain binders made in Cleveland by the Davis Platform Binder Company (1889). Like a Currier & Ives picture, the composition integrates picturesque elements by envisioning a home across

Figure 3.7. Romanticized farm scene from the cover of the 1889 *Davis Platform Binder Company Catalog*. Courtesy of Iowa State University Library Special Collections and University Archives.

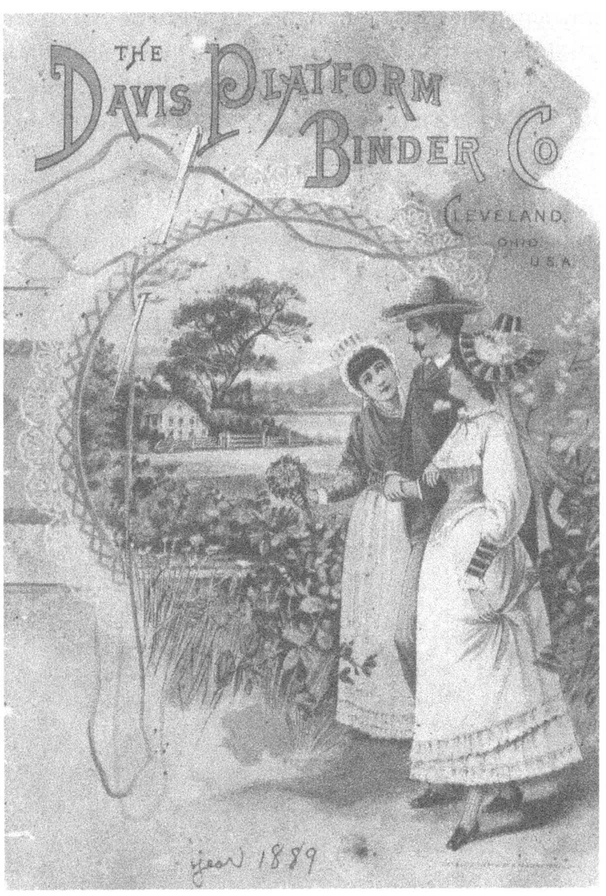

from a field of grain, framed by trees and a rough-hewn fence, with a pond in the foreground embellished with flowers. This ideal farmstead is framed with needlework that binds the threads, feminizing and domesticating the scene. The figures in the foreground enhance the picturesque setting, with a young woman (probably a bride) on the arm of a well-dressed man, while another woman stands close by him, suggesting an emotional attachment and further reinforcing the binding theme. Together the landscape, handmade needlework, and romantic figures create a sentimental effect aimed at eliciting a strong emotional response from potential customers.

Sentimentalism for commercial purposes was also visualized with images of children. Romanticism celebrated the innocence and imagination of childhood, epitomized in poems by William Blake and William Wordsworth and in the paintings of German artist Philipp Otto Runge.[10] These sentimental figures appear in figure 3.8, a promotional piece for axle grease where children navigate a wagon on a dirt road that winds through a picturesque landscape, with dwellings, fences, and trees checkering the distant fields. The youngsters appear to be relishing their country jaunt:

Figure 3.8. Children promoting Paragon axle grease by riding and pulling a wagon (Meriam & Morgan Paraffine Company 1880). Courtesy of the Library of Congress, Prints and Photographs Division [LC-DIG-pga-03987].

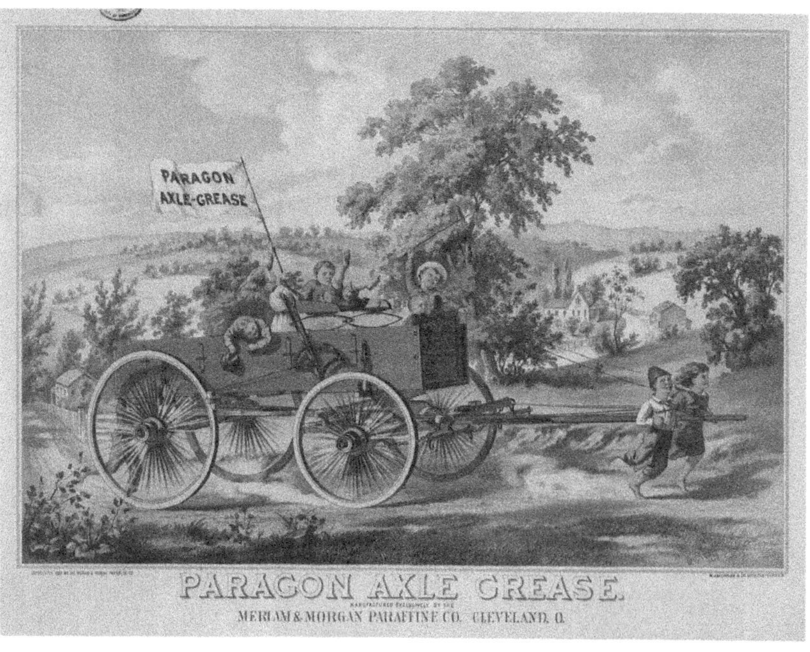

one of them holds the company flag, others frolic in the wagon, and two barefooted children pull it, a bit of hyperbole illustrating the effectiveness of the grease. Although enlisting children to promote axle grease might seem far-fetched today, visualizing children was common in the Victorian era, fostering both nostalgia and levity. Adults could sentimentally recollect similar scenes from their own childhoods as well as experience a humorous diversion from the harsh conditions of rural life.[11]

Visualizing Domestic Life: Sentimental Scenes from the Home

The home also provided a setting for visualizing sentimental values for persuasive purposes. Images of women figure prominently in Victorian visual culture, especially toward the end of the century, with domestic spaces arousing emotions of security and comfort and insulating women from the competition and corruption of the commercial and industrial world outside the home (see Tange 2004). Figure 3.9, a sales poster for

Figure 3.9. Promotional piece for the Standard Wax Company that shows a domestic interior from the Victorian era (ca. 1880s). Courtesy of the Library of Congress, Prints and Photographs Division, Popular Graphic Arts Collection [LC-DIG-pga-11390].

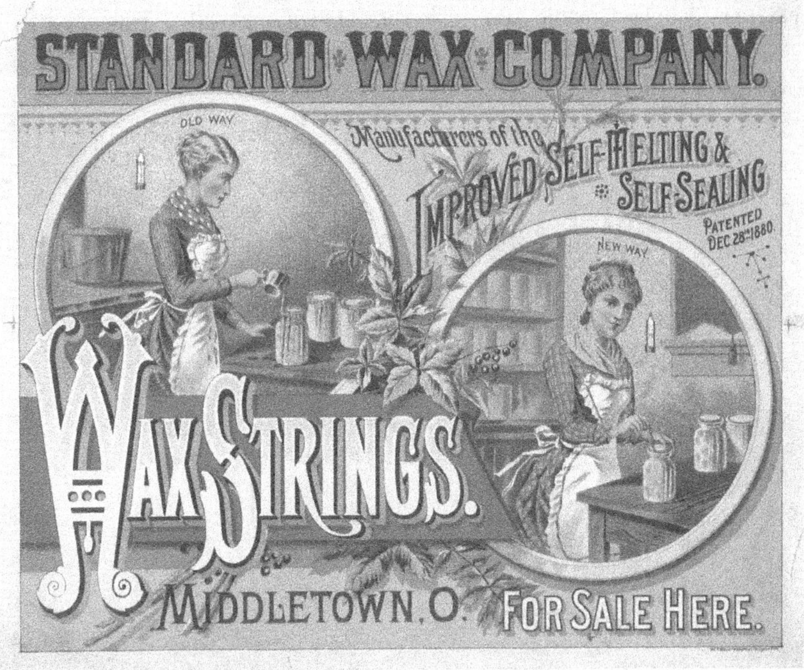

the Standard Wax Company, shows a woman in her kitchen using wax strings to seal canning jars, both the old version of the wax strings (upper left) and the new and improved version (lower right), which seems more pleasing to her. Embellished with vegetation and slab letters (with thick serifs), this image combines sentiment and typographical punch. As Sears, Roebuck and Company and other commercial enterprises increased the demand for consumer products, images like this became a staple of late Victorian culture, blending business with domestic life.

Domestic sentimentality is envisioned both textually and graphically in the popular *Home Sweet Home* lithograph by Currier & Ives (1874) shown in figure 3.10. The profusion of flowers and vegetation connects the home to nature and to the moral sensibility associated with handwriting capable of executing this detailed and exquisite design. As I mentioned earlier, handmade designs were touted by the Arts and Crafts movement, which championed the moral and social value of labor and renounced the mass-produced goods of industrialism. Displaying this sign in the home—near a door or in a parlor or dining room—aroused the senti-

Figure 3.10. Exquisite penwork of a Currier & Ives (1874) lithograph celebrating ideal domestic life. Courtesy of the Library of Congress, Prints and Photographs Division, Popular Graphic Arts Collection [LC-DIG-pga-09100].

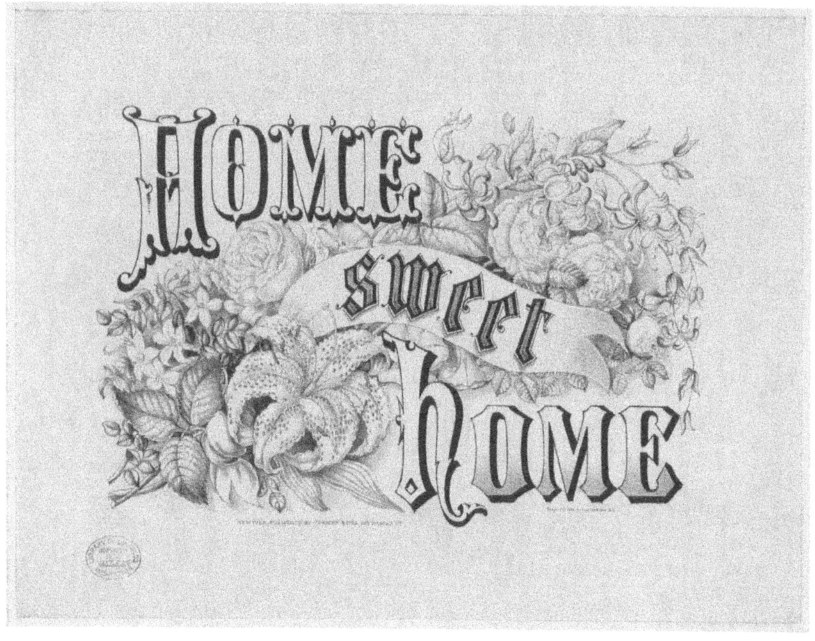

ment of a safe, blissful, and morally principled dwelling. The epitome of textual sentiment, this sign still hangs today in many homes, conjuring up a nostalgic image, however faithful to reality, of domestic life from the Victorian past.

Picturing Women at Work: Sentimental Agents

As the age of sentimentalism unfolded, women also appeared outside the home in agrarian settings to promote agricultural products. Figure 3.11 shows a woman named "Minnie" handling bundles of grain, while

Figure 3.11. Promotional piece for the Minneapolis Harvester Works (ca. 1882) showing "Minnie" handling bundles of grain in a picturesque setting. Courtesy of Iowa State University Library Special Collections and University Archives.

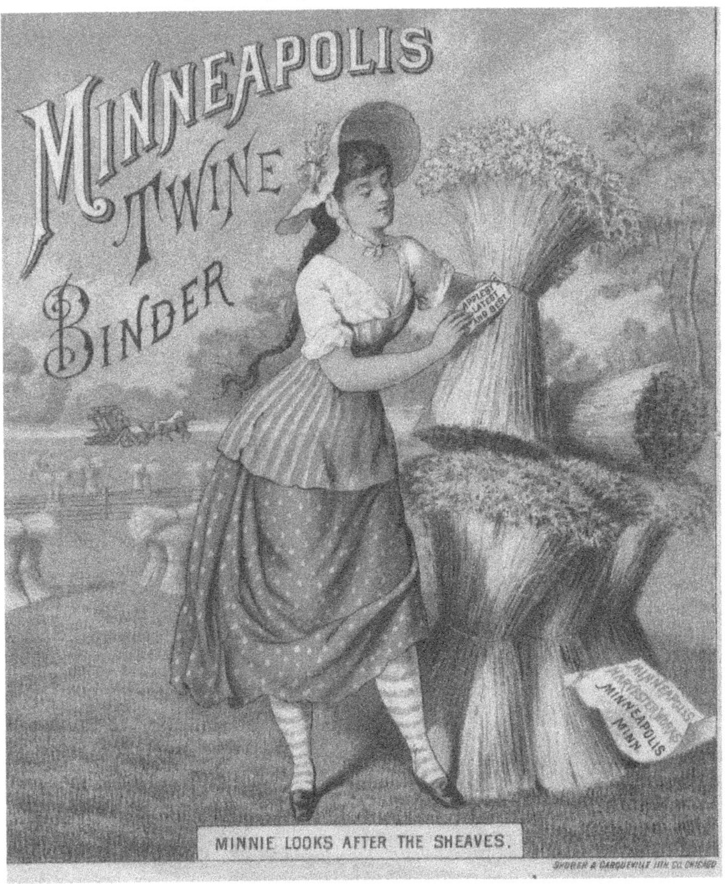

another figure harvests and binds it in the background amid a picturesque landscape that blends crop fields and woods. Representing the Minneapolis Harvester Works, which manufactures the twine binder, Minnie is an agrarian Rosie the Riveter who works hard and *produces*, the main selling point of this promotional piece. However, beyond her functional efficiency, Minnie also persuades with emotion by injecting a sentimental domestic appeal. Sporting long braided hair, a colorful skirt, fancy hat, and striped stockings, this hardworking farmer hasn't relinquished any of her femininity, something that both male and female audiences might find equally persuasive.

Women were also portrayed erotically for commercial purposes, inspired by Pre-Raphaelite painters like Dante Gabriel Rossetti and John Everett Millais, whose neo-medieval style followed in the stream of the Gothic Revival, as did Art Nouveau design, which followed in its wake. Achieving both aesthetic renown and public visibility, Pre-Raphaelite (and Art Nouveau) pictures typically featured women with long flowing hair and gowns, elongated necks, and distant, longing expressions. The woman in figure 3.12 promoting McCormick harvesting machines (1901)

Figure 3.12. Cover of the *McCormick Harvesting Machine Company 1901 Catalog*. Courtesy of Iowa State University Library Special Collections and University Archives.

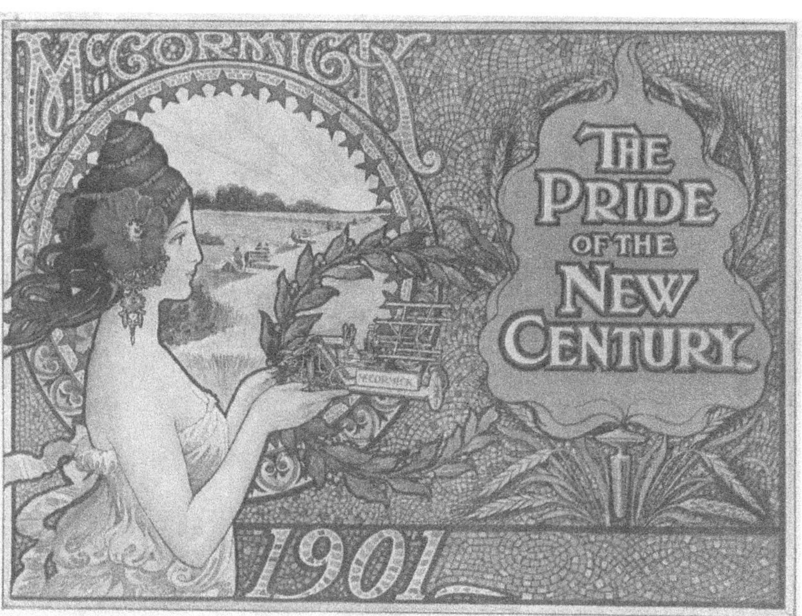

possesses these alluring features, while the flowers and jewels in her hair add a sentimental element, as do her intricate earrings, the toy harvester in her hands, the pastel mosaic around her, and the other profuse details that typify late Victorian design. Alluring women like this—simultaneously sentimental and erotic—appear in many other promotional materials of the time, not only for agricultural equipment but also consumer goods ranging from clothing and soap to food and tobacco.

In a variety of ways, then, nineteenth-century document design projected emotional appeals allied to larger cultural forces—the picturesque, the Gothic, the sentimental, and other Romantic values—by activating memory, creating affecting pictures of nature, and idealizing domestic life. The profuse decoration embodied in Victorian design—intricate lettering, flowery borders, detailed patterns, striking colors—provided sensory stimulation that further heightened emotional engagement. Do practical communications still draw on these design elements to foster emotional appeals? Indeed they do: Victorian designs are constantly redeployed and repurposed in retro designs for restaurants, hotels, antique malls, theme parks, and other enterprises that try to evoke the memory of a simpler, more authentic past and its lingering sentimentality. However, part of the answer to this question can also be found in examining the tension between modernism and the sentimental past: specifically, the intent of modernism to subdue the emotions aroused by the perceived excesses of nineteenth-century design.

Modernism: The Sentimental Rebuked with a Fresh Start

The late nineteenth century marked the zenith of sentimental values, with post-Romantic nostalgia at its height and the Arts and Crafts movement fully mobilized, both in its practice and in the public consciousness. At the same time, the forces of early modernism began to coalesce around a new aesthetic that broke away from sentimentalism (M. Bell 2000, 160) and that valued instead machine-age functionalism (Banham 1960), scientific objectivity, or "*Sachlichkeit*" (Kinross 1985, 26), and minimalist design forms.[12] In the fine arts, the push-back against sentimentalism occurred on several fronts: the French Fauve painters (the "wild beasts"), the Italian Futurists, and the Bauhaus in Germany, among several other modernist innovators and movements in the Netherlands, England, Russia, and Austria. The sharpest, most aggressive attack against sentimental values came from the Futurist Filippo Marinetti, who in "The Founding and Manifesto of Futurism" ([1909] 1972) sang the praises of industrialism and its power, speed, and machine-age dynamism. Pursuing this revolutionary Futurist program, Marinetti argued in his "Manifesto," required a complete rejection of the past: "Do you, then, wish to waste all of your best powers in this eternal and futile worship of the past, from which you emerge fatally exhausted, shrunken, beaten down?" (43). The Bauhaus had its own forward-looking mantra, "starting from zero" (Wolfe 1981, 12). The conventions of the past didn't resonate any more, and after the horrors of World War I, designers searched for international forms that unified cultures in lieu of traditional forms that divided them.

This new lean minimalism is exemplified in the 1930s WPA poster in figure 3.13 with its sans serif typefaces arranged in an arch and at

Figure 3.13. Federal Art Project poster by Russell West (1936–1938) announcing an exhibition in Boston. Courtesy of the Library of Congress, Prints and Photographs Division [LC-USZC2-975].

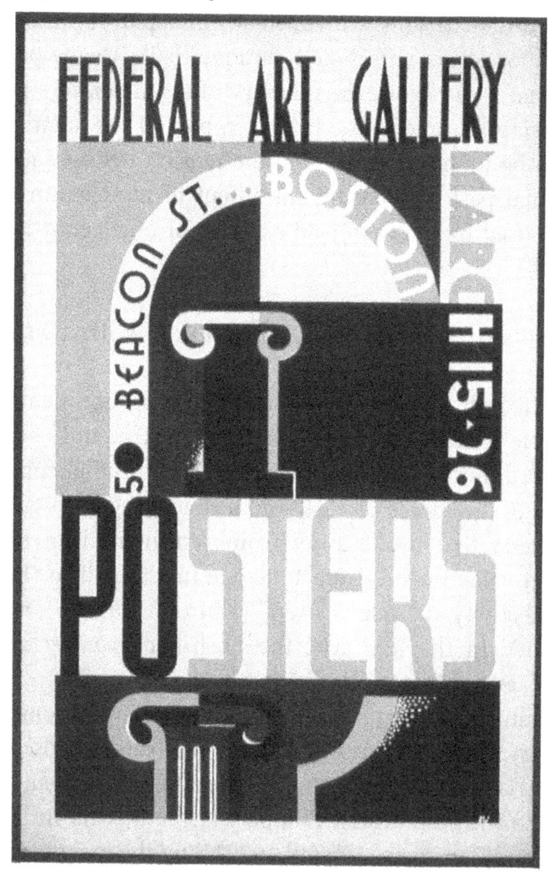

right angles and its abstract, high-contrast graphic elements that create a perceptually dynamic design. The classical ionic columns recall the Federal period of early America, but they trigger that memory stoically, sans sentimentalism, with scarcely any appeal to emotion. The same effect occurs in figure 3.14, a pictograph from *On Relief: General Relief Program* (U.S. Federal Emergency Relief Administration 1935), which charted the economics of the Great Depression, here the numbers of men and women receiving government assistance. The simple, high-contrast icons echo the Isotype system of Otto Neurath (1936), which like other forms of modernism had its roots in perceptual principles (Lupton 1986). In the *On Relief* chart, men appear in bib overalls, women in work dresses, both of them faceless and anonymous, with the subdued emotion attributable to their bleak, jobless lives as well as the minimalist design principles that shaped their appearance. These were lean, hard times, and this chart captures that reality by both stating the grim facts and recognizing their human toll but doing so without resorting to the heightened emotion of Victorian designs.

Figure 3.14. Modernist chart from the 1935 U.S. Federal Emergency Relief Administration's *On Relief: General Relief Program* (fig. XVI).

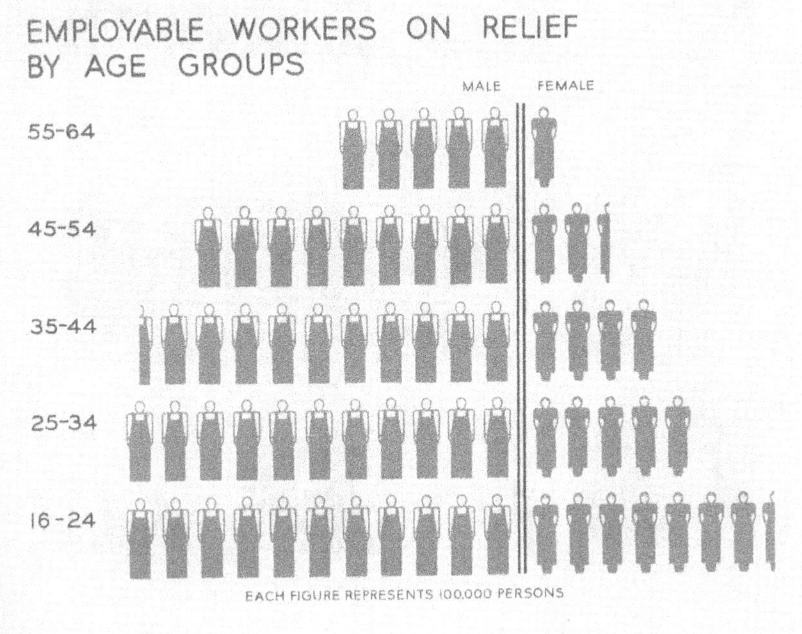

Modernist design was capable of subduing emotion even in the private domestic realm, where close personal relationships and passion typically go hand in hand. The guide to saving household electricity in figure 3.15, which appeared in a 1941 issue of the U.S. Department of Agriculture's *Consumers'*

Figure 3.15. Energy-saving guide with an abstract modernist design that appeared in a 1941 issue of the U.S. Department of Agriculture's *Consumers' Guide* (5).

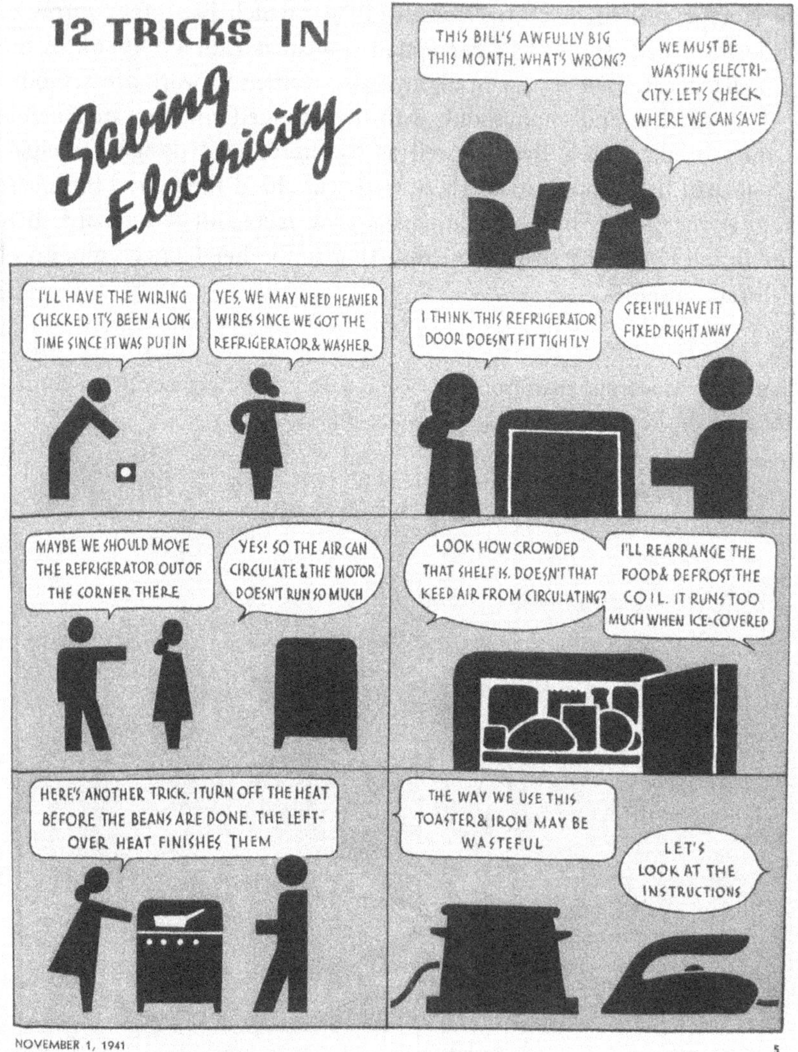

Guide, embodies the modernist style with its high-contrast geometrical forms that are abstract, impersonal, and matter-of-fact. The anonymous husband and wife share ideas with each other—and the audience—about how to save energy in their home by, for example, ventilating the refrigerator, repairing its door, and turning off the burner on the stove. The two characters appear to agree on the problems and the remedies, though blame seems to be implied in places. The muted, abstract figures, however, visually subdue the husband-and-wife banter, enabling the audience to absorb the advice without getting sidetracked emotionally.

That same suppression of emotion occurs in most warnings that feature highly stylized modernist figures. These flat, high-contrast figures serve multiple purposes by warning their audiences against being electrocuted, getting their fingers mangled in machines, or falling on slippery floors, off steep cliffs, or into abandoned mines—all highly charged emotional situations that threaten life and limb. However, the abstract figures that appear in these kinds of warnings, one of which we saw in figure 2.25, depersonalize their distress, enabling us to feel the pain vicariously without having to empathize with a specific, identifiable individual, reducing the proximity (the "connexion," as George Campbell put it) between us and the figure. We vicariously experience the danger of falling or the pain of mangled hands wondering if it's someone we know.

Modernism still has its fingerprints on most contemporary designs: sans serif typefaces that dominate instructions, fillable forms, timetables, and the internet; abstract corporate logos and sleek icons; and high-contrast figures on public information signs in airports, stadiums, museums, and parks. Like any major cultural movement, modernism had many threads and off-shoots—Futurism, De Stijl, Dada, Art Deco, and post–World War II branches, including Swiss typography and the invention of Helvetica and modular systems like Univers (see Kinross 1985, 27). Most of these manifestations of modernism dwell in the present and the future, rather than the past and the memories it evokes, and in that sense modernist design is timeless and universal—or at least aspires to be. Nonetheless, ironically, today even modernist designs can trigger memory and emotion when audiences contextualize them culturally and breathe new life into aging forms. With the distance of time, even hard-edged figures that roll off the assembly line, like the jobless men and women in the *On Relief* chart, can later incite nostalgia.

Today the tension and interplay between modernist values on the one hand, and memory and sentiment on the other, continue to play out in everyday practical designs. However, the domination modernism enjoyed in twentieth-century design was tempered by the postmodern turn. Robert Venturi's *Complexity and Contradiction in Architecture* (1966) acknowledged the importance of considering historical context in designing buildings and urban spaces—of taking the past seriously and legitimizing the value of memory. From a postmodern perspective, planting a steel-and-glass skyscraper in the middle of a traditional urban neighborhood might not be an acceptable option. Rather, designers should consider spatial and temporal context, mindful of how a building relates to its nearby predecessors and the values those designs still embody.

When we think about design this way—in relation to spatial and temporal context—we are explicitly reminded how much the past continually shapes visual language. Indeed, even the rejection of the past by early modernists acknowledges its influence, however obliquely. Of course, buildings and other works of fine art have a stable, long-term visual presence that differs from ephemeral print and digital artifacts that are readily absorbed by time. Still, visual culture extends to all forms of visual language in whatever media, with broader cultural influences steering all of them, like a stream with many tributaries that all flow from the same source.

Rhetro Emotion: The Past Recovered in the Present

Retro design isn't a recent invention. As we saw earlier, eighteenth-century landscape gardens were embellished with artificial edifices (mock ruins or "follies") that evoked emotion by recalling the distant past. Today, however, retro design generally doesn't encompass landscapes and private parks. So how do visual invocations of the past arouse audience emotion today? What rhetorical purposes are served by "rhetro" designs—the rhetoric of evoking the past—in order to warrant the "h" in "rhetro"? Why do designers and their audiences long to go back in time, and to what end? To answer these questions, I'll consider the role of technology in nostalgic design, then examine how it's evoked in several different forms of visual language: handwriting, typography, icons, pictures, maps, data design, and color. In all of these modes, rhetro design provides a compelling window to the past; at the same time, it also offers insights into the future. William Kurlinkus (2018) points out the importance of understanding the values associated with nostalgia, as "designers need to know a culture's longing, pride, loss, and desire to know what's to come"

(30). In other words, understanding the values and emotions that drive rhetro design gives us insights into those that will shape future design.

Revival of Older Technologies

The design technologies we use today at work and at home have undergone major shifts—for many of us in our own lifetimes. Graph paper for designing data displays, electric typewriters for composing letters and reports, mimeograph machines for duplicating documents (in purple ink)—these are among the dinosaurs of not-too-distant technologies. Looking back at these earlier technologies makes us grateful for our present capabilities. However, as reluctant as we might be to trade our laptops for typewriters or our laser printers for mimeograph machines, we yearn nostalgically for old technologies. More to the point, we long for the visual language that simulates old-school methods.

Some of that nostalgia might be triggered by the burdens of the new. Democratizing information design with technology has both benefits and drawbacks: We need to learn (and constantly relearn) how to use digital tools as they evolve, and we need to acquire design knowledge to use them effectively. Just as early adapters of desktop publishing had to learn about typography, page composition, and picture editing—tasks they previously had little responsibility for—information designers today increasingly need to know something about screen and web design, interactivity, generative AI, and online accessibility. These changing dynamics of design technologies remind us that they are constantly in flux, creating tighter learning curves. Longing for simpler technologies (though largely unwilling to revive them), we embrace the nostalgia of their vestigial visual language, a phenomenon I'll explore in the next several sections.

The Recovery of Handwritten Emotion

Handwriting has long played a key role in practical communication, as I discussed in chapter 1, mainly serving functional roles of creating clear, legible, and credible text appropriate for a given genre and situation. For a very long time—centuries, in fact—penmanship was serious business, and writers and audiences valued it because their livelihoods and reputations depended on it. Today, however, handwritten text is mostly valued for its affect, for its personal and sentimental value, which originates in a performative act hidden from the audience but visibly present in the text. Handwritten messages are formed by the unique bodily gestures of a sentient individual, replete with all of the idiosyncratic nuances they reveal—the size of the text, its slant, the shape of the letterforms, the thickness of the strokes, the color and quality of the ink, the blots and erasures. A handmade image or object will always attract our attention, usually add value to it, and remind us that another human being touched it and put some thought into its creation, unlike the product of a mechanical process and the repetitive, impersonal kinetics that manufactured it.

Handwritten text, then, affords an alternative to the machine-age aesthetic of modernism that emphasized industrial precision and efficiency, epitomized graphically in sans serif typefaces with their geometrical forms and clean, uniform lines, as if extruded from a machine. To the extent that modernism wrung out emotion from design, especially any sentimentality or subservience to the past, today handwritten text partly fills that rhetorical vacuum, whether transcribed by hand, machine, or AI, or digitally simulated by a typeface. We routinely encounter "handwritten" envelopes and sales messages that arrive on our doorstep almost daily, inviting us to contribute to a charity, replace windows or gutters, or sell property for cash. The seemingly handwritten text of these messages creates an informal, personal tone that harkens back to a simpler time when people penned messages to family, friends, and customers. Some contemporary variations of these messages combine script *and* print, marking a generational shift away from cursive, which has largely been supplanted by print letterforms. For example, figure 3.16 (on the left) shows a message I created using the Bradley Hand Bold typeface. So, while "handwritten" messages like this one arouse emotion through nostalgia, they also embody an au courant visual language, heightening their credibility with younger audiences. AI will likely make mechanical handwriting even more accessible and authentic looking. The unique style of an individual's handwriting will be programmed to generate text (similar to voice cloning software), personalizing it and creating a new frontier for rhetro design.

Of course, nothing is more personal, emotionally engaging, and persuasive than the real thing. Figure 3.16 (on the right) shows a postcard created by Letter Friend (2025), a company that (literally) handwrites messages for its clients, including corporations, small businesses, nonprofits, and private individuals. Even though the message is written by a surrogate writer employed by Letter Friend, it bears none of the calculated, prosaic regularity of machine-generated text. The authentic appearance achieved by a genuinely handwritten message—by the physical and rhetorical gesture

Figure 3.16. On the left, a fictitious message I created using the Bradley Hand Bold typeface; on the right, a handwritten message created by Letter Friend (2025), a company that handwrites letters for its clients. Handwritten message courtesy of Letter Friend.

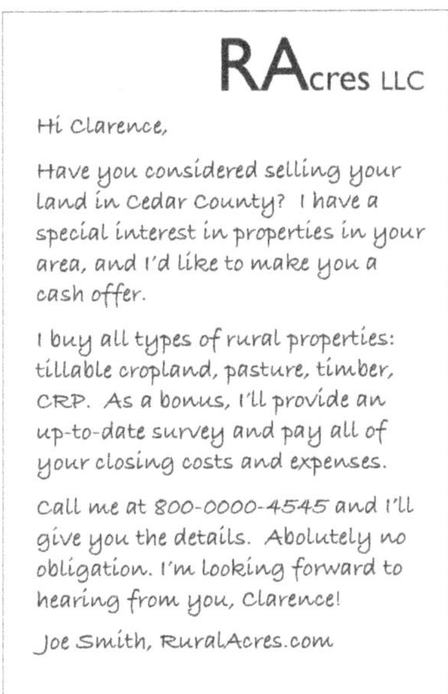

RAcres LLC

Hi Clarence,

Have you considered selling your land in Cedar County? I have a special interest in properties in your area, and I'd like to make you a cash offer.

I buy all types of rural properties: tillable cropland, pasture, timber, CRP. As a bonus, I'll provide an up-to-date survey and pay all of your closing costs and expenses.

Call me at 800-0000-4545 and I'll give you the details. Abolutely no obligation. I'm looking forward to hearing from you, Clarence!

Joe Smith, RuralAcres.com

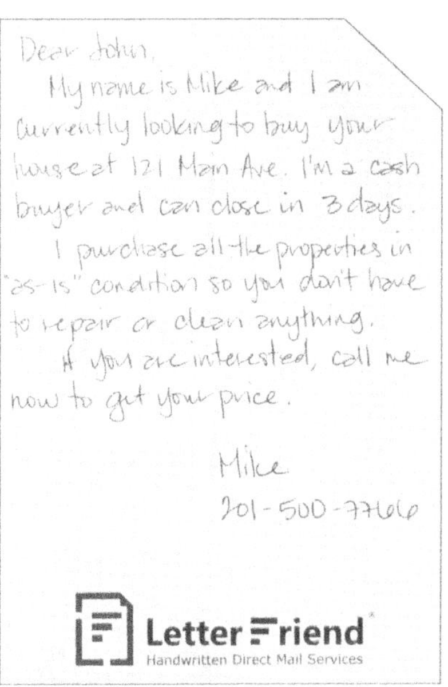

Dear John,

My name is Mike and I am currently looking to buy your house at 121 Main Ave. I'm a cash buyer and can close in 3 days.

I purchase all the properties in "as-is" condition so you don't have to repair or clean anything.

If you are interested, call me now to get your price.

Mike

201-500-7766

of the human hand—amplifies its personal appeal as well as evokes the nostalgia of notes and letters received in envelopes bearing five-cent stamps.

Rhetro writing comes in many other (less personal) forms, some of them within reach on our desktops. Figure 3.17, which I created, shows a couple of examples, one a formal script typeface (Edwardian) that's used for epideictic documents like wedding and graduation invitations and that has many other variations, including Zaph Chancery, Park Avenue, and Snell Roundhand, named after the early eighteenth-century penman Charles Snell. The other the kind of script typeface (Rastanty Cortex) mimics the look of ordinary handwriting that appears, for example, in sales letters we routinely receive, beginning with our address on the envelope. This more contemporary and relaxed script typeface also has many counterparts, ranging from Comic Sans to carefree forms like Mistral and Chalkduster. How effectively these typefaces stimulate the desired emotional responses is hard to gauge; however, many of them are surely intended to generate nostalgia.

Figure 3.17. Nostalgic typefaces that evoke the near and distant past.

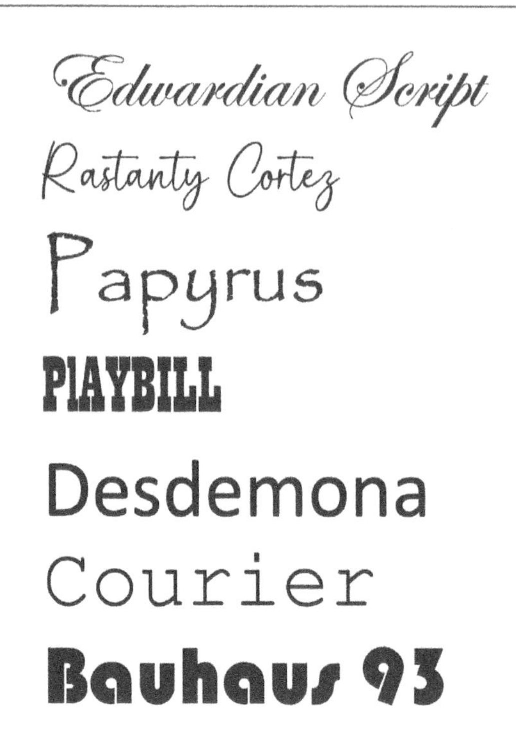

Typewriting and Nostalgic Text

The typewriter was initially—and has always been—a writing machine, a series of keystrokes and mechanical letterforms that typists replicated from one message to another. As such, its practical, utilitarian style and uniform functionality allied it with modernist principles of simplicity, efficiency, and minimalism (S. Walker 2018). By the early twentieth century, despite its impersonal design, typewritten text quickly inundated business and the professions, and the less efficient practice of handwriting dwindled. Nonetheless, as I explained in chapter 1, the conventions of the handwritten letter (address, salutation, spacing) were readily adapted to the typewriter, maintaining visual and emotional continuity for audiences familiar with these long-standing conventions.

Today most text or page design programs include typewriter fonts like Courier (fig. 3.17) or American Typewriter, a testament to the once complete domination of typewriter fonts in the workplace—as well as to their lasting perceptual and cultural legacy. Although typewriters are now all but extinct (with the exceptions I discussed in chapter 1), most people who have never even used a typewriter associate these fonts—with their uniform lines, large serifs, and mono spacing between characters—with the old technology. For those that have experienced using a typewriter, the nostalgia is further heightened. Figure 3.18 shows a document that

Figure 3.18. Sales letter (June 2024) for the *Mayo Clinic Health Letter* with typewriter text and a handwritten note (Mayo Foundation for Medical Education and Research). From: Mayo Clinic Press. *Mayo Clinic Health Letter* June 2024. Rochester, MN: Mayo Clinic, June 2024; used with permission of Mayo Foundation for Medical Education and Research, all rights reserved.

Mayo Clinic® Health Letter
Rochester, MN 55905

From the desk of:
Daniel Roberts, M.D.
Medical Editor
Mayo Clinic Health Letter

June 2024

Ms. Cla
705 SW
Boone,

MAYO
CLINIC

Dear Ms

Astro ... 77.
Poet Ma ... al of
Freedom ... at age 87.
Frank L ... age 91.
Dr. Wil ... age 70.

Willi ... rk)
blasted

What ... Why do
some st ... gical age?

Most ... just
regular ... ness and
inactiv

And t ... ed to
help th ... ler to
achieve

It's ... in this
letter, ... news on
better ... and more.

Have you ever wondered why some people are so robust and energetic in their 70s, 80s and beyond, while others appear to be losing their vitality while still in their 50s and 60s? Mayo Clinic doctors answer questions like this every day. In this letter, we'll share some of our top healthy aging findings with you.

Read to the end and you'll also learn how to claim a valuable free gift to help you sleep better.

So please read on!

Regards,

Daniel Roberts, MD

MAYO CLINIC HEALTH LETTER – written to help you improve your health, maintain your lifestyle, reduce your medical expenses or even save your life – is still only $1.97 a copy.

These days, you don't usually get much for $1.97. But that's the price for what could be one of the most valuable purchases you'll ever make.

Sixteen times a year, readers of MAYO CLINIC HEALTH LETTER get eight pages of reliable medical news for healthy aging, and not a word of third party advertising.

(over, please)

uses both typewritten text and a handwritten sticky note to inform its audience about aging and health and persuade them to subscribe to a newsletter from the Mayo Clinic. The typewriter text appears on several pages of lined paper typically used for handwritten text, further reinforcing the personal tone of the sticky note attached to the first page. Both the note and typewritten text deploy nostalgic visual language that the target audience (based on its demographic) could easily recognize and recall, fostering a sentimental response and heightening the emotional appeal of the letter. For most audiences, regardless of age, we now see typewritten text through the lens of time—not as bland and impersonal (as it appears in the Orville Wright telegram in fig. 1.5) but with a nostalgic yearning for its naive simplicity, along with empathy for its fragility. This creates a temporal-rhetorical paradox: Visual language once viewed as matter-of-fact and utilitarian has transitioned into the nostalgic and sentimental.

Contemporary text design encompasses many other rhetro elements whose emotional appeals rely on their connection to the near or distant past. Rhetro typefaces that evoke ancient cultures are available in the pull-down menus of most word processing software. The Papyrus typeface (fig. 3.17), for example, emulates the look of ancient text on parchment, and other typefaces recall ancient Roman, Greek, and Mideastern cultures. Many traditional typefaces embody an Egyptian design: thick "slab" serifs and tall letter forms, the kind used in the nineteenth century for advertisements, promotional pieces, and "wanted" posters in the West. These characteristics are visible in Playbill (fig. 3.17), a typeface still widely used today for theater programs and posters, as well as in Desdemona (fig. 3.17), also a headline typeface but lighter and more elegant. In the Victorian era Desdemona was also known as "Quaint" (Hardwig 2025), explicitly suggesting its sentimental origins, which are still visible today.

Even modern typefaces have contemporary rhetro versions. For example, Bauhaus 93 replicates the clean minimalist geometry of the original typeface created by typographer Herbert Bayer, who used a few repeatable strokes to create the entire letter set, paring down the design to only its most essential parts, a Bauhaus trademark. Stencil, a 1930s typeface, creates a similar machine-age look with its repeatable segments while also creating a slightly distressed appearance. Numerous distressed typefaces appear on desktops, designed to look old and worn and ravaged by time, an "antiqued" effect that also applies today to furniture, cabinets, and picture frames. Other typefaces evoke specific styles and periods: Art Deco from the 1930s, the "neon" look of the 1950s, and the "hippie" look of the 1960s. These are only a few of the hundreds of rhetro typefaces available today that capture the look and feel of bygone eras.

All of these typefaces do little to enhance the clarity or structure of the text—in fact, most of them degrade legibility by today's standards; instead, they are deployed largely to evoke emotion. To do so, rhetro typefaces embody subjective qualities—what Gerrit Willem Ovink (1938) called "atmosphere-value"—that create a certain mood or ambience, engendered here by time and memory.[13] For example, restaurant menus frequently use rhetro typefaces to create a nostalgic mood that matches the food and decor of the business. Figure 3.19 shows two pages from the online menu of a popular restaurant, Hickory Park. The slab headings throughout the menu create a nostalgic theme that harkens back to the nineteenth-century Egyptian style, and the graphical elements at the bottom of the boxed text further enhance the traditional theme, as do the stars and the watermark of the American flag. The menu complements the interior of the restaurant, its walls decorated with antique farm tools and other mementos of rural life that recall a simpler time and a slower

Figure 3.19. Menu for Hickory Park Restaurant (2025) in Ames, Iowa, that integrates nostalgic elements. Printed with the permission of Hickory Park Restaurant.

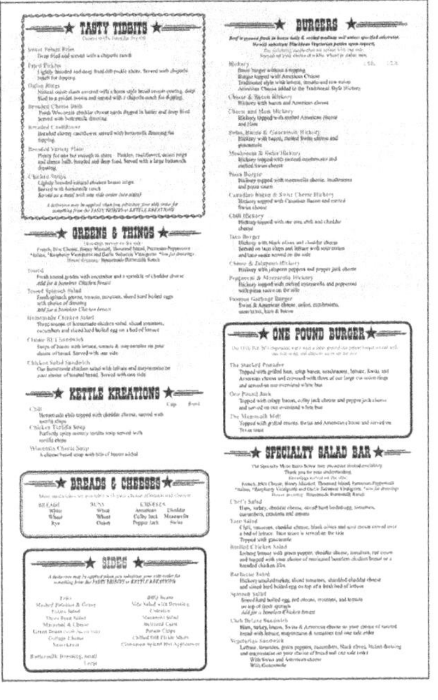

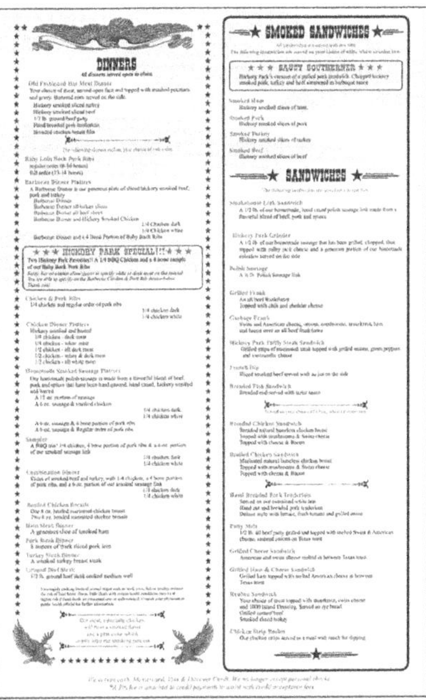

pace. All of these visual elements reinforce the message—that customers get plenty of wholesome food at a reasonable price, just as they did in the old days.

Iconic Anachronisms

Other sentimental associations with the past can be found in icons that envision artifacts that, despite their age, still have a useful life communicating information. These visual anachronisms, some of which appear in the display of common digital icons that I assembled in figure 3.20, arouse sentiment while retaining a surprisingly tenacious clarity. For example, even though contemporary audiences have little if any experience using a rotary telephone, icons of it still widely appear, either as the whole device or just the receiver. Why do we still use this traditional icon when mobile phones dominate, traditional landlines are disappearing, and the real thing can be spotted only in reruns of movies and sitcoms? Apparently, icon conventions have too much grip—and sentimental value—for users to give them up. Another resilient icon, which has experienced a prolonged digital life, is the traditional curbside mailbox (fig. 3.20). Complete with a swinging door, moveable flag, and oblong container with a rounded top, these mailboxes are still widely used in rural areas and suburbs, though they lie outside the experience of most urban audiences. Still, this traditional icon represents what mail delivery *should* look like—that is, in an idealized rural setting—in the minds at least of some audiences. As such, in an era where mail arrives more in digital form than paper, the icon makes a sentimental gesture that humanizes new media, despite whatever cultural baggage that gesture might bear. Digital interfaces are populated with flocks of icons—trash cans, file folders, clipboards, paper clips, and the like—some of the many skeumorphic images (common, recognizable objects that appear on digital interfaces) that users can connect to their real-world experiences (see Norman 2013, 159). However, as users become further removed from file folders and paper clips, these images will also become less directly relatable and increasingly rhetro.

Many other anachronistic icons populate digital and print media, among them hourglasses, pointing hands, and scissors (fig. 3.20). Hourglass icons serve the generic function of marking the passage of time, telling audiences to wait for something to happen (an egg to cook, medicine to work) or to remain patient during a delay—in loading an online image or retrieving data from a customer's account. An ancient technology, the

Figure 3.20. Digital icons that arouse sentiment and nostalgia.

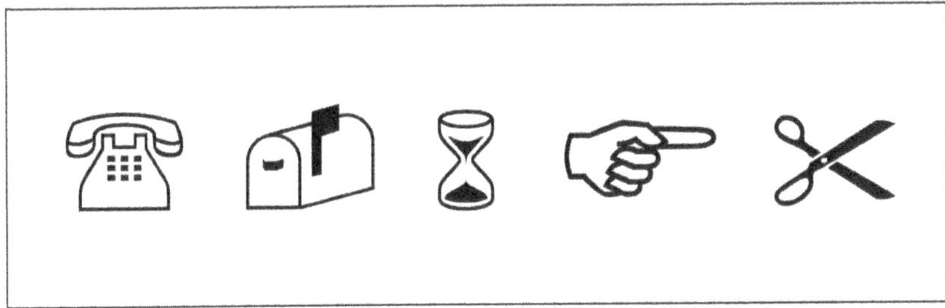

hourglass is a nonthreatening way to subdue the audience's anxiety about having to wait, a form of visual therapy. Of even greater sentimental value, the pointing hand appears on communications where something important needs to be "pointed out" or emphasized: a warning, an item on a restaurant menu, a restroom in a tavern, a "return to sender" stamp on an envelope, and so on. In the nineteenth century the pointing hand appeared in advertisements, on posters, and on a host of other printed materials, providing emphasis to direct the audience's eyes and assert the gravity of a message. Pointing hands were almost universally male hands, with stiff cuffs, tightly clenched fingers, and an authoritarian tone. Although it retains its functional goal of guiding and emphasizing, today's pointing hand has acquired sentimental value, its severity has diminished, and we associate it with a folksy tone and light-hearted prodding, a nostalgic wink rather than a stern gaze. The scissors icon evokes a more recent past—a paper-based world some of us once inhabited and, more universally, our experiences in early childhood education. Although we may occasionally still use scissors as adults, we're more likely these days to do digital cutting and pasting, with vague sentimental recollections of the real thing.

Pictures as Nostalgic Time Travel

Victorian pictures like those I've previously discussed continue to appear in private and public spaces, as well as in practical documents, arousing sentiment by invoking the past. Aging technical pictures can also be similarly deployed: Pictures of early engineering inventions (fig. 1.12), architectural drawings (fig. 1.14), or even drawings of industrial machines (fig. 1.15) can all evoke curiosity and nostalgia. Retrofitting (or repurposing) older pictures offers another method for eliciting emotion. Appearing in promotional materials, blogs, and memes, famous pictures like Leonardo da Vinci's *Mona Lisa* (1509–1519) Grant Wood's *American Gothic* (1930), or Edvard Munch's *The Scream* (1893) can be repurposed to persuade audiences to buy a product, attend an event, or consider a new idea or course of action. Repurposed pictures require the audience's knowledge of the original picture to achieve their rhetorical effect (irony, hyperbole, humor) by juxtaposing past and present—the picture the artist originally conceived versus its repurposed version. How does this repurposing arouse emotion? Retrofitting *Mona Lisa*, for example, can evoke humor and levity by showing someone in a quasi-serious pose like Mona Lisa's, or the comparison might have a darker side by satirizing someone with too much hubris and provoking the audience's anger or jealousy. A similar range of emotions might be sparked by Munch's *The Scream*: empathy for someone experiencing extreme anxiety or surprise (or dismay) at something unusual happening.

Rhetro images often incorporate visual elements that connect audiences emotionally with an organization's past, arousing sentiment and nostalgia. Figure 3.21 shows the logo for Grand Hotel (2025), a landmark structure built in the late nineteenth century on Mackinac Island in northern Michigan. Transportation on Mackinac Island is limited to feet, bicycles, and horse-drawn carriages, which carry visitors from the dock up to the hotel, its sprawling porch (the longest in the world) lined with pillars that are visible from the ferries that shuttle back and forth from

Figure 3.21. Logo for Grand Hotel, Mackinac Island, Michigan (2025). Printed with permission from Grand Hotel.

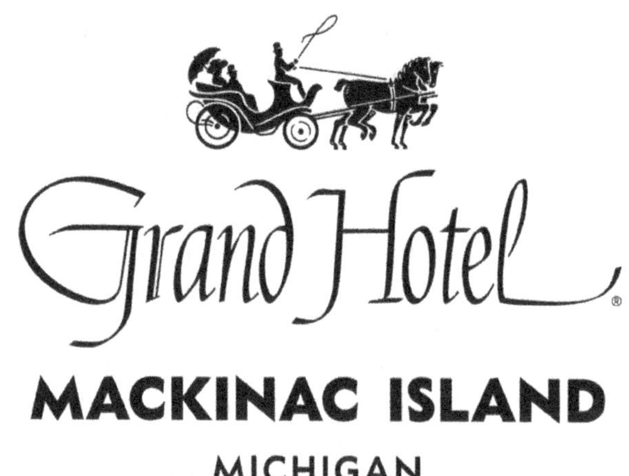

the mainland. Like most businesses on the island, Grand Hotel thrives on tradition and its linkage to the past. The horse-drawn carriage on the logo takes visitors back in time to the hotel's storied past, arousing nostalgia for the era in which Grand Hotel first opened its doors and which it continues to preserve.

Some company logos evolve continually over time but still retain their connections to the past, evoking memory and emotion. For example, the John Deere Company began in the nineteenth century by producing plows in a small Illinois town and today is an international company that manufactures farm, construction, and lawn and garden equipment. Early versions of the trademark logo show clumps of prairie grass and a log over which a deer leaps (John Deere 2025; Kayla F. 2020). Deere's plow enabled farmers to break the prairie sod, so those details envision the company's original purpose and implicitly evoke the agrarian culture in which the Deere plow thrived. This rural scene eventually gave way to the flat, high-contrast modernist design that, in today's green and gold version, appears on its products around the world—from cornfields and construction sites to golf courses and suburban lawns.

Nevertheless, the leaping deer continues to symbolize the company's past, connecting audiences to its rural roots of breaking sod and creating an emotional attachment to that pioneer past and the values it represents. Logos and trademarks like John Deere's trigger memory and emotion through their longevity, while those of more recent companies often look shallow and fleeting, especially when absorbed into larger entities with fresh images and identities that decouple their audiences from the past, fueling their appetite for nostalgia.

Images with rhetro effects arousing emotion through memory or nostalgia appear in several other forms:

- Comics used for instructions, warnings, and textbooks have a sentimental, nostalgic effect, a vestige of twentieth-century popular culture when comics had a wide following.

- Rhetro stock photos—of buildings, cars, clothing, sports, and domestic life—routinely appear in promotional materials, posters, social media, and invitations to events.

- Pixelated images and sawtooth text spur audiences to recall the early days of digital design with antiquated technologies like low-resolution monitors and Pac-Man games.

- Digital filters in Photoshop and other programs can age or "distress" images, transform them into paintings and Polaroid pictures, or add grainy texture or embossing.

We don't have to look very far in our professional and personal lives to find images like these, which rely on nostalgia to intensify their rhetorical force.

Maps as Visual Traces of the Past

Maps meet our expectations as users by locating things in the here and now: roads, rivers, lakes, cities, towns, parks, counties, and so on. But if we look closely, we can also see things as they *were*: traces of earlier routes, settlements, events, and other significant places from the past. Most maps contain symbols with "points of interest" that identify (with red dots and icons) important historical sites: houses, castles, monuments, ruins, battlefields, former state capitals, birthplaces of presidents and other famous people—all places that evoke the past and that travelers might be attracted to. Some maps also show old abandoned routes like the Santa Fe Trail, Oregon Trail, and Mormon Trail as well as more recent routes like the Lincoln Highway and Route 66. In these ways, maps serve as geographical archives and time capsules of memory, arousing both curiosity and nostalgia. All maps are rhetro to some degree, given that they are superimposed on older maps (and those superimposed on yet older ones) whose vestiges still remain whether or not we take the time to recognize them (see Barton and Barton 1993, "Ideology").

The opposite also occurs, with rhetro maps designed specifically to layer the past on the present—for example, in large cities that have long histories and have experienced massive changes in their urban landscapes. These rhetro maps enable users to engage in sentimental journeys at their own pace, experiencing vicariously historical events, places, and lore that might delight, inspire, or terrify them. Rhetro maps that guide tourists through cultural epicenters like London, Paris, Rome, and Old Jerusalem evoke even deeper, more reflective emotions from their users.

The Rediscovery of Emotion in Data Design

With the labor-saving benefits of digital technology, several data design genres from the past have been recovered and enhanced. For example, rectilinear area charts (mosaics) that were popular in the late nineteenth century are reincarnated today as treemaps (fig. 1.7), though audiences won't connect them to their past lives, and therefore they won't generate nostalgia. In some cases, however, contemporary designers have re-created famous historical charts and graphs. For example, Charles Joseph Minard's *Carte Figurative* of Napoleon's Russian campaign (shown in fig. 2.26) was re-created by Sam Dragga and Dan Voss (2001), who added icons to the chart to elicit empathy for those who lost their lives (270). Makeovers like this trigger memory by retelling stories that have long been dormant, in this instance by adding graphical elements to arouse emotion.

On a far larger scale, contemporary data design injects emotion through innovation and graphical richness experienced during its "golden age" in the later nineteenth century (Funkhouser 1937, 330; Friendly 2008). Creative designs and sensory stimulation through color, lavish detail, and sometimes animation and multimodal elements infuse digital charts and graphs. It's hard to imagine these recent transformations in data design—and the rekindling of emotion they foster—without the innovative exemplars of the "golden age." Moreover, interactive data design allows audiences to customize charts according to their own needs and preferences, heightening their emotional appeal. Before digital design was in full swing, Edward Tufte (1990) explained how well-designed static charts enable audiences to explore data at both the "macro" and the "micro" levels (43–51), with the "micro" level granularity allowing them to "personalize" their interactions with the display (50). Today interactive digital displays intensify this effect by allowing virtually any audience to visualize data in close proximity to themselves, which creates a "connexion" with them, as George Campbell argued, that stimulates emotion (Kostelnick 2016, "Re-Emergence"). Advances in digital technology, then, have revived the emotionally charged data designs of the Victorian "golden age." These digital wonders aren't retro per se, but they seethe with the vestigial emotion of their predecessors.

The Reminiscent Power of Color

The expressive power of color was thoroughly exploited by Victorians with their profusely decorated house façades, along with the wallpaper, furniture, and domestic objects inside. Color was also used extensively in data design (Kostelnick 2016, "Mosaics")—for example, in the late nineteenth-century *Statistical Atlases of the United States* (figs. 1.6, 1.8, and 2.21), Du Bois's charts of African Americans (fig. 1.11), and the poverty maps of Charles

Booth (Kimball 2006). Saturating the senses with color was a Victorian trademark, and the technology of chromolithography greatly enhanced its proliferation. Modernism altered the calculus, favoring the perceptual purity of primary colors, the use of monochrome colors (fig. 3.14), or the elimination of color altogether to objectify and simplify information design (fig. 3.15), exemplified in the 1914 and 1925 *Statistical Atlases of the United States* (Sloane), which relied heavily on high-contrast black-and-white charts and maps. So the color pendulum swung from the Victorian arousal of emotion through sensory stimulation to subduing these effects with the advent of modernism (Kostelnick 2004, 235–38).

With its powerful cultural and political associations and fidelity to place and time, color also appeals to the emotions by creating identity and maintaining tribal cohesion. Nations adopt color palettes that evoke strong passions in their citizenry: red, white, and blue; the French and Italian tricolors, Ukrainian blue and yellow, Dutch orange, Chinese red—all create identification with audiences based on their collective interpretations from one generation to the next. Color becomes embedded in their cultural and political DNA, fostering long-term emotional affinities with the past. Often our emotional associations with color, however, lie closer to home, creating spatial and temporal proximity as well as a strong "connexion" with friends, family members, teammates, congregations, and coworkers. A school color can invoke a lifetime of sentimental attachment in its alumni, as can colors for sports teams (e.g., Dodger blue) and other organizations we've affiliated ourselves with, long after we've parted ways. How individuals navigate these color associations can be highly personal and idiosyncratic, but they all share the power to elicit emotional responses, largely based on time and memory.

Color also arouses emotion through nostalgia, which connects us to the near and distant past with rhetro forms of visual language. Sepia-colored images, for example, create a nostalgic look, harkening back to the early days of photography in the nineteenth century. When we see a photograph or graphical element colored in sepia, we associate it with age, and even though the photo might have been taken yesterday in an antique photography studio, processed digitally, or edited on the desktop, the rhetorical effect is sentimental and nostalgic. The same emotional responses are aroused by colors that mimic parchment surfaces: antique white, beige, tan, and gold. All strongly suggest past eras, heightening the serious tone and ethos of the message: it's old, it's fragile, it deserves respect. Not surprisingly, reproductions of iconic documents (like the

Declaration of Independence) are printed on golden parchment, aimed at simulating the aged look and feel of the original.

Close-Ups: Memory and Sentiment Within a Lifespan

As George Campbell claimed, the close proximity of time will create a strong emotional effect, so by implication, design elements that appear in our own lifetimes will resonate more strongly than those at a longer temporal distance. For example, our awareness of the temporal variations of color is most closely tied to our personal experiences, which both filter and enhance emotion. In the 1960s turquoise became a popular color for homes appliances, which today for many Baby Boomers arouses memories from their childhoods. Most younger consumers today, however, don't have those associations with turquoise, and for them the emotions they associate with it are pure and immediate, unfiltered by prior experience. The same applies to rhetro images from the 1970s, 1980s, and subsequent decades, which induce different emotional responses. Although everyone experiences nostalgia, for those that lived with those images when they first became popular, the emotion runs deeper.

On the other hand, sometimes we emotionally rehabilitate artifacts from that past that have developed personal connections. For example, operating instructions for appliances and equipment might lie around for years before being consulted again (or for even the first time), stored in a basement or garage with other family mementos. Figure 3.22, for example, shows two pages from a manual for a Montgomery Ward gas stove (Roberts ca. 1950) that my parents purchased when they moved into their first house, a small brick ranch house in the Chicago suburbs. My parents kept the manual when we later moved to another house nearby, and the stove came along and found a home in the basement. The operating manual survived as well—now a personal memento rather than a utilitarian document. The pictures in this manual were directed toward housewives who had just acquired one of these modern gas stoves. Figure 3.22 shows a couple of pages from the manual, with a woman cheerfully imagining all the things she can cook for her family on her new stove and the instructions below guiding her. The images and text (some handwritten) paint a picture of domestic bliss that recaptures the charm and innocence of the 1950s, with memory submerging the era's social constraints, inequities, and threats of nuclear war. The manual that originally enabled homeowners

Figure 3.22. Pages from a cooking manual for a Montgomery Ward gas stove (Roberts ca. 1950, 14–15). Reprinted with permission from Montgomery Ward.

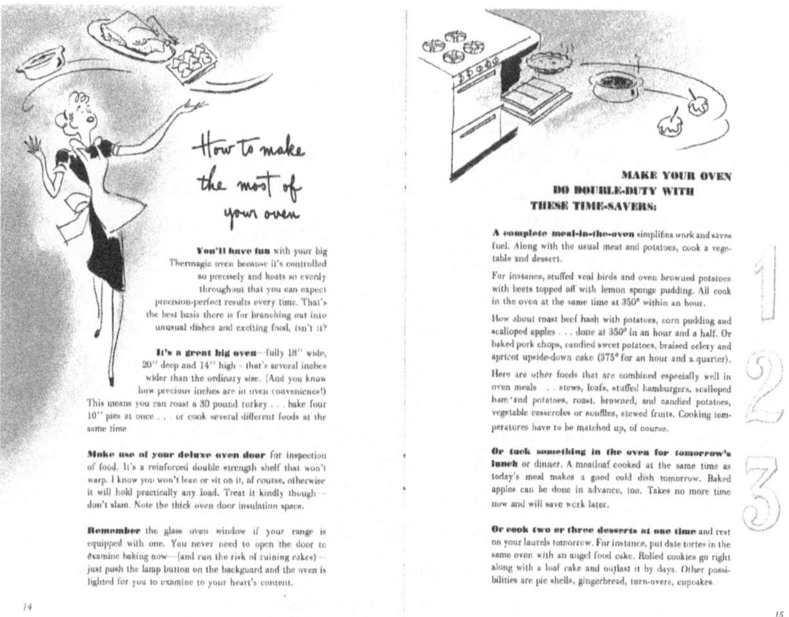

How to make the most of your oven

You'll have fun with your big Thermagic oven because it's controlled so precisely and heats so evenly throughout that you can expect precision-perfect results every time. That's the best basis there is for branching out into unusual dishes and exciting food, isn't it?

It's a great big oven—fully 18″ wide, 20″ deep and 14″ high—that's several inches wider than the ordinary size. (And you know how precious inches are in oven convenience!) This means you can roast a 30 pound turkey . . . bake four 10″ pies at once . . . or cook several different foods at the same time.

Make use of your deluxe oven door for inspection of food. It's a reinforced double strength shelf that won't warp. I know you won't lean or sit on it, of course, otherwise it will hold practically any load. Treat it kindly though—don't slam. Note the thick oven door insulation space.

Remember the glass oven window if your range is equipped with one. You never need to open the door to examine baking now—(and run the risk of ruining cakes)—just push the lamp button on the backguard and the oven is lighted for you to examine to your heart's content.

14

MAKE YOUR OVEN DO DOUBLE-DUTY WITH THESE TIME-SAVERS:

1

A complete meal-in-the-oven simplifies work and saves fuel. Along with the usual meat and potatoes, cook a vegetable and dessert.

For instance, stuffed veal birds and oven browned potatoes with beets topped off with lemon sponge pudding. All cook in the oven at the same time at 350° within an hour.

How about roast beef hash with potatoes, corn pudding and scalloped apples . . . done at 350° in an hour and a half. Or baked pork chops, candied sweet potatoes, braised celery and apricot upside-down cake (375° for an hour and a quarter).

Here are other foods that are combined especially well in oven meals . . . stews, loafs, stuffed hamburgers, scalloped ham and potatoes, roast, browned, and candied potatoes, vegetable casseroles or souffles, stewed fruits. Cooking temperatures have to be matched up, of course.

2

Or tuck something in the oven for tomorrow's lunch or dinner. A meatloaf cooked at the same time as today's meal makes a good cold dish tomorrow. Baked apples can be done in advance, too. Takes no more time now and will save work later.

3

Or cook two or three desserts at one time and rest on your laurels tomorrow. For instance, put date tortes in the same oven with an angel food cake. Rolled cookies go right along with a loaf cake and outlast it by days. Other possibilities are pie shells, gingerbread, turn-overs, cupcakes.

15

to operate and maintain their Montgomery Ward gas stoves now serves as a nostalgic memento of post–World War II suburban life experienced by many of the Baby Boom generation.

Other practical documents that survive the ravages of time become nostalgic reminders of the past. Even the most functional ones like catalogs, bills of sale, receipts, letters, contracts, annual reports, building plans, meeting minutes and agendas, and the like take on new lives after their utilitarian ones have long since passed. The same transformation occurs with brochures, tickets, museum guides, park maps, and other mementos from vacations, which initially enable us to make decisions or navigate unfamiliar spaces but which later acquire sentimental value. Like retired license plates hanging on a garage wall, their practical purposes morph into pure affect. Although some of these artifacts were designed to have second lives as keepsakes, most never had that aim. Either way, if they survive very long after their initial use, their audiences will feel their rhetorical impact, generated mostly by memory and nostalgia.

However much their emotional appeal, many aging documents still possess functional value as well. For example, a user's manual for an antique sewing machine might supply vital information for restoring it to working order, or the original drawings for a building might need to be reviewed to make plans for a remodel, addition, or demolition. Or to complete a sales transaction, a land survey or a property abstract might need dusting off to provide legal information about location, size, or ownership. The rhetors who created these documents might be long gone, but they continue to speak from their graves, their designs kindling a sentimental glow while still having practical value after a long hiatus.

Conclusion and Implications for Design

Emotion and memory constantly filter our perception of the world around us in everyday encounters, especially with artifacts that are specifically designed to evoke them (monuments, statues, memorials) as well as through interactions with practical information designs. Arousing emotion through visual stimuli was actively cultivated in the eighteenth century where feeling, sentiment, and memory coalesced aesthetically, intellectually, and rhetorically in the sentimental, the picturesque, and the Gothic. These developments infused nineteenth-century design through idealized pictures of rural and domestic life and through profuse graphical and textual decoration that heightened sensory stimulation and emotional engagement. In contrast, modernism attempted to subdue the emotions fostered by the perceived excesses of nineteenth-century design. The tension between the Victorian sentimental and modernist objectivity continues to play out today with rhetro designs that appeal to memory and emotion and with digital technology that regenerates traditional forms and heightens user interactivity.

These developments underpin our interpretive experiences with information designs that evoke memory. Even if we don't recognize the synergy between memory, design, and emotion—or the intellectual and aesthetic forces that shape these affinities—this rhetorical process is constantly unfolding all around us in everyday interactions, hidden conceptually but plainly visible and continually felt. How can contemporary designers elicit emotion with nostalgia, memory, and sentimentality? Designers have a variety of strategies to choose from, though implementing them begins with considering the various, and sometimes contrary, perspectives audiences have about the past.

Recognizing the Multiplicity of Pasts and Nostalgias

Audiences differ in their responses to nostalgic designs: to some, the visual language of the Victorian age will appear charming and sentimental; to others, it will evoke associations with stifling social mores and rampant colonialism. Consequently, designers must recognize what William Kurlinkus (2018) identifies as "conflicting nostalgias" (4). Negotiating these perspectives presents an ever-present rhetorical challenge for rhetro designers, but as Kurlinkus argues, it also offers opportunities to recognize diverse nostalgias that can democratize design as well as alter audiences' perspectives on the future (4, 30–31). Here are some suggestions for implementing these ideas:

- Identify the varying perspectives of audiences on what they find nostalgic and how to visualize it effectively. Also identify what they *don't* find nostalgic and why. These analyses can provide a conceptual and rhetorical baseline for rhetro design.
- Diversify appeals to nostalgia by, for example, interweaving traditional Hispanic, African American, and Native American images and other inclusive visual elements into the rhetro design.
- After creating a prototype rhetro design, seek feedback from the intended audiences to gauge what resonates with them and what's missing or off the mark.

These may seem like commonsense suggestions, but sometimes designers have limited awareness of how others interpret the past and which facets of it arouse nostalgic yearning—and which don't.

Imitating Old Technology for Emotional Appeals

The designs we create and interpret are time stamped with the technology used to generate them. By simulating older technologies, designers can use time rhetorically to create emotional appeals based on nostalgia. Old-school technology can be invoked in several forms:

- Textual: typewriter, script, and pixelated typefaces; the Papyrus typeface simulating text on ancient scrolls; Egyptian slab typefaces evoking the handset type of the Old West.

- Graphical: vegetal scrolls, flourishes, and decorative borders that simulate penwork; images or icons that appear to be stenciled or stamped.
- Pictorial: freehand drawings; photos aged with sepia coloring, a grainy texture, or a Polaroid effect; scrapbook frames or shadows around pictures in a newsletter.

The efficacy of these design elements depends, of course, on the rhetorical situation—whether they're used persuasively in sales materials, restaurant menus, or social media or more formally in graduation diplomas, professional certifications, awards, and other epideictic genres that connect audiences with past achievements or recognitions.

Visualizing Handmade Nostalgia

Because we're so immersed in technology, machine-age production, and lean modernist aesthetics, hand-formed designs immediately transport audiences back in time, melding memory and emotion. Inspired initially by the Arts and Crafts movement in the later nineteenth century, handmade designs created today generate emotional appeals based on nostalgia and sentimentality. Although actual handwritten messages (to customers, donors, volunteers) are still highly regarded in professional communication, designers have a repertoire of "handmade" digital designs readily available at their fingertips. Script typefaces, from calligraphic to chalkboard; hand-drawn clip art and pen-and-ink decorations; simulated woven, wood, and parchment surfaces—all of these aim to look like they were produced by a human hand, the epitome of sentimental design that, in one form or another, resonates across cultures and generations in our global high-tech era.

Like any form of visual language intended to arouse an audience's emotions, however, designers have to tailor rhetro design to the rhetorical situation (with the "h" in rhetro guiding the way). The nostalgic affect generated by a calligraphic typeface that would be appropriate for an invitation to a graduation ceremony or the program of a classical concert would be a mismatch for a technical report, letter of application, or set of instructions. A beige woven texture on a flyer for an antique mall might resonate emotionally with its audience but not on a bar chart in a board room. Like other emotional appeals, nostalgia requires a snug rhetorical fit.

Chapter 4

Visual Storytelling

Temporal Elasticity and the Interplay Between Micro- and Macro-Narratives

We're all familiar with the storytelling genres that saturate contemporary culture—novels, films, mysteries, short stories, plays, TV documentaries—as well as storytelling genres of the distant past—epics, parables, fables, and romances. Storytelling also occurs in literary illustrations and graphic novels as well as in the fine arts—painting, sculpture, and public memorials. These forms of storytelling serve to instruct and entertain, record and celebrate past events, cultivate collective memory, and promote cultural values. Narrative also plays an important role in practical communications: to tell stories about how things work, how to perform a task, or how something has changed over time. Like literary and artistic forms of storytelling, these also narrate actions and events, whether they extend for a few minutes or a few millennia.

Narratives in literature and the fine arts typically experience long lives, extended indefinitely on library shelves, in multiple editions, in museums and public spaces, and through digital media.[1] The lifespans of many forms of *practical* storytelling, however, are often acutely short-lived: timetables change, data are constantly refreshed, and procedures come and go. This ephemeral aspect of practical designs makes them far more vulnerable to loss and erasure—in recycling bins and digital trash cans and to other forms of disintegration or neglect. Given these tenuous conditions, it's easy to overlook the wide array of functional purposes that practical storytelling performs in our everyday lives—and its rhetorical power to engage, inform, and persuade us.

The power of *written* narrative in professional communication has been widely acknowledged over the past several decades, early on through the lens of theoretical constructs from other fields (Perkins and Blyler 1999; Barton and Barton 1988). Written storytelling occurs in a wide array of professional communications, including annual reports, proposals, manuals, social media posts, and meeting minutes and agendas. Large organizations have a huge stake in representing themselves through compelling and persuasive written narratives, often through the stories of their leaders, which scholars like Janis Forman (2013) have studied in depth. Several scholars have also examined *visual* narratives: in text, diagrams, and networks (Rosenberg and Grafton 2010; Lima 2011), illustrations (Eisner 2008; McCloud 1993; Kostelnick 2019, *Humanizing*, 97–130), and charts, graphs, and maps (Barton and Barton 1993, "Ideology," "Modes of Power"; Brasseur 2003; Tufte 1990, 96–119; Wainer 1997, 2005; Tominski et al. 2017). With concepts from semiotics, narratology, rhetoric, and art history providing springboards for analysis, these and other studies showcase a multitude of captivating examples, both historical and contemporary.

Indeed, practical visual narratives occur in so many disparate forms over such a long stretch of time that I can't possibly account for all of them in a single chapter. Rather, my goals here are threefold: (1) to assess how past narrative practices have influenced the present—their vestigial forms—through genres and conventions with long-term currency; (2) to define how those conventional practices have evolved over time, particularly in response to rhetorical exigencies, technology, and culture; and (3) to show how our everyday micro-narratives are embedded in macro-level cultural narratives that have become so ingrained and taken for granted that we often don't recognize them.

To achieve these goals, I'll explore the relationship between narratives in the fine and applied arts, the temporal elasticity of narratives, and the socializing effects of their visual conventions. I'll then examine narrative modes for storytelling—textual, pictorial, and graphical—at a wide range of temporal spans, ranging from timetables and decision trees to how-to instructions, comics, and data visualization. I'll then explore how several narratives in pictures and data displays intertwine with larger national, cultural, and scientific macro-narratives like westward migration, economic prosperity, civil rights, and climate change. Finally, I'll examine the relationship between those larger macro-narratives and the narratives we encounter every day. Although I can't claim to connect practical and cultural narratives with a high degree of precision or authority, these synergies infuse everyday information design, both narrative and non-narrative, and become clearer with the passage of time.

Visual Storytelling: Genres, Temporal Range, and Conventions

Practical storytelling occurs primarily in three visual modes: textual, pictorial, and graphical. Many of these forms have coalesced into genres with conventional visual language that combines these three modes, one of which usually dominates:

- Textual. Timelines that annotate events along a linear path; timetables for transportation, theater performances, mortgage payments, or academic courses; tables showing data over time; decision trees showing a series of steps with contingencies; lists sequencing tasks or events.
- Pictorial. Series of pictures explaining a process or teaching audiences how to perform a task (assemble a product, perform CPR, cook a meal); comics telling a practical story; a single picture describing an important event (a ribbon-cutting, an awards ceremony); a training video recording how-to instructions.
- Graphical. Data displays like line, radar, or area charts or other forms of data design (static or interactive) that plot time in units ranging from seconds to years; animated maps and bubble charts that show variations from past to present; diagrams or flow charts describing a process, method, or series of steps.

In any of these three modes, time can be visualized in the past, present, and future. Here are a few examples:

- Past. A line graph showing how well a business performed last quarter or how population changed over the last decade;

a map showing the geographical evolution of nations and empires; a picture showing how something used to be done or used to look.

- Present. A timetable showing when a bus arrives at a given corner; an interactive dashboard that shows current business operations; a stock or commodity chart that shows minute-by-minute trading; video surveillance that shows activity in a store, office, or plant.
- Future. A bar chart in an annual report showing future sales, production, or carbon emissions; an online digital table showing the arrival times of incoming flights; pictures visualizing the potential effects of sea level rise or a tsunami on land, buildings, and people.

Sometimes more than one temporal mode is visualized simultaneously—for example, a bar chart in an annual report that shows sales over the past five years, sales in the current year, and aspirational sales for the next few years; a GPS mapping system that shows current location (driving, walking, flying) relative to past and future locations. The order of the temporal sequence can also vary. Although we typically think of time moving linearly from past to present, sometimes the narrative is reversed. For example, in online blogs and social media the most recent posts appear first (Venditti, Piredda, and Mattana 2017, S275), in email chains we have to scroll down to see the sequence of responses to the original message, and in charts and graphs (especially in annual reports), sometimes the most recent year appears first and earlier years appear in reverse order.

Temporal Elasticity: Narratives from Mini-Micro to Super

Practical narratives vary immensely in their durations, ranging from a few seconds to millennia. On the very short temporal end of the spectrum, figure 4.1 shows a simple set of instructions for using a charger in a car, a task that takes only a few seconds. Pictures in the form of simple line drawings tell the story—plug the charger into the port in the car; then plug in the device—which probably never needs repeating once the user completes the process. We experience these mini-micro-level narratives every day: from instructions and warnings on equipment that we use to digital maps and schedules that guide us to our destinations. These

Figure 4.1. Simple instructional narrative about how to use an Anker car charger for electronic devices (Anker Innovations 2023). Reprinted with the permission of Anker Innovations Limited.

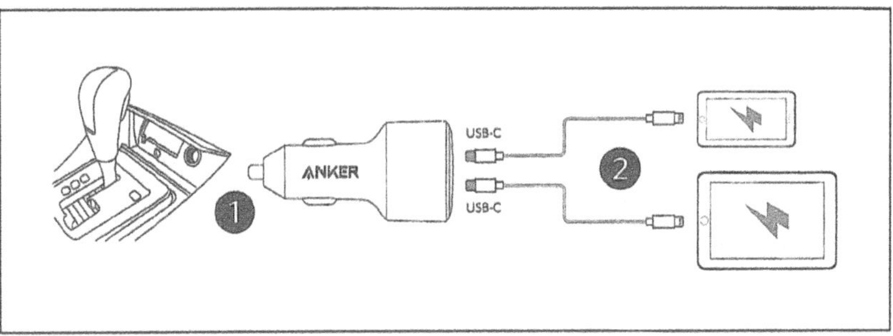

transient little narratives are so interwoven in our daily lives that we often overlook their storytelling power.

At the other end of the temporal spectrum, some visual narratives enable us to see time in panoramic views, like standing on a mountaintop and gazing far and wide to get a bird's-eye view. Other large-scale visualizations give us distant views into the future by prodding us to ponder its likely effects on the world we currently inhabit. These super narratives extend far beyond the myopia of daily life, allowing us to travel long distances, both backwards and forwards. Beginning in the eighteenth century, for example, wide-spanning historical timelines began to appear in several ingenious forms (S. Davis 2017), a storytelling genre that became increasingly popular as modern historiography emerged. Especially notable among these visual narratives were Joseph Priestley's *A Chart of Biography* (1765), which graphically plotted the lifespans of historical figures from antiquity to modern times, and his *New Chart of History* (1769), which diagramed empires and major events over the past three thousand years.[2]

Perhaps the most spectacular and elaborate of these historical timelines was Sebastian Adams's *Synchronological Chart of Universal History* (1881). This gigantic chart (over twenty feet wide) plots the history of the world from Adam and Eve to the present, recording nearly six thousand years of empires, rulers, biblical figures, battles, monuments, artifacts, and key discoveries with text, images, and graphical flow diagrams. The segment of the chart in figure 4.2 covers six hundred years, with biblical

Figure 4.2. A segment of Sebastian Adams's mammoth graphical and pictorial timeline of world history (Adams 1881, 6). Image credit: David Rumsey Map Collection, David Rumsey Map Center, Stanford Libraries.

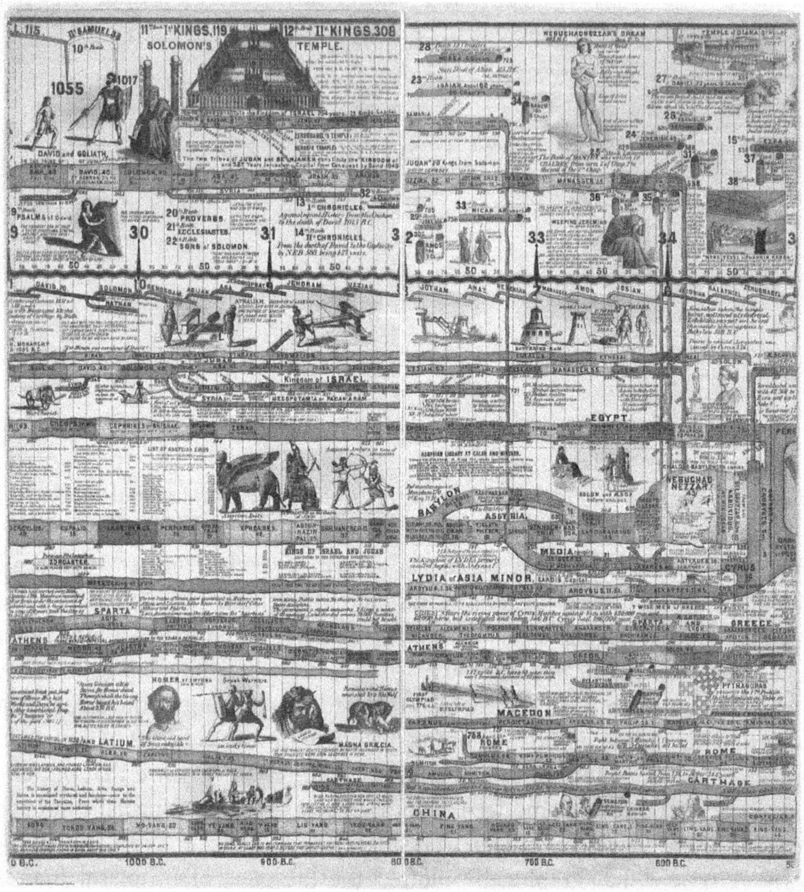

history occupying most of the top (David and Goliath to the Babylonian captivity) and beneath that the history of the major centers of power, from Egypt, Greece, and Babylon to Rome, Carthage, and China. Kings, queens, inventors, military heroes, authors, philosophers, prophets—all find their places on this mammoth timeline. The grandiosity of Adams's super narrative of humanity lies almost beyond comprehension, given its graphical richness and the sheer scale and the complexity of the stories it interweaves.

Been There, Seen That: The Socializing Effects of Conventions

Practical storytelling is a very flexible and creative activity, but like other forms of information design it relies on visual conventions—time capsules that constantly import the past into the present. Highly social constructs, conventions are adopted by visual discourse communities, and audiences within those communities come to understand and expect them (Kostelnick and Hassett 2003, 82–96). For example, members of professions learn the conventions of their discipline: heart doctors learn to read EKGs, electrical engineers learn to read circuit diagrams, and architects learn to read construction drawings. Virtually all professions have some visual conventions, however expansive or minute, specific to their discipline. On a broader, more general level, most of us learn the visual language of popular genres like pie charts, newsletters, social media, and TV schedules directly through lived experience.

Some conventions of practical storytelling parallel those used in literature and the fine arts, and some differ sharply (Kostelnick 2019, *Humanizing*, 98–103). Practical narratives using primarily text—flow charts, decision trees, timetables—parallel literary storytelling by sequencing words in temporal order. However, these practical narratives also use graphical and spatial elements—lines, arrows, boxes, lists, matrices, networks—to conjoin their text segments syntactically and temporally. Graphical narratives of data on Cartesian grids (line graphs, bar charts) have virtually no equivalent in literary or fine arts narratives, though data narratives embody aesthetic elements that draw from the same wellspring of cultural codes: Victorian charts and graphs use striking colors, complex graphical elements, and titles with heavy serifs; modernist charts, on the other hand, streamline graphical coding with the stark minimalism of a steel-and-glass façade or a poem by William Carlos Williams (1986).[3]

Although literary texts can create free-flowing narratives, and graphic novels and illustrations can vividly visualize those stories, storytelling in the fine arts has traditionally occurred one temporal moment at a time. In the eighteenth century, Gotthold Lessing ([1766] 1874) distinguished between visual and verbal narration using the story of Laocoön, contrasting the literary version in Virgil's *Aeneid* (2016, 70–71, bk. 2, lines 190–217) with the classical sculpture of *Laocoön and His Sons* (40–30 BC), which captures a single poignant moment of the story. Paintings also typically envision a single, temporally static image, though they can, as Wendy

Steiner (1988, 12–18) claims, possess narrative qualities if they meet certain conditions (Steiner 2004).

Narrative conventions for practical communications, on the other hand, don't bear these temporal constraints and enable designers to tell stories that are both complex and elastic. These can range from a series of data points on a graph that "tells a story," as Edward Tufte (1983, 30) puts it, to a series of pictures describing how to perform a task or how something happened, generating narratives with widely divergent forms and temporal spans (as we saw in figs. 4.1 and 4.2). Designers of these visual narratives draw on conventions to tell their stories, visual vestiges from the near and distant past that enable audiences to make decisions, get things done, and understand the world around them. The historical trajectories of these conventions are also shaped by the available technology and its affordances for innovation and usability, as we'll see with the narrative forms I'll explore in the next section.

Micro-Narratives: Text, Graphics, and Pictures

Most practical narratives we encounter every day tell modest stories with relatively short time spans and specific short-term goals—getting us places, helping us make decisions, or telling us how to do something. Through text, graphical elements, and pictures, time is visualized in small, discrete units that usually flow in a linear pattern from start to finish, though sometimes we can jump into the middle of the story (*in medias res*) to locate a stop on a bus schedule, find a data point in a time series, or complete a step in a set of pictorial instructions. All of these narrative forms rely on genre conventions for their success, though some of these conventions are being redefined by digital technology with interactive features that allow users to customize the narrative for their individual needs.

Traveling Timetables: From Static to Dynamic

Visualizing time with text has long been one of the most common and useful kinds of information design, enabling audiences to compare data in tables from one point in time to the next, row by row and column by column. Timetables are one of the inescapable necessities of modern life and routinely encountered by most people on the planet. Not surprisingly, scholars have analyzed their rhetorical features (Kinross 1985) as well as their usability and aesthetics (Tufte 1990, 101–5). Since the invention of mass transportation in the nineteenth century, timetables and schedules have enabled people to get across towns and cities and whole continents, telling audiences when and where they can travel—by stagecoach, train, bus, subway, ship, or plane. As they do with other kinds of tables, audiences "look up" values—time and place of departure and arrival—that relate most directly to their situations and ignore most other times within their purview. Like other kinds of tables, the arrangement of these values holds the key to their usability, taking the form of lists or matrices with rows, columns, and headings—visual conventions that have served audiences for nearly two centuries.

Figure 4.3 shows an early railway timetable (1849) for freight and passenger trains between Boston and Worcester (outward), with the stops along the way listed as row headings on the left; the table below (inward) lists them in the opposite direction. Both tables intermix passenger and freight times with a steamboat option (and a run to Milford), creatively blending multiple modes of transportation. The table at the bottom displays the times for the Newton trains, with the Brookline times tacked on the right in separate columns. Like most contemporary timetables, the Boston

Figure 4.3. An 1849 timetable for the Boston & Worcester Railroad. Courtesy of the Library of Congress, Rare Book and Special Collections Division, Printed Ephemera Collection [2021769436].

To take Effect] **BOSTON & WORCESTER RAIL ROAD.** [Oct. 8, 1849.

RUNNING TIME FOR THE PASSENGER AND FREIGHT TRAINS.

OUTWARD.

To Leave.	1st Freight.	1st Pass.	2d Pass.	Milford	Worcest. & Way Fr'ght.	Milford Fr't	Third Passenger.	Fourth Passenger.	Milford.	Steamboat.	Night Fr't.
Boston,	4 30 A. M.	7 30 A. M.	8 00 A. M	9 00 A. M.	12 00 M.	12 30 P. M.	2 30 P. M.	4 00 P. M.	4 30 P. M.	5 00 P. M.	5 15 P. M.
Brighton,	4 50	7 40 m. N'n	8 10 m. Mil	9 15 m. P.	12 20 m. F.	12 55	2 41 m. Mil	4 10	4 45	5 10 m. Nt.	5 35
West Newton,	5 10	7 47 m. Milf	8 19	9 27	12 30 P. M.	1 10	2 52	4 19	4 57 m. Nt.	5 17	5 55
West Needham	6 00	7 57	8 32 m. P.	9 42	1 00	1 45	3 05 m. F.	4 32 m. P.	5 12	5 27 m. P.	6 30
Natick,	6 20	8 03	8 40 m. Mil F	9 50	1 10	2 00	3 12	4 38	5 20	5 33	6 40
Framingham,	6 40	8 15 m. P.	8 52	10 00	1 30 m. F.	2 25 m. F.	3 23	4 50	5 30	5 45	7 05
Ashland,	6 55	8 21	9 05	10 15 Holl.	1 45	2 55 Hol.	3 31	4 56	5 45 Holl.	5 51	7 20
Southboro',	7 20	8 29	9 12 m. F.	10 35 Mil.	2 05	3 30 Milf.	3 40	5 07	6 06 Milf.	5 59	7 40
Westboro',	7 45 m. P.	8 35 m. F.	9 22		2 25		3 50	5 17		6 06	8 05
Grafton,	8 55 m. F.	8 47	9 36		2 50		4 03 m. P.	5 30		6 18	8 40
Reach Worces.	9 25	9 00	9 53		3 15		4 20	5 50		6 35	9 15

The regular Trains will start promptly from Boston, though expected Trains have not arrived. *In going out*, give the road to any Train or Engine on the Providence Railroad, taking the greatest care, always.

* First Ft. Outward must keep out of the way of 1st and 2d Pass. Outward. ☞ See 13th Rule. Outward Trains must not exceed 15 miles the hour west of 43d mile post. Great care must be used in approaching and leaving the Boston Station—not to exceed four miles the hour between the new Freight House and Station.

INWARD.

To Leave.	Milford	Milford Fr't	First Passenger.	1st Freight.	Steamboat.	Second Passenger.	2d Freight.	Milford.	3d pass.	Fourth Passenger.	Night Fr't.
Worcester,			7 30 A. M.	8 00 A. M.		10 30 A. M.	12 00 M.		4 10 P. M.	10 00 P. M.	11 00 P. M.
Grafton,			7 45	8 30 m. P.	May be expected	10 46	12 30 P. M.		4 30		11 30
Westboro',	Leave Milford		7 56	9 00 m. P.	to leave Worces-	11 00	1 00	Leave Milford	4 42		12 00
Southboro',	6 45 A. M.	Milf. 7 20 A. M.	8 04	9 20	ter early every	11 08	1 20	1 00 P. M.	4 50		12 20 A. M.
Ashland,	7 05 Holl	Hol 8 00	8 12 m. P.	9 25	morning, except	11 16	1 40	1 20	4 58 m. P.		12 40
Framingham,	7 20	Fra. 8 30 m. P.	8 25	9 55	Monday. It will	11 27	2 05	1 35	5 10	10 50	1 00
Natick,	7 28	Nat 8 50	8 33 m. P.	10 05	avoid all regular	11 35	2 25	1 45	5 18 m Mil		1 15
West Needham	7 35	Nee. 9 10 m. P.	8 46	10 20	trains.	11 49	2 40 m. P.	1 50	5 25		1 30
West Newton,	7 50	New. 9 45	8 53	10 40		11 55	3 05	2 05	5 37 m. P.		1 50
Brighton,	8 04 m. Nt.	Brig. 10 10	9 05 m. Mil.	10 55		12 07 P. M.	3 35 m. Nt.	2 17 m. Bkl.	5 50		2 10
Reach Boston,	8 20	Bos 10 30	9 20	11 15		12 20 m. F.	4 05	2 35 m. Nt.	6 05 m. Nt.	11 35	2 30

First Inward Freight will wait two and a half hours for New York Express Freight—taking care to keep out of the way of 2d Inward Passenger. ☞ See 13th Rule.

Night Freight will wait two hours for W. R. R. Freight.

The "Second" and "Third" Passenger Trains will each wait for the West. R. R. Trains, one hour. The "2nd Freight" will wait till 7 P. M. for W. R. R. Freight. All Trains leaving Worcester will halt above the W. R. R. Switch to see if all is right; also pass the switches at Newton and Brookline Junctions with great care.

NEWTON TRAINS.

Leave	1st Inward. A. M.	1st Outward. A. M.	2d Inward. A. M.	2d Outward. P. M.	3d Inward. P. M.	3d Outward. P. M.	4th Inward. P. M.	4th Outward. P. M.	5th Inward. P. M.	5th Outward. P. M.
Boston,		7 45		12 30		3 30		5 00 m. P.		9 30
Brighton,	7 07 m. P.	8 00 m. Mil.	9 24	12 45	2 53	3 45 m. Mil.	5 08 m. P.	6 15	8 38	9 45
Newt. Corner,	7 01	8 06	9 16	12 51	2 46	3 51	5 02	6 21	8 31	9 51
West Newton,	6 55	8 12	9 11	12 57 m. P.	2 40 m. P.	3 57 m. F.	4 55 m. Mil	6 27	8 25	9 57
Lower Falls,	6 45	———	9 00	———	2 30	———	4 45	———	8 15	———

BROOKLINE.

Leave Brookline.	Leave Boston.
7¾ A. M.	[illegible] 10 A. M.
8¾ "	10 "
10¾ "	12¼ P. M.
1¾ P. M.	2½ "
3¾ "	4½ "
5¾ "	6½ "
9 "	10¼ "

Newton and Brookline trains must avoid all others, and each other, while entering on and leaving the main tracks. Trains on the main track will look out for signals in approaching the junction of either branch, and stop if signal is given.

Engineers will allow no person to ride upon the engine without express authority.

and Worcester design encompasses a single twenty-four-hour day. Although this early timetable isn't nearly as clean, streamlined, and perceptually efficient as we'd expect today—U.S. passenger trains had existed for only a couple of decades when this document appeared—its structure creates a prototype for the genre and the flocks of timetables that followed it.

A similar temporal pattern appears in the current Metra train timetable for northeast Illinois (fig. 4.4), which shows the times for destinations between Fox Lake and Union Station in downtown Chicago, a trip that takes 80–90 minutes. On some runs, the train doesn't stop at some locations (indicated with a dash), which slightly speeds up the trip. However, the visual language of this timetable differs markedly from its Boston ancestor. The sans serif typeface is clean and legible, and the grayscales across every other row enhance the display perceptually by guiding the audience's eyes across the lines. For places along the route undergoing construction, icons appear below the columns to notify passengers of possible delays during work hours. Overall, this lean, efficient design is legible and structurally transparent, a child of modernist functionalism.

With the growth of modern transportation, creating usable timetables like the one in figure 4.4 has been an ongoing project in the quest for leaner and more perceptually transparent displays.[4] Although timetables are designed primarily for their clarity and utility, their typography can also serve a persuasive purpose, as Robin Kinross (1985) argues, that

Figure 4.4. Timetable for the Metra rail system in northeast Illinois (2024). Reprinted with permission. The Metra logo is a service mark of, and the inbound weekday schedule for Metra's Milwaukee District North Line dated, July 15, 2024, is copyrighted by, both the Commuter Rail Division of the Regional Transportation Authority and the Northeast Illinois Regional Commuter Rail Corporation.

Milwaukee District North Line

Metra

The below weekday schedule is effective July 15, 2024. Please check metra.com for updates and service alerts.

WEEKDAY SCHEDULE

Zone	FOX LAKE TO CHICAGO	2102	2104	2106	2108	2110	2112	2116	2118	2120	2124	2126	2128	2130	2132	2134	2136	2138	2140	2142	2144	2146	2148	2150	2152	2156	2158	2160
		AM	AM	AM	AM	AM	AM	AM	AM	AM	AM	AM	AM	AM	AM	PM	PM	PM	PM	PM	PM	PM	PM	PM	PM	PM	PM	PM
4	Fox Lake	4:40	5:15	5:40	5:55	6:14	6:25	6:45	7:01		7:28		8:46		10:46		12:46		2:46			4:25					8:37	10:30
4	Ingleside	f4:44	f5:19	—	f5:59	—	f6:29	f6:49	—		f7:32		f8:50		f10:50		f12:50		f2:50			f4:29					f8:41	f10:34
4	Long Lake	f4:47	f5:22	—	f6:02	—	f6:32	f6:52	—		f7:35		f8:53		f10:53		f12:53		f2:53			f4:32					f8:44	f10:37
4	Round Lake	4:51	5:26	—	6:06	—	6:36	6:56	—		7:39		8:57		10:57		12:57		2:57			4:36					8:48	10:41
4	Grayslake	4:56	5:31	—	6:11	—	6:41	7:01	—		7:44		9:02	10:02	11:02	12:02	1:02	2:02	3:02	3:51		4:41				7:33	8:58	10:46
4	Prairie Crossing	5:00	5:35	—	6:15	—	6:45	7:05	—		7:48		9:06	10:06	11:06	12:06	1:06	2:06	3:06	3:55		4:45				7:37	9:02	10:50
4	Libertyville	5:06	5:41	—	6:21	—	6:51	7:11	—		7:54		9:12	10:12	11:12	12:12	1:12	2:12	3:12	4:01		4:56			6:51	7:48	9:08	11:01
4	Lake Forest	5:17	5:52	6:08	6:32	6:42	7:02	7:22	7:29	7:47	8:05	8:27	9:23	10:23	11:23	12:23	1:23	2:23	3:23	4:12		5:07	5:27	6:02	7:02	7:59	9:19	11:12
4	Deerfield	5:23	5:58	6:14	6:38	6:48	7:08	7:28	7:35	7:53	8:11	8:33	9:29	10:29	11:29	12:29	1:29	2:29	3:29	4:18	4:43	5:13	5:33	6:08	7:08	8:05	9:25	11:18
3	Lake Cook Rd.	5:26	—	6:17	—	6:51	—	—	7:38	—	8:14	8:36	9:32	10:32	11:32	12:32	1:32	2:32	3:32	4:21	4:46	5:16	5:36	6:11	7:11	8:08	9:28	11:21
3	Northbrook	5:30	6:03	6:21	6:43	6:55	7:13	7:33	7:42	7:58	8:18	8:40	9:36	10:36	11:36	12:36	1:36	2:36	3:36	4:25	4:50	5:20	5:40	6:15	7:15	8:12	9:32	11:25
3	Glen/N. Glenview	5:34	—	6:25	—	6:59	—	—	7:46	—	8:22	8:44	9:40	10:40	11:40	12:40	1:40	2:40	3:40	4:29	4:54	5:24	5:44	6:19	7:19	8:16	9:36	11:29
3	Glenview	5:38	6:09	6:29	6:49	7:03	7:19	7:39	7:50	8:04	8:26	8:48	9:44	10:44	11:44	12:44	1:44	2:44	3:44	4:33	4:58	5:28	5:48	6:23	7:23	8:20	9:40	11:33
3	Golf	5:41	—	6:32	—	7:06	—	—	7:53	—	—	8:51	9:47	10:47	11:47	12:47	1:47	2:47	3:47	4:36	—	5:31	5:51	6:26	7:26	8:23	9:43	11:36
2	Morton Grove	5:45	6:14	6:36	6:54	7:10	7:24	7:44	7:57	8:09	8:31	8:55	9:51	10:51	11:51	12:51	1:51	2:51	3:51	4:40	5:03	5:35	5:55	6:30	7:30	8:27	9:47	11:40
2	Edgebrook	5:50	—	6:41	—	7:15	—	—	8:02	—	—	9:00	9:56	10:56	11:56	12:56	1:56	2:56	3:56	4:45	—	5:40	6:00	6:35	7:35	8:32	9:52	11:45
2	Forest Glen	5:53	—	6:44	—	7:18	—	—	8:05	—	—	9:03	9:59	10:59	11:59	12:59	1:59	2:59	3:59	4:48	—	5:43	6:03	6:38	7:38	8:35	9:55	11:48
2	Mayfair	5:57	—	6:48	—	7:22	—	—	8:09	—	—	9:07	10:03	11:03	12:03	1:03	2:03	3:03	4:03	4:52	—	5:47	6:07	6:42	7:42	8:39	9:59	11:52
2	Grayland	6:00	—	6:51	—	7:25	—	—	8:12	—	—	9:10	10:06	11:06	12:06	1:06	2:06	3:06	4:06	4:55	—	5:50	6:10	6:45	7:45	8:42	10:02	11:55
2	Healy	6:04	—	6:55	—	7:29	—	—	8:16	—	—	9:14	10:10	11:10	12:10	1:10	2:10	3:10	4:10	4:59	—	5:54	6:14	6:49	7:49	8:46	10:06	11:59
2	Western Ave.	6:10	6:32	7:01	7:12	7:35	—	8:02	8:22	8:27	8:49	9:20	10:16	11:16	12:16	1:16	2:16	3:16	4:16	5:05	5:21	6:00	6:20	6:55	7:55	8:52	10:12	12:05
1	Chicago (Union Station)	6:22	6:44	7:13	7:24	7:47	7:55	8:14	8:34	8:39	9:01	9:32	10:28	11:28	12:28	1:28	2:28	3:28	4:28	5:17	5:33	6:12	6:32	7:07	8:07	9:04	10:24	12:17

makes them more rhetorically potent than users might realize. This persuasive function, however, was far more explicit in earlier nineteenth century timetables when (unlike the public corporate entities that provide transportation services today), the *business* of travel—its marketing, customer service, economics, and adventure—was still an integral part of the rhetorical process, with competing services vying for travelers' dollars.

Figure 4.5, for example, shows a timetable from 1857 that announces the summer schedule for trains leaving Buffalo, New York, and traveling to various destinations in the Midwest. The daily departure times are listed for the four "express" trains, with detailed itineraries below, including stops and mealtimes. The large and varied typefaces, the graphical embellishments (including pointing hands), the textual border proclaiming the low fares, and the drawing of the steam engine and passenger cars—all entice audiences to travel with *this* company, the New York Central and Lake Shore and Michigan Southern Rail Road Line, and *none* other. Today this persuasive approach has shifted primarily to tourist travel like cruises, which entice potential travelers (both in print and online) with luxurious itineraries and who, once aboard, often receive daily schedules itemizing activities along the route—musical events, crafts, games, exercise, and social gatherings. These leisure/holiday schedules have an entirely different purpose and pace than modern commuter schedules and their Boston and Worcester ancestor.

Although the textual mode dominates schedules, graphical methods have also been used to visualize time, epitomized by Etienne-Jules Marey's chart of trains running between Paris and Lyon (fig. 4.6). Mirroring previous methods of graphical display (Wainer, Harik, and Neter 2013), Marey visualized time on a rectilinear grid with slanted lines connecting locations—with the steeper the line, the faster the train. This ingenious display allowed users to escape the granular confines of tables by giving them a macro-level perspective of how fast trains travel from one destination to the next. In the contemporary world, graphical methods similar to Marey's have been applied digitally to show movement over time. For example, Mike Barry and Brian Card (2014) designed an interactive visualization of Boston's daily commuter trains that uses sloping lines to show the movement of trains from station to station and that links the graphical display to a map. Like Marey's timetable, their design enables users to see movement at a macro level but with the added benefit of interactive features that provide precise granular views of time and place along the way.

With the increase in interactive digital design, schedules have been radically transformed to accommodate individual users, employing both textual and graphical elements to guide them to their destinations,

Figure 4.5. Railroad schedule for summer 1857 for trains going from Buffalo, New York, to the Midwest (New York Central and Lake Shore and Michigan Southern Railroad). Courtesy of the Library of Congress, Rare Book and Special Collections Division, Printed Ephemera Collection [2021768598].

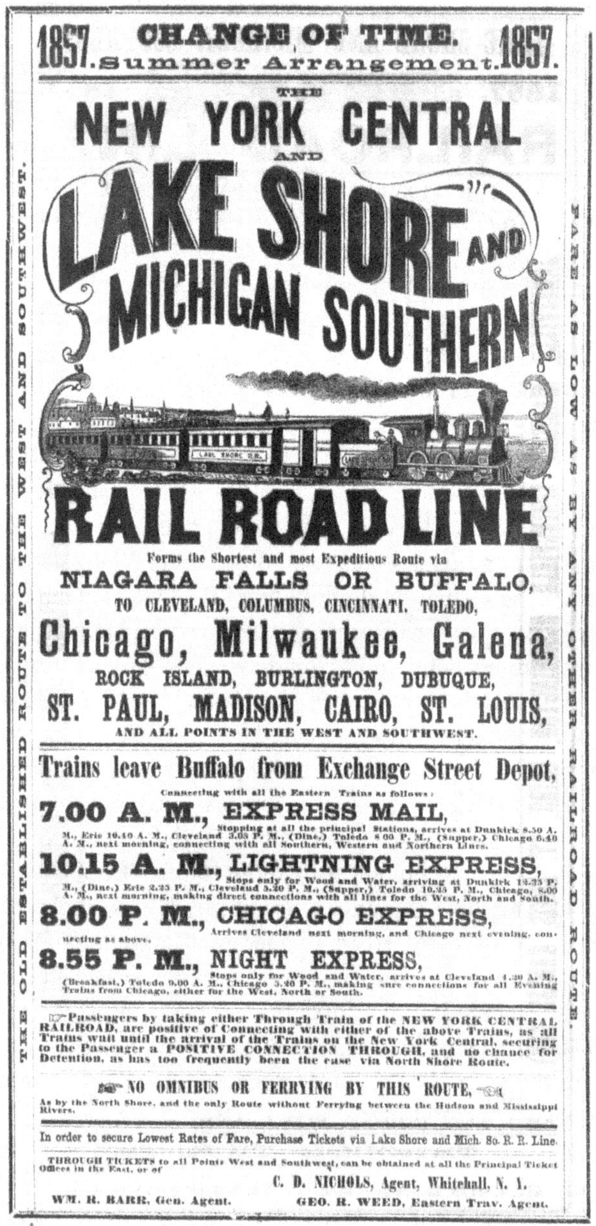

Figure 4.6. Graphical timetable by Etienne-Jules Marey (modeled after a design by Ibry) that charts trains traveling between Paris and Lyon (Marey 1878, 20, fig. 7).

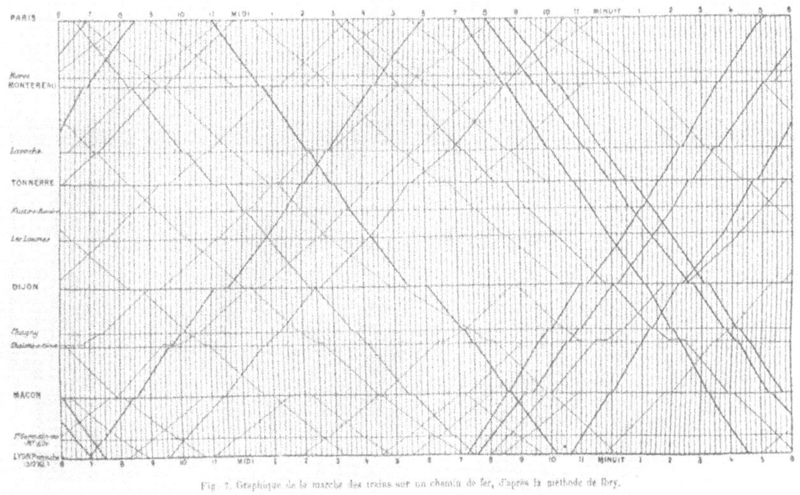

with mapping tools enabling users to navigate a schedule one stop at a time (at the micro level) or to see the whole route on an interactive map. Many users prefer to customize their travel plans, so they use the interactive display to choose where they want to go. For example, figure 4.7 shows an example of a live interactive display for the CyRide local bus service in Ames, Iowa, which provides transportation on the Iowa State University campus and around the surrounding Ames community. Like most interactive schedules, users (primarily students) choose from among several routes on the interface, which displays real-time updates at a given location—here the current location of the next bus on the route and the bus's estimated time of arrival at the Memorial Union on campus. Users can also access this information on a Google map, which reveals additional details—for example, the percentage of passengers occupying a given bus. In these ways, by showing these micro-level details, the schedule and accompanying map enable users to "personalize data," as Tufte (1990, 50) describes the experience, by showing them only what they want to see based on their travel plans at a given moment, an interactive process that creates an emotional connection with each of them.

Many other interactive applications (for airlines, subways, trains) follow similar patterns of prompting users to customize their schedules

Figure 4.7. Interactive display showing the schedule for the CyRide (2025) bus service on the Iowa State University campus. © GMV. Reprinted with permission from CyRide.

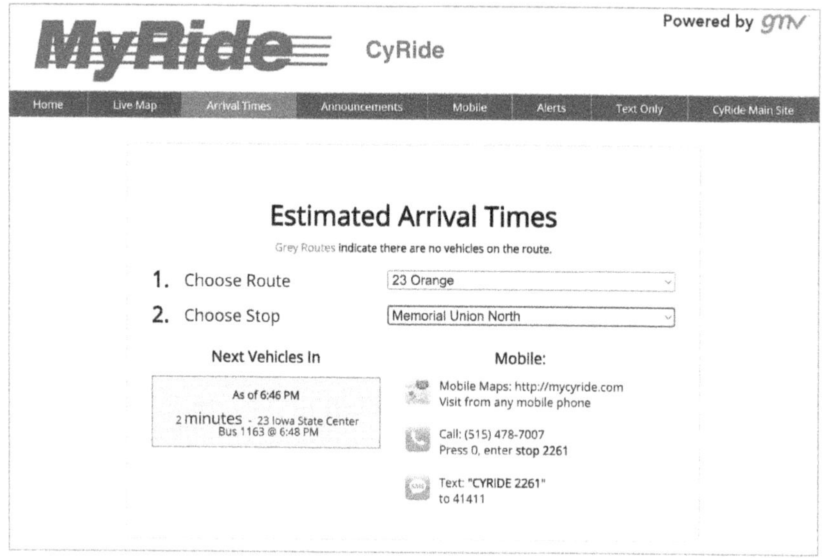

for particular times and routes. As a result, these micro-narratives play out differently for each traveler, pinpointing the search for particular data values that traditional timetables display across heavily populated rows and columns. Interactive schedules eliminate this noise by enabling users to focus only on the pertinent details, creating an efficient and customized path to a given destination and compressing the narrative into a single kairotic story of the moment. This more intimate and focused interaction might be particularly important with urban mass transportation systems whose complex print/PDF schedules account for every time and route, most of which are irrelevant to a given traveler. However, interactive schedules can also become rather claustrophobic, heightening the "analytic" view at the expense of the "synoptic" (Barton and Barton 1993, "Modes of Power," 142–49), though luckily for CyRide users the interactive map gives their route spatial and temporal context on campus and around Ames. These digital designs continually reshape the visual conventions of the timetable, which were initiated by prototypes like the Boston and Worcester schedule (fig. 4.3) and which have served generations of users, for whom timetable conventions remain part of their daily lives.

Decision Trees: Stories Users Customize and Control

Audiences can also customize narratives with branching diagrams—most notably decision trees, which can shorten or extend the narrative depending on the audience's circumstances. The visual conventions of decision trees descend from displays like *Encyclopédie*'s genealogy chart of French royalty in figure 2.5, with its branches and vegetation invoking the tree metaphor, and from Linnaean taxonomies that visualize hierarchies in nature, which originated earlier in the eighteenth century. A modern variation of the branching diagram, decision trees encompass diverse fields—law, criminal justice, medicine, statistics, business, construction—enabling its users to follow a path of decisions or actions to achieve a certain outcome, which can take days or months. Figure 4.8, for example, shows the

Figure 4.8. Decision tree for how to respond to patients needing surgery during the COVID-19 pandemic (General Surgeons Australia 2020). © 2020 General Surgeons Australia. Used with permission.

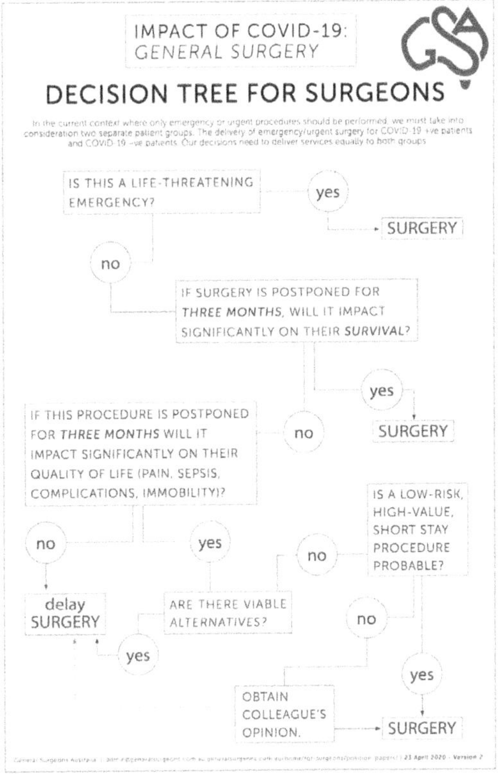

decision-making process for performing surgeries during the COVID-19 pandemic, balancing the needs of COVID-19 patients with those needing surgeries for various conditions, a dilemma that affected millions of people around the globe needing medical care. The branching tree configuration and yes/no nodes enable surgeons to find their way through the decision-making quickly and confidently, though some information gathering or consultations may be needed along the way to decide which path to take.

Another complex narrative unfolds in figure 4.9 where decision-makers need to assess threats to schoolchildren. Intended for an international audience, the tree relies heavily on icons and pictures to guide the audience through a series of decisions based on the kind of threat (an explosion, fire, flood, or tsunami) and the respective outcomes of these threats, ultimately leading to a resumption of classes or returning students to their families. The diagram uses conventions of the decision tree genre, with action flowing from the top down, arrows directing the traffic, and diamonds indicating decision-making moments in response to questions. The chart also embodies other conventions: triangles to signify warnings and spot color (red and green) to signal positive or negative responses to the questions. The complexity of this decision tree could especially benefit teachers and administrators during staff training sessions so that they were prepared to act quickly and decisively once a threat has been observed, with the diagram then serving as a reminder of how to respond based on an agreed-upon policy.

Today, decision trees are designed in both static and interactive forms, with digital versions often allowing users to zoom in on only the most immediately relevant parts of the display. Much like contemporary digital timetables, these displays conceal the parts of the display unessential to an immediate contingency, which greatly benefits users navigating their way through detailed, complex processes. However, users also miss seeing the bigger picture and are left wondering what might have been if they decided on a different path through the narrative—for example, online consumers who "build" their cars on a manufacturer's website, gamers who create role-playing characters, or singles looking for compatible soulmates. Tightly focused interactive displays like this are both efficient and confining, their benefits and drawbacks the inverse of traditional static displays that give users the big picture at the cost of visualizing more than they need to know.

Figure 4.9. Decision tree for responding to school emergencies across the globe (International Federation of Red Cross and Red Crescent Societies 2018, 135). Reprinted with permission from the IFRC. Copyright: International Federation of Red Cross and Red Crescent Societies.

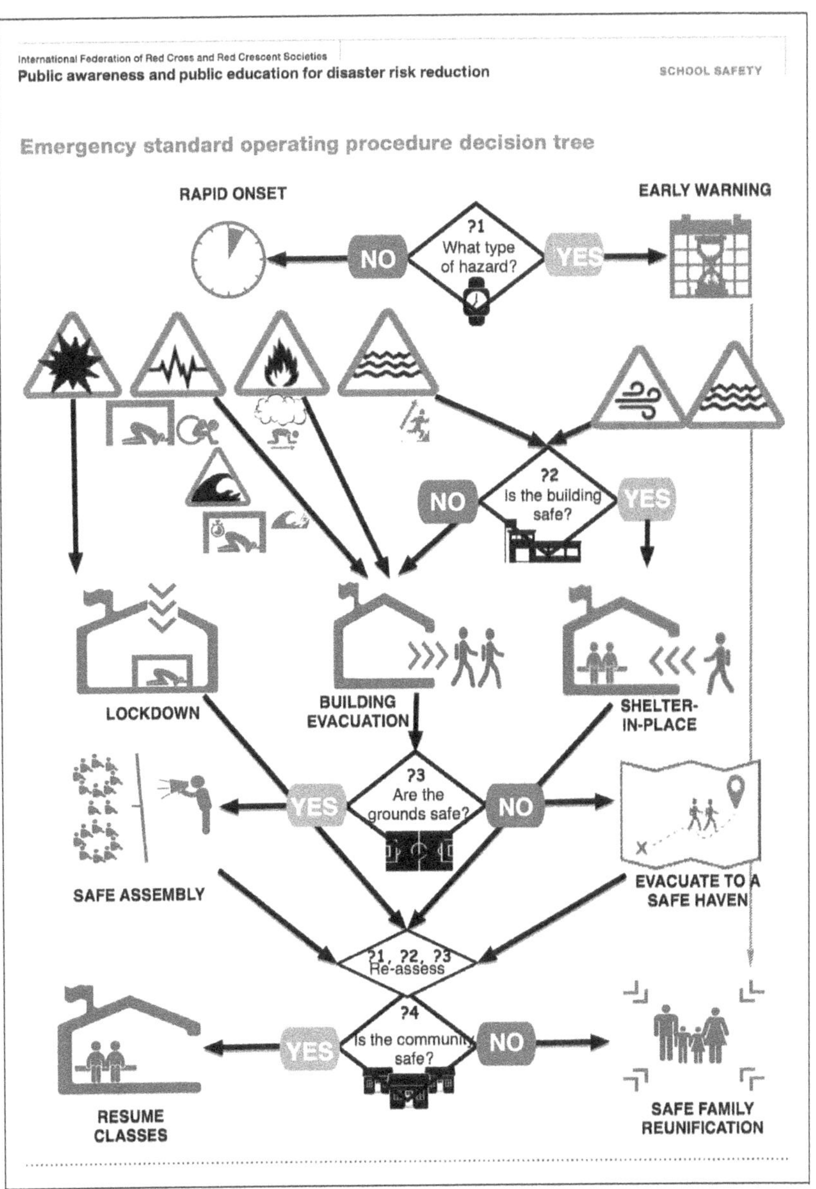

Pictorial Time: How-to Narratives and Practical Comics

Our world teems with how-to designs in the form of pictorial instructions and comic-like narratives with identifiable characters. As we saw in previous examples—the IKEA wordless instructions for assembling furniture in figure 1.18, the choking poster in figure 1.20, and the car charger instructions in figure 4.1—how-to displays rely heavily on pictures. You're likely to find any number of these narratives in the workplace, at home, traveling, or engaging in leisure activities.

Archetypes for these practical narratives appeared in early mining, gardening, and engineering books and coalesced most famously in the plates (the *Recueil de Planches*) for the eighteenth-century French *Encyclopédie* (Diderot and d'Alembert 1751–1765). Many of the *Encyclopédie*'s illustrations narrate methods of building and manufacturing things, from seafaring ships and masonry walls to textiles and glass. Figure 4.10 shows the process for making artificial pearls through enameling. Each woman in the picture does her part in making the jewelry—extracting color from fish scales, applying the color to the fake pearls, filling them with wax—and in these ways the narrative unfolds in a single picture, step by

Figure 4.10. The process for making artificial pearls illustrated in the *Recueil de Planches* (Diderot and d'Alembert 1762–1772, vol. 4, "Emailleur, a la Lampe, Perles Fausses," plate III, top) that accompanied the *Encyclopédie*. Courtesy of Iowa State University Library Special Collections and University Archives.

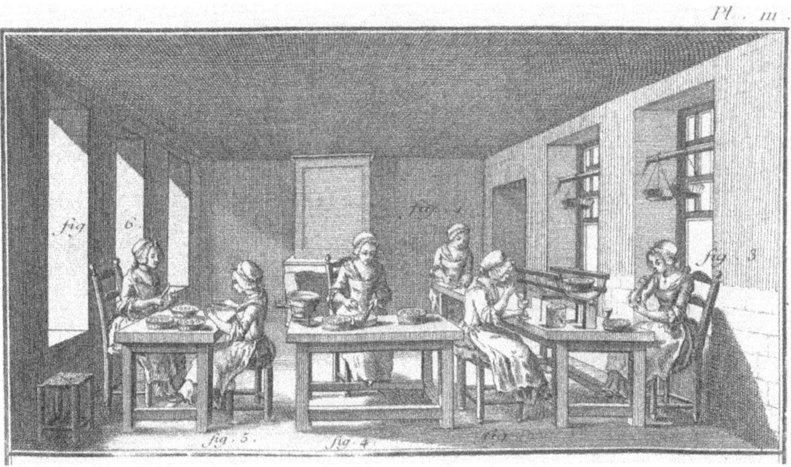

step, as the activities circulate around the room. Some audience members might already engage in these activities, or they might consult the picture to learn how to do them; others might buy, sell, or wear artificial pearls. For most audiences, however, this plate's purpose is primarily epistemological, enabling the audience to understand how this trade is performed. For whatever reasons audiences engage with this picture, the diligent pearl-makers remain in perpetual motion, endlessly repeating their tasks in their well-orchestrated ensemble.

Modern pictorial narratives in the late nineteenth and early twentieth centuries had narrower and more quotidian purposes—to instruct consumers in how to use mass-produced machines ranging from sewing machines and typewriters to farm equipment (see Tebeaux 2010). Fueled by burgeoning consumerism (Schriver 1997, 17–26), these narratives featured photographs and detailed drawings, legacies of the realism that dominated Victorian aesthetics. These were followed by leaner, more abstract narratives that reflected the values of modernist minimalism—exemplified by the high-contrast images in figure 3.15 that visualize a conversation about home energy efficiency.

Popular culture also made its impact on instructional materials with the surge in comics from the late 1930s to the 1950s. As detective stories and superheroes began to inundate popular culture, practical applications of comics also found their way into the mainstream (Yu 2015), partly because of the imprimatur of the U.S. government, which included comic-style pictures in their publications to tell stories about labor unions, military activities, agricultural practices, and domestic life. Comics had the double benefit of telling practical stories in an informal and riveting style while at the same time legitimizing them, transforming the commonplace into the heroic.

As a narrative technique, comics are composed of series of pictures in what comics illustrator Will Eisner defined as "sequential art," a concept Scott McCloud (1993, 10–17) elaborated on using historical examples like Egyptian paintings, the *Bayeux Tapestry*, and William Hogarth's pictorial narratives (e.g., *A Rake's Progress*, a segment of which appears in fig. 2.23).[5] Combining pictures and words, comics engage audiences in stories with abstract, conversational drawings arranged in a tight, linear structure, one frame after another, with characters expressing their thoughts in word bubbles. An example of a practical workplace comic appears in figure 4.11, which narrates a contract negotiation between a union and a company. In this sequence of pictures from a 1947 issue of the U.S Department of

Figure 4.11. Comic-style narrative from the 1947 *Labor Information Bulletin* that explains how union contracts are negotiated with the help of a conciliator (U.S. Department of Labor, Bureau of Labor Statistics 1947).

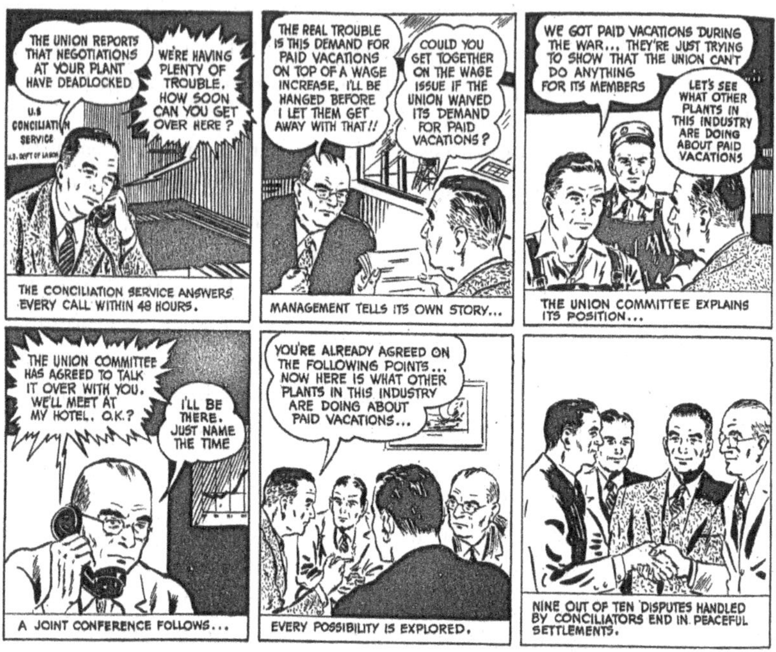

Labor's *Labor Information Bulletin*, a representative of the U.S. Department of Labor (Conciliation Services) acts as a mediator between a company and its union workers. At issue are wages and possible paid vacations, which the conciliation agent tries to get the company and the union to discuss together. Appearing in each picture, the conciliator plays the role of active agent, meeting separately with the company manager and the union representatives, then bringing them to the table together. After he negotiates between the two parties, an agreement is likely reached, though how long this takes isn't specified. The story simply narrates how the process unfolds, ostensibly to persuade audiences (both labor and industry) that they will benefit from having a conciliation agent act as their mediator.

Numerous other comic-style narratives appeared in the *Labor Information Bulletin* in the 1940s and 1950s, as well as in the U.S. Army's *PS Magazine: The Preventive Maintenance Monthly*, which since the early 1950s has used comics to instruct soldiers on how to maintain military equipment. Illustrated initially by comics pioneer Will Eisner, *PS Magazine* featured a regular cast of characters, including the omniscient Sgt. Half-Mast McCanick, the intelligent and alluring Connie Rodd, and the well-meaning but inept Joe Dope and his accomplice Private Fogsnoff (U.S. Department of the Army 1951–1971). Figure 4.12 illustrates one of

Figure 4.12. Comics narrative from the U.S. Army's *PS Magazine* in which Joe Dope tries to persuade soldiers not to hoard spare parts (U.S. Department of the Army 1954, 783). Courtesy of Special Collections and Archives, Virginia Commonwealth University Libraries.

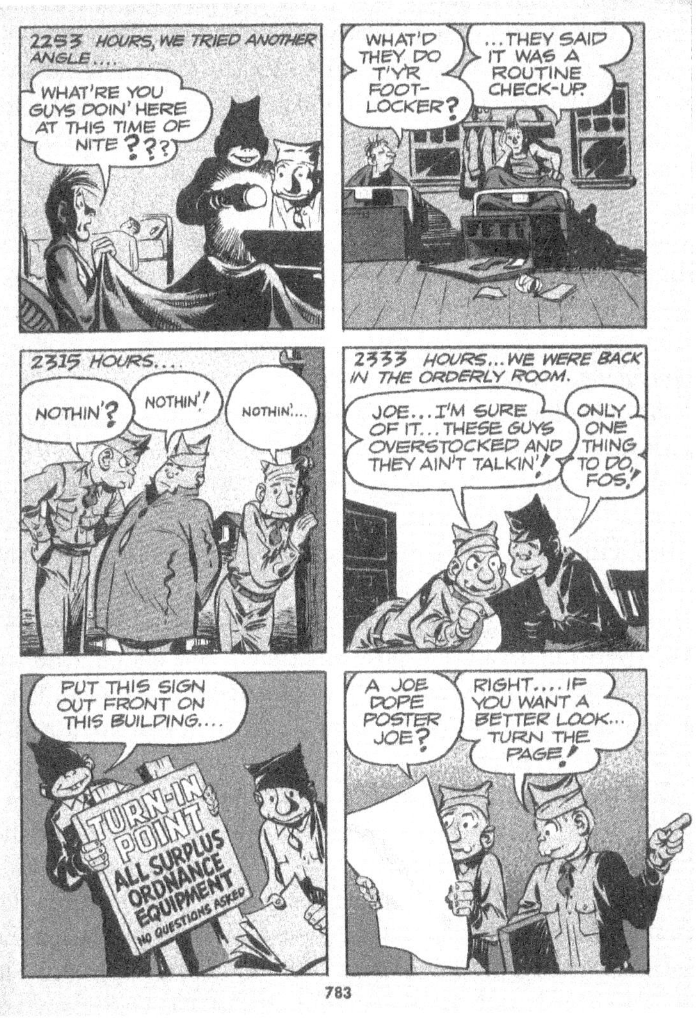

Joe Dope's adventures, here searching in the middle of the night for overstocked parts, presumably to prevent waste and supply chain problems. The hoarders of the parts play the villains while Joe and his accomplice search for the goods and a way to convince their fellow GIs to turn them in.

Comic-style narratives that tell practical stories have proliferated in recent decades: Sid Jacobson and Ernie Colón's graphics version of *The 9/11 Report* (2006), Scott McCloud's comic narrating the development of Google Chrome (2009), Robert Sikoryak's rendition of Apple's "terms and conditions" (2017), and Ray Doty's popular collection of graphical stories explaining projects around the home (1996). As a medium to tell practical stories, comics have also been used for teaching and learning—in textbooks and other forms of instructional design, including those for technical communication (e.g., see Yu 2015; Watkins and Lindsley 2020). Today how-to pictorial narratives inundate our personal and professional lives, appearing on equipment (printers, projectors), in airline safety cards, in online help menus, in manuals for vehicles and machinery, and with virtually any product that arrives at the doorstep. Like comics, the pictures are typically line drawings at varying levels of abstraction; they appear in sequential order, step by step; and unlike characters in *PS Magazine* or a graphic novel, the people (or parts of people) they visualize are typically an anonymous "Everyperson" without a specific identity (Kostelnick 2019, *Humanizing*, 42–46, 69–71).

The ubiquitous Everyperson appears in the how-to narrative in figure 4.13, which illustrates the procedure for a self-test for COVID-19, a task performed globally during the pandemic. Audiences typically complete this task in their dwelling space with the test kit (swab, tube, tube holder, test strip) and a series of pictures and text to guide them. Test takers are anxious about the results but highly motivated to complete the process, which takes about 12–15 minutes. When the process is completed, the user discovers the negative or positive results—one ending with a sigh of relief, the other with isolation and possible medical treatment. Regardless of the results, retesting probably follows as well, at that point making the instructions a reminder of what to do rather than a step-by-step guide.

When we use pictorial narratives to complete tasks like these, we're conscious of the time it takes relative to our expectations (or the time estimated in the instructions). If we pay attention to the warnings (Step 3A in the COVID-19 test instructions), we might proceed more slowly to avoid mistakes, or we might dwell on some steps more than others because of their complexity or less familiar conventions. Some steps, for example,

Figure 4.13. Some of the steps for completing a Quidel COVID-19 test (Quidel Ortho Corporation 2021). QuickVue® At-Home OTC COVID-19 Test, QuidelOrtho Corporation ©2023 QuidelOrtho Corporation. Used with permission.

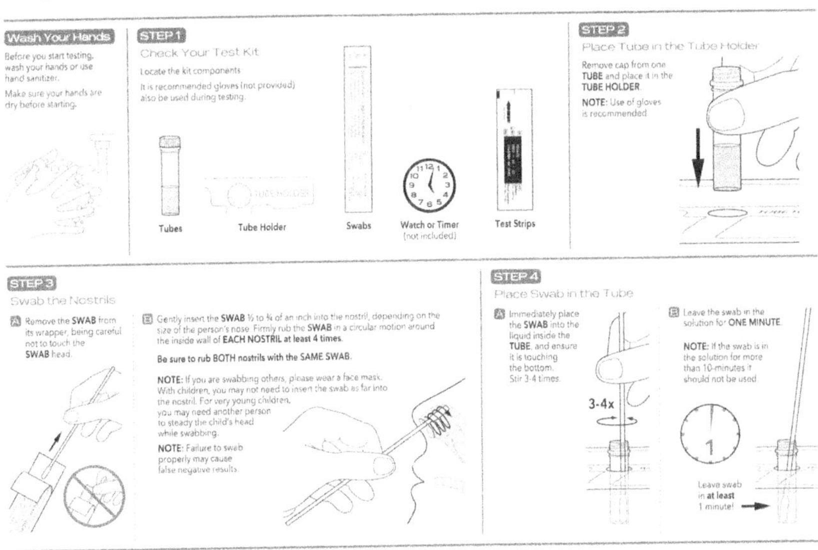

embed micro-narratives by using arrows to show motion: in Steps 2 and 3A to show the direction for moving the tube and swab and in Steps 3B and 4 to show the circular motion of the swab. These drawing conventions, among others, are familiar to most audiences for visualizing motion in micro-narratives, though a usability study would provide insights into how clearly, comfortably, and punctually audiences interpret these and other pictorial elements.[6] The actual time on task for many kinds of pictorial instructions will be more fluid than fixed, with some moving at a faster pace than the COVID-19 test, others much slower, with steps that require time lapses—glue that has to dry, food that has to cook, plants that have to germinate.[7] These and other hiatuses (checking social media, getting a sandwich, going to the store for supplies) can extend the narrative for minutes, hours, or days.

Moreover, many practical narratives like the labor negotiation in figure 4.11 envision an ideal story—here by streamlining the negotiating process, skipping the tense back-and-forth interactions, threats of a strike, and breakdowns. This idealization of the story has persuasive ends—for example, persuading consumers that a product is worth buying because it's fast and easy to assemble, or persuading reluctant do-it-yourselfers to attempt tasks they find intimidating. At some level, then, these stories are fictions because the actual task performed by audiences rarely unfolds exactly the way it's envisioned: managers and union bosses dig in their heals, parts for assembly are missing, cell phones ring, mistakes are made. On the other hand, users who have little choice in doing a task don't need prodding, however anxious they might be about successfully completing it. The COVID-19 test instructions in figure 4.13 repeatedly warn users about miscues that could derail the test, resulting in inaccurate results or retesting, but users are highly motivated to conduct a valid test, however reluctant they might be to do it in the first place.

Have practical comic-style narratives run their course in the digital age? Despite waning interest in popular comics, lavishly illustrated graphic novels have filled that gap in popular culture, further fueling the appetite for storytelling with pictures, including practical narratives using the visual conventions of comics. At the same time, digital media has opened many new avenues for practical storytelling. For example, animations have become increasingly popular for illustrating medical and safety procedures, for envisioning how mechanical devices operate (O'Neal 2024; Szczepaniak 2025), and for explaining countless other practical subjects. In addition, self-made videos (by amateurs and professionals) can quickly and cheaply visualize a task and be posted instantly on social media. Nonetheless, static pictorial narratives continue to benefit from access and control because not everyone has the technology to view online videos while they perform a task. Moreover, simple line drawings, including those done in comic style, give designers a great deal of control over content, eliminating the potentially noisy distractions of video. This mix of static, interactive, and digital media will continue to foster pictorial storytelling for the foreseeable future, complementing and competing with each other to meet the needs of diverse audiences and situations.

Data Narratives: In Buckets and Teaspoons

Data narratives in graphical form have a long history that spans the past four centuries, beginning in the later seventeenth century with charts that visualized weather (barometer, temperature, precipitation) and later transitioned to displays of statistical data about population, health, and economics (see Tufte 1990, 96–119). Michael Friendly and Daniel Denis's website Milestones in the History of Thematic Cartography, Statistical

Graphics, and Data Visualization (2001) provides an interactive timeline of this long and complex history. Much of this history entailed various forms of narrative, ranging from line graphs and bar charts to area charts, wind roses, data maps, and a host of other genres that flourished during the "golden age" (Funkhouser 1937, 330; Friendly 2008) of data design in the later nineteenth century as statistics emerged as a discipline. Today, data designers have a wide array of spatial and graphical options for plotting visual narratives (Tominski et al. 2017), with "storytelling" recognized as a key element of data visualization that's worthy of scrutiny (Kosara and Mackinlay 2013).

Until the last few decades, charts visualized time in graphical forms that were static and immutably fixed on the page. As a result, a chart that effectively "tells a story" (Tufte 1983, 30) needed to be as data rich and clear as possible, with its narrative spanning months and years, if not longer. Playfair's trade chart (fig. 2.20), for example, visualizes its story over a century, as does the rank population chart from *Scribner's Statistical Atlas* (fig. 2.21). This protracted temporal clock was due chiefly to the "statistical" nature of the data, collected by nation-states only periodically, often once a decade. Data collection methods determine the kinds of narratives that are even possible to visualize: the more data, either collected directly (weather observations) or from sources (the U.S. Census), the more possibilities for visualization. In creating his now famous chart of Napoleon's invasion of Russia (fig. 2.26), how did Charles Joseph Minard know the size of the French army at a given time and location on the map? How did he know the temperatures on certain days? An accomplished civil engineer who knew something about terrain and logistics, Minard obviously did his homework.

Although static charts with longer temporal horizons will always play an important role in narrating data over time, the technology of digital displays has transformed the way we envision time—about money, health, sports, weather, and personal activities—and consequently our ability to *act* on what we *see*. In our age of "big data" we think about data and time in much smaller, more immediate increments, with mini-micro-narratives narrowing the temporal scope and refining time in ever smaller granules. As a result, we now consume time in teaspoons rather than buckets.

Figure 4.14, for example, shows another data display from StockCharts.com, an online charting company that focuses on stock trading. Earlier I showed a StockCharts.com area chart (fig. 1.7) that captured daily stock prices on the S&P 500. Figure 4.14 includes multiple narratives about

Figure 4.14. StockCharts.com chart showing the daily performance of the Dow Jones Industrial Average in multiple time series and chart genres (2025, "Dow Jones"). Copyright StockCharts.com. Chart courtesy of StockCharts.com.

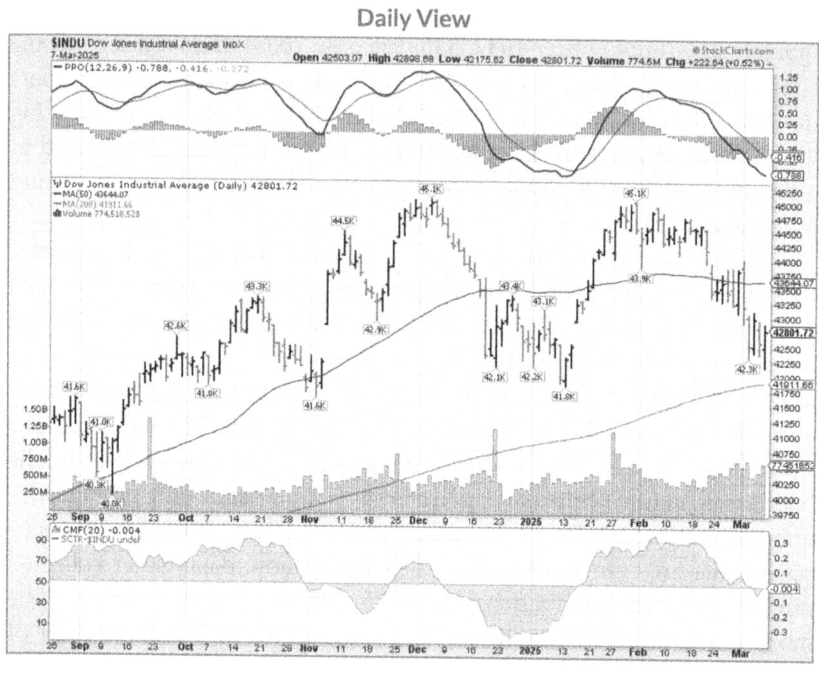

six months of stock trading in the Dow Jones Industrial index, with the main plot frame showing high/low fluctuations on a given day, the trading volume shown as a bar chart (histogram) at the bottom of the plot frame, and lines showing various other indexes and technical metrics. By clicking on this chart, the user can access multiple controls to customize the time period, stocks, and other categories, as well as the user's preferred genre.

The density of this chart reflects the "big data" trend and the needs and expectations of users, who can slice and dice time on command within the parameters of the chart. Many other types of charts have hyper micro-level features that show time in miniscule increments in near real time: charts plotting sporting events, maps displaying the movement of storms and traffic, apps showing popular times in restaurants. Teaspoon by teaspoon we consume a data rich diet.

This proliferation of mini-micro-narratives has occurred in many forms and genres, as I've shown previously, leaving us to wonder if technology itself has caused this shift or if our frenetic lifestyles, cognitive strain from information overload, or other social or cultural factors have fostered our appetite for temporal immediacy. Probably all of these contribute. In the next section we'll explore longer-term social and cultural forces that shape practical narratives, past and present.

Macro-Narratives: Visualizations of Social and Cultural Values

Language and narrative are deeply intertwined with culture, as Jerome Bruner (1990) argued, and this affinity is epitomized by macro-narratives that represent broad social and cultural trends that permeate entire eras. Several theorists have invented models that illustrate macro-narratives and how they have shaped and predicted communication and behavior, both on a large, "grand" scale as well as in smaller, more local increments (Lyotard 1984, 60; see Chandler and Munday 2020). On whatever scale, these macro-narratives provide the temporal and cultural cohesion that enables us to make sense out of the events and "stories" in which we are immersed, in whatever media we encounter them. Although these theoretical constructs give us insights into literary and fine arts narratives, they inform practical narratives as well. Information designs that visualize time are often embedded in larger collective narratives, which provide interpretive lenses for audiences that share (or tacitly acquiesce to) their values. In this way, visual narratives—even mini-micro-narratives—have the capacity to tell much larger stories that galvanize audiences by connecting them with the social, cultural, and political milieu of a given era.

For example, Sebastian Adams's massive chart of human history in figure 4.2 is itself embedded in a Western macro-narrative, with biblical figures foregrounded at the top, heroes and artifacts featured in each segment, and all of these woven together to create a compelling story about human progress. As sure-footed as this historical narrative might have been to its Victorian audience, macro-narratives inevitably change, perhaps because of new discoveries—artifacts, archeological excavations, remote mapping—but also because of changes in culture, politics, and ideology. Given these potent forces for designing and interpreting, today's audiences might challenge Adams's narrative: Why were certain people and events chosen and others disregarded? How do we distinguish the heroes from the villains? If this chart were created by someone from a non-Western culture (or from a country formerly colonized by Europeans), what would it look like? Questions like these might not have occurred to audiences in the Victorian era, when Western empires extended to every corner of the globe and historical progress reigned supreme, fueled by advances in science and technology. Inevitably, however, the macro-narrative changes: glory morphs into shame; the obscure becomes mainstream; new stories emerge and old ones fade.

Manifest Destiny: Visualizing the Narrative of Westward Expansion

The influence of macro-narratives extends to many other kinds of practical visualizations that are designed and interpreted within these larger temporal domains. The movement of European immigrants across the U.S. in the nineteenth century provided a compelling national narrative that was celebrated in literature, painting, and visual culture. During the creation of the *Statistical Atlases of the United States* in the late nineteenth century, the epic concept of Manifest Destiny and the decree "Go West, young man" (often accredited to Horace Greeley), occupied the national consciousness. In the mosaic chart in figure 1.6 (from the 1874 *Statistical Atlas*) the rectangles displaying the outflows of population for each state represent primarily the mass migration from the East to the West.

The westward movement of the nation's population was also shown *geographically* in the *Statistical Atlases*. In the 1874 *Atlas* (F. Walker and the U.S. Census Office), the first "centre of population" map (based on a map by J. E. Hilgard) appeared in a section titled "The Progress of the Nation—1790–1870" (6), with dark stars on the map charting the westward movement with each census. Also in the 1874 *Atlas*, an individual population map for each of the nine U.S. censuses showed its respective center of population (plates 16–19).[8] Updated "Center of Population" maps appeared in the *Statistical Atlases* of 1883 (Hewes and Gannett, xl), 1898 (Gannett and the U.S. Census Office, 10), and 1903 (Gannett, plate 16). Figure 4.15 shows the map from the 1898 *Atlas*, which like the earlier maps marks the geographical centers of population with dark stars, beginning on the East Coast and with each new census working their way across the Blue Ridge Mountains in Virginia, into Ohio and the Midwest, and all the way into southeastern Indiana with the 1890 census. Moving at a rate of about fifty miles per decade, the stars travel almost due west, following the tide of immigration across the prairie, plains, and mountains to the West Coast. The trajectory of that movement has only one direction

Figure 4.15. Map from the 1898 *Statistical Atlas of the United States* showing the geographical centers of the U.S. population from each census (Gannett and the U.S. Census Office 1898, 10). Courtesy of the Library of Congress, Geography and Map Division [07019233].

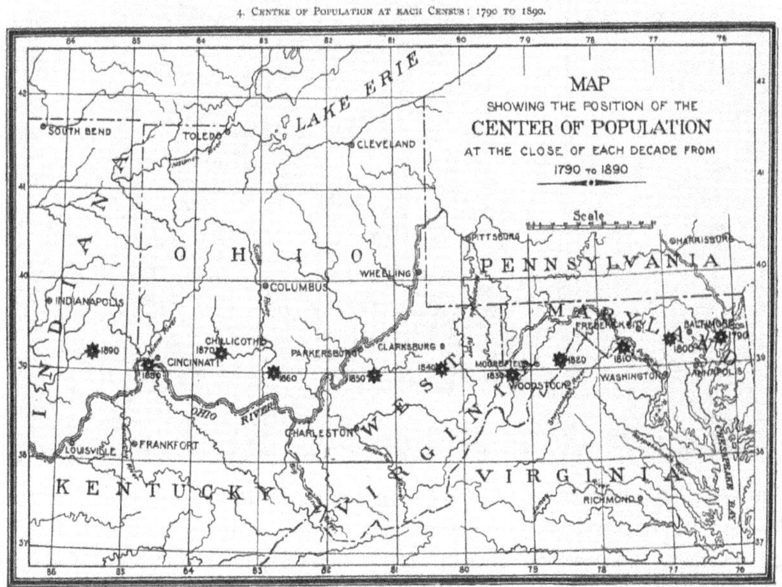

to go, with fate driving it relentlessly westward. Audiences viewing this data map would have found its story a source of pride and optimism, a narrative that they (or people they knew) participated in.

American audiences interpreted these time series maps through the lens of the national narrative of Manifest Destiny, which was visualized in two mythic pictures: Emanuel Leutze's *Westward the Course of Empire Takes Its Way*, a mural painted for the U.S. Capitol Building in 1861–1862, and George Gast's *American Progress*, painted in 1872 and reproduced by George Crofutt. Leutze's mural (fig. 4.16) envisions the march westward with a stream of wagons pulled by oxen wending their way across rugged terrain, the figures undaunted by hardship. Men strain to move their wagons, mothers embrace their children, a man atop the highest rock holds an American flag, and everyone looks, points, or leans forward, determined to push ahead. Although few Americans may have been able to visit the U.S. Capitol to view Leutze's grand mural, it vividly embodied the nation's

Figure 4.16. Mural in the U.S. Capitol Building by Emanuel Leutze (1861–1862) titled *Westward the Course of Empire Takes Its Way*. Credit: Architect of the Capitol. Used with permission.

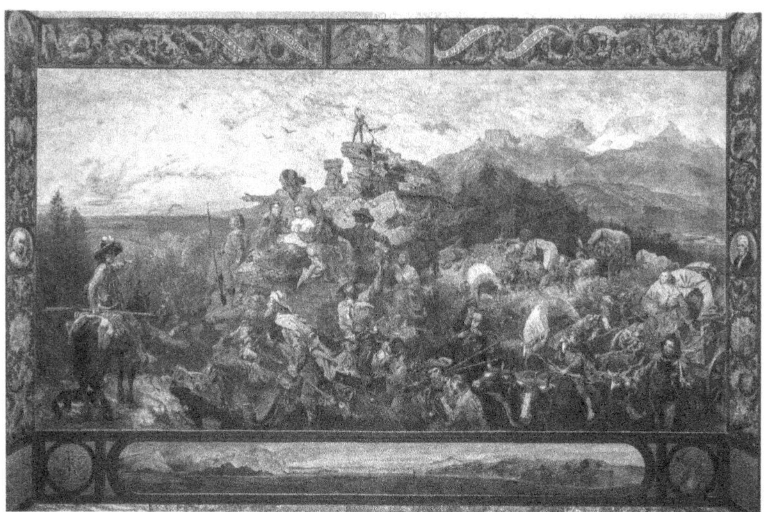

Figure 4.17. Picture, based on an 1872 painting by John Gast, titled *American Progress* (or *Westward the Course of Destiny*) that visualizes the movement of settlers across the continent (Crofutt 1873). Courtesy of the Library of Congress, Prints and Photographs Division [LC-USZC2-1332].

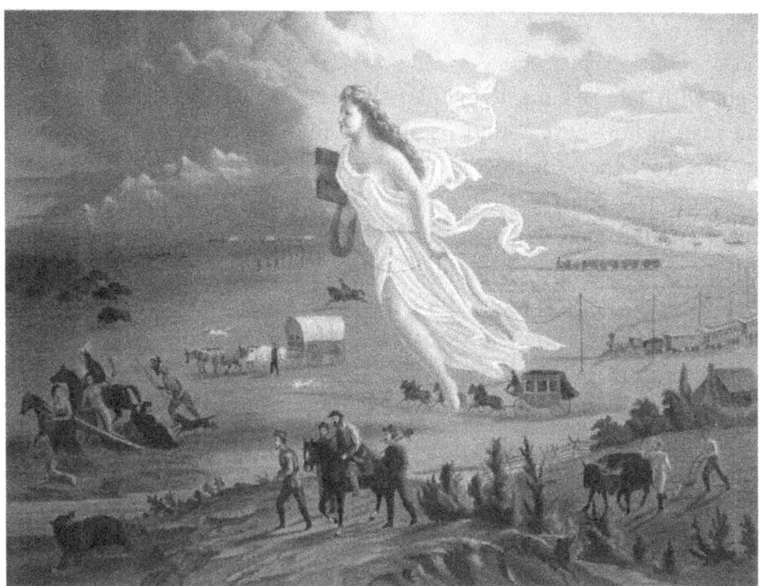

conception of Manifest Destiny. Gast's picture (fig. 4.17), which was widely circulated in color prints, envisions westward migration in equally epic terms, with a mythic woman, Columbia, overseeing the journey from above, with a schoolbook in one hand and a telegraph line in another and a panoramic landscape beneath her from the sunlit east to the dark, foreboding west. In their westward trek, pioneers transport themselves by every available means—on horse, on foot, in wagons, stagecoaches, and railroads—killing the buffalo and plowing up the soil, with Native Americans fleeing ahead of them. This epic process of westward movement portrayed in the pictures by Leutze and Gast captivated the nation, as pioneers collectively nudged the stars on the "Center of Population" maps of the *Statistical Atlases* farther west.

In 2010, the U.S. Census updated the center of population map (fig. 4.18), which by then had moved from southeast Indiana, across southern Illinois and into south central Missouri, taking a decidedly southwesterly

Figure 4.18. U.S. Census Bureau map showing the geographical centers of the U.S. population over 220 years (U.S. Department of Commerce, U.S. Census Bureau, Geography Division 2022). Courtesy of the U.S. Census Bureau.

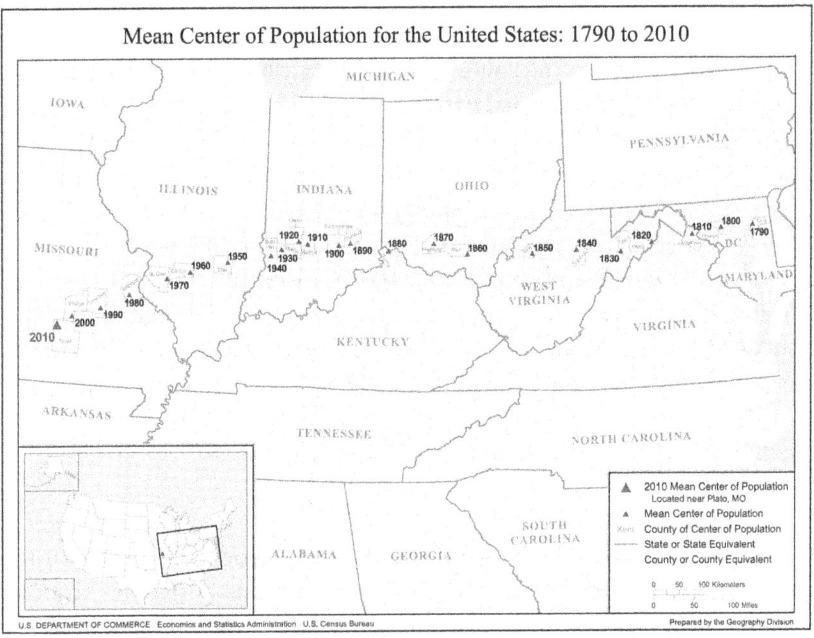

trajectory as people moved increasingly to the "Sun Belt" in the 1960s and 1970s. By the time this map was created in 2010, however, Manifest Destiny had long ceased to capture the national imagination, and, moreover, historians increasingly questioned the merit and legitimacy of that concept, especially its impact on Native Americans and the natural environment, epitomized by the near extinction of the buffalo. Culture, science, and the age of the U.S. population had changed dramatically, and along with those changes created very different interpretive lenses, both of the 2010 map and the 1898 version. Americans began thinking about retiring in the warm Southwest, fleeing "Rust Belt" cities for a fresh economic start, and seeking opportunities for outdoor recreation in the open spaces—and later worrying about degrading those spaces as populations exploded in fragile environments. Given these new priorities, the 2010 map definitely does not look and feel like the map of its forebears. Mediated rhetorically by the world that produced it, the 1898 map and the macro-narrative that shaped its design and interpretation has long since faded for many contemporary viewers.

The "Center of Population" map from the 1898 *Statistical Atlas* (fig. 4.15) showing westward migration illustrates, first and foremost, how graphical displays have the capacity to tell powerful stories, which scholars like Edward Tufte (1983), Howard Wainer (2005), Lee Brasseur (2005), and Miles Kimball (2016), have eloquently explained with numerous examples from the past and present. The map also illustrates how audiences interpret visual language within the context of the era in which they live—here with the map invoking the national narrative about migration across a continent. Geographical visualizations over such broad swaths of time, like the 1898 map, both reflect and bond themselves to large-scale narratives that galvanize their audiences. Maps envisioning migrations, wars, empires, plagues, and other major human events create a rhetorical nexus of the temporal, spatial, and cultural, and macro-narratives like westward migration provide a framework for them to congeal. However, as time passes and macro-narratives change, so do the collective interpretations of audiences, an iterative process that shapes and reshapes the graphical narratives they encounter.

American Dreams: Charting the Narrative of Economic Prosperity

As business, industry, and capitalism generated goods and wealth in the eighteenth and nineteenth centuries, narratives that supported this vision of prosperity sprang up, both in Britain and later the U.S. William Playfair's historical charts of England's trade (fig. 2.20) and public finances in his *Commercial and Political Atlas* (1801) made national economic data widely accessible as the British Empire grew and flourished, an epideictic tale that embodied the macro-narrative about economic progress and prosperity. Although the battle of the "ancients and moderns" raged decades earlier, pitting the classical world against modern achievements, by the end of the eighteenth century, "progress" firmly controlled the macro-narrative, in Britain and the U.S. at least, fortified by advances in science, technology, and industry.

These advances provided the fuel for a growing middle class and its expectations for economic prosperity. Figure 4.19 visualizes the "Present Prosperity" of the U.S. economy with a direct argument and call to action: to vote for Theodore Roosevelt for U.S. president (Patterson 1904). All of the indicators in the graphical timeline show favorable economic results during the seven years under Republican presidents McKinley and Roosevelt (who became president when McKinley was assassinated), with exports, bank deposits, tax revenues, and other categories increasing and

Figure 4.19. Chart showing key economic indicators in the decade before the U.S. presidential election of 1904 (Patterson 1904). Courtesy of the Library of Congress, Rare Book and Special Collections Division, Printed Ephemera Collection [2020779948].

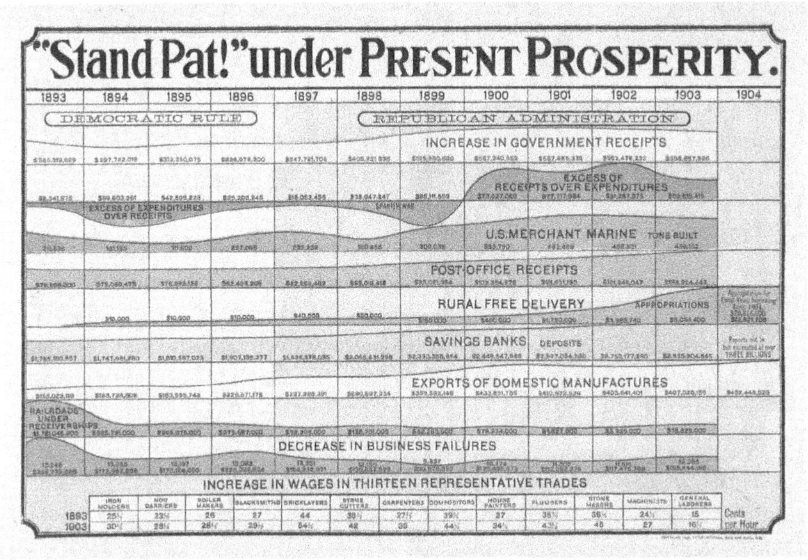

business failures decreasing. Although the scales vary from one category to the next, hampering exact comparisons, the positive trends are consistent across the board, contrasting with the same indicators during the previous administration. The chart argues implicitly that a vote for Roosevelt will continue that positive trend into the future—that is, progress toward even greater prosperity. The narrative must have resonated with voters: Roosevelt won by a huge landslide.

Other visualizations in the late nineteenth century engaged audiences in the macro-narrative of economic progress and prosperity through Horatio Alger stories lauding individuals whose hard work, ingenuity, and steadfast commitment begat their success. A quintessential success story appears in figure 4.20, a brochure for the McCormick Harvesting Machine Company printed in color with chromolithography, the cutting-edge technology of the era for telling persuasive corporate stories. Figure 4.20 shows the two inside panels of an 1886 brochure created by Cosack & Company that narrates a

Figure 4.20. Promotional piece for the McCormick Harvesting Machine Company telling the story of its growth (McCormick Harvesting Machine Company 1886). Courtesy of Iowa State University Library Special Collections and University Archives.

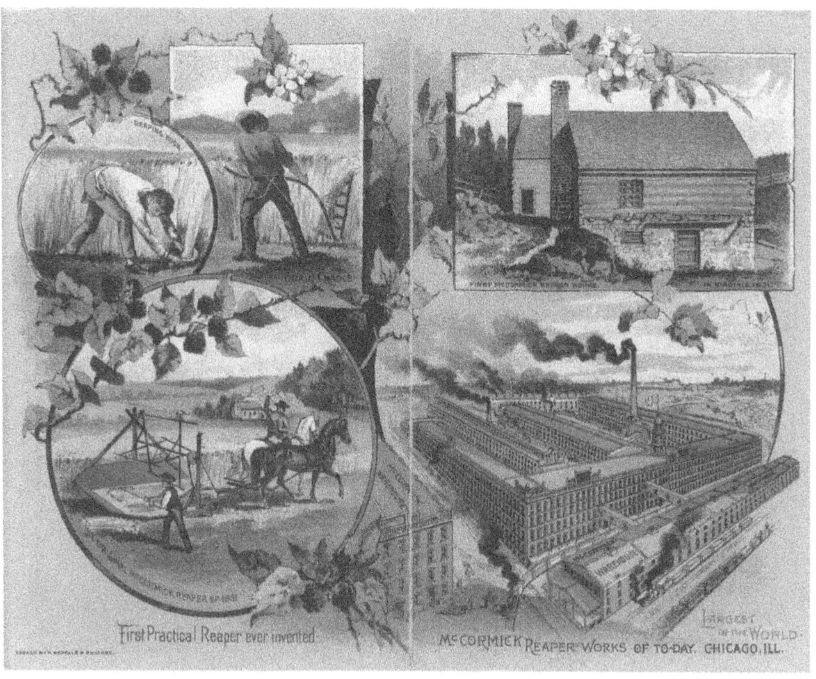

pictorial history of Cyrus McCormick's reaper: from the hand tools used before its invention (upper left), to the invention of the first reaper (lower left) and its first manufacture (upper right) to its production at McCormick's massive Chicago plant (lower right). This pictorial story juxtaposes McCormick's humble beginnings in a log building in rural Virginia and his sprawling factory half a century later. His "rags to riches" story must have enthralled and gratified audiences at the time, further persuading them that McCormick's products were tried and true, having achieved such a high level of production. Viewed through the macro-narrative lens of economic progress and prosperity, audiences would have interpreted McCormick's story in a larger technical and cultural context, the pictorial montage representing not only his story but those of many others (like John Deere) that began as small rural operations and followed industrialism into the city.

The McCormick success story is also envisioned graphically in figure 4.21, with progress heightened by a bit of hyperbole.[9] Using the shape of a pyramid, the chart in figure 4.21 from the McCormick Harvesting Machine Company's 1888 annual catalogue shows the sales growth of agricultural equipment from 1844 until 1887. The chart uses the volume of the pyramid to show progress, with time beginning at the top representing the modest sales in the company's early days, with the most recent year (in red) visualized at the bottom. However, the proportions of the most recent year's sales exceed the actual numbers, making this feat seem even more impressive—an amplification that enhances the narrative of a company that's growing by leaps and bounds, its agricultural machines distributed by rail and ship "to all parts of the Earth." McCormick has grown from its cottage industry days (the point at the top) to becoming a global operation and "The Wonder of the Age," an achievement explicitly compared to the ancient pyramids, further heightening the hyperbole. Interest in Egypt, aroused earlier in the century by Napoleon's expedition, flourished in the 1880s with archeological discoveries, and soon all things Egyptian—typefaces, graphics, architecture, interior design—resonated with audiences enthralled with the ancient culture and its revival. McCormick's pyramidal chart of progress was au courant, and its audience knew it.

The macro-narrative of economic prosperity, however, can be weakened or disrupted when conditions change, casting a shadow over micro-narratives that try to paint a rosy picture, something most annual reports do even in challenging times. Figure 4.22 shows a graph from the 1938 annual report of General Motors that visualizes financial data from 1919 to 1938, a twenty-year window that includes the prosperous 1920s, the Great Depression, and the long, lethargic recovery. Entitled

Figure 4.21. Chart from the *57th Annual Catalogue of the McCormick Machines* (1888, 2) that uses the volume of a pyramid to show sales growth from the 1840s to the 1880s. Courtesy of Iowa State University Library Special Collections and University Archives.

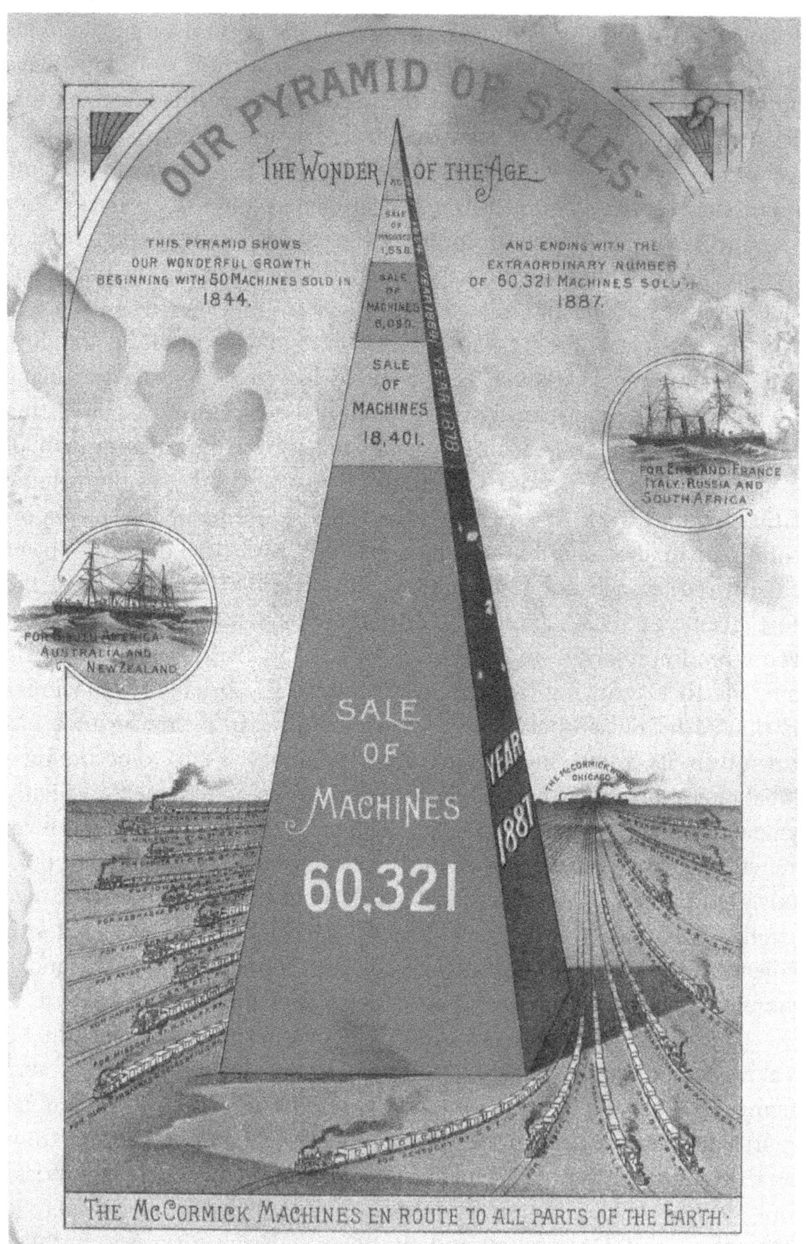

Figure 4.22. Chart from the 1938 annual report of the General Motors Corporation that interweaves several narratives (General Motors Corporation 1939). Reprinted with permission of General Motors 2025. Image courtesy McGill University Libraries.

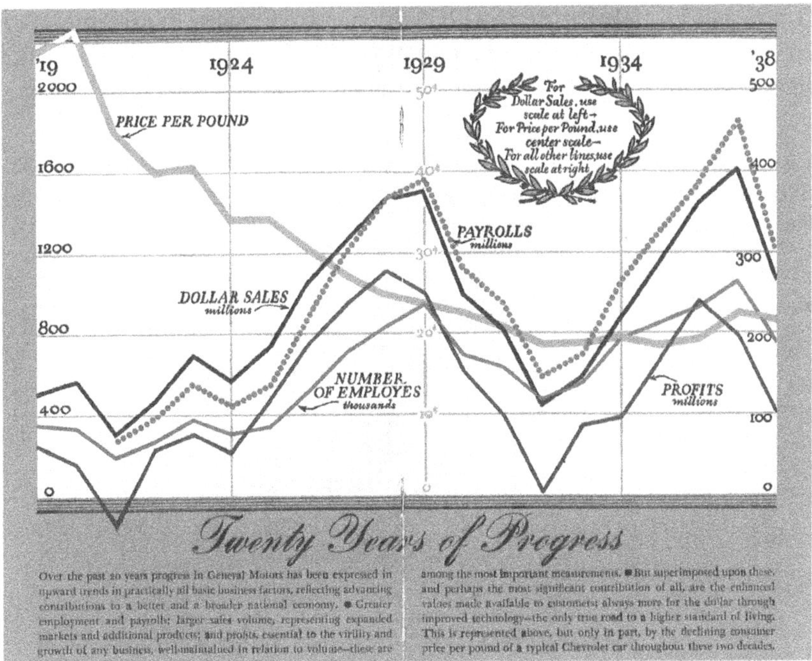

"Twenty Years of Progress," the chart visualizes data about sales, payroll, number of employees, profits, and the "price per pound" of a GM car. As expected in an annual report, the chart tells the GM story for investors by visualizing sales and profits. However, it also appeals to employees and consumers by emphasizing the decreasing cost of cars per pound, which in two decades dropped by over half. To manage such a complex story, the chart uses three different scales (left, center, and right), a risky strategy that succeeds here in creating a more compelling narrative, not only about GM but also about the national economy, the long recovery from the Depression, and advances in technology that raised the standard of living. In addition to supplying a conventional element of the annual report, then, the line graph enables audiences to visualize "progress" in broader, more relatable ways and to connect the data to the longer-term narrative of economic recovery and a return to prosperity.

For many large, successful organizations like the General Motors Company and McCormick Harvesting Machine Company the story of economic prosperity extends over long stretches of time. Contemporary digital technology can heighten the depth and detail of stories like these as well as enhance their accessibility. For example, the Principal Financial Group tells its 150-year story with an interactive timeline in increments of 20–25 years, beginning with its founding in 1879 (fig. 4.23). The timeline shows the steady growth of the enterprise, with milestones and images

Figure 4.23. Interactive timeline from the Principal Financial Group website (2025) showing the history of the company. Reprinted with permission from Principal Financial Services, Inc. ©1879–2025.

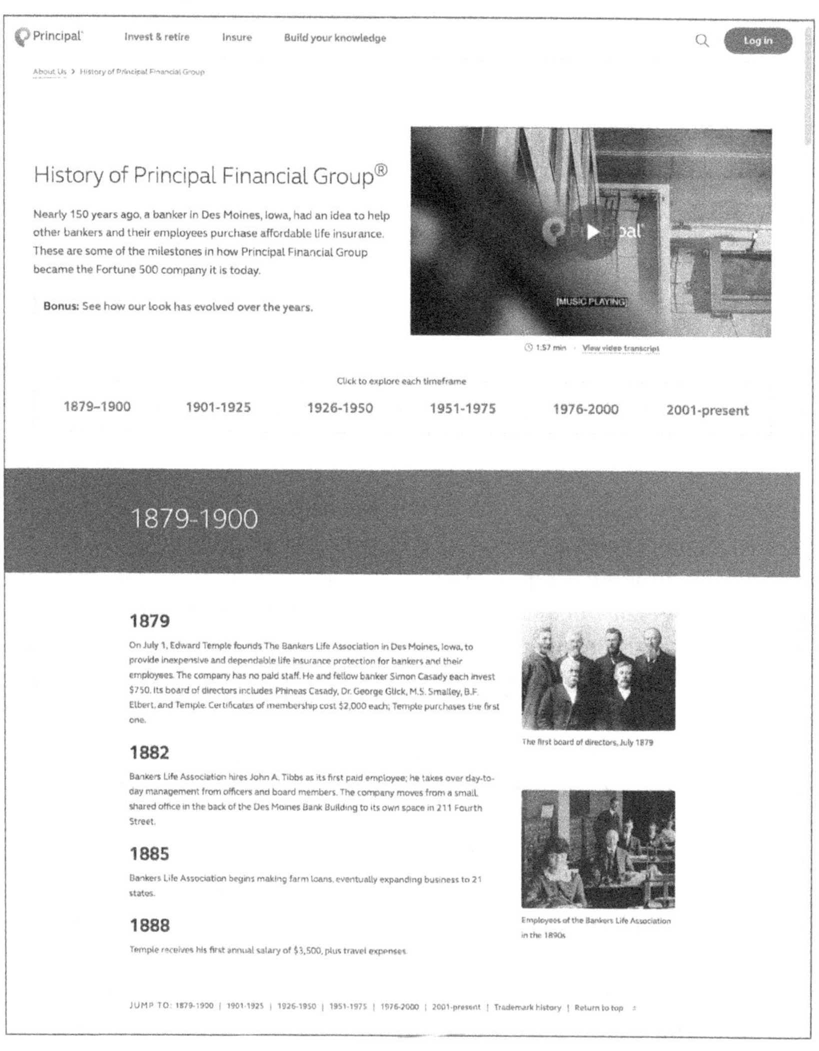

documenting its slow rise to a world financial powerhouse. Users can navigate their way through this history by choosing which time period to explore, with a collage of Principal logos at the end providing a visual history of its trademark.

The narrative of economic prosperity, of course, had a global dimension, especially after World War II when European and Asian countries rebuilt their postwar economies. A story of recovery and prosperity appears in figure 4.24, which shows a timeline for the Toyota Motor Corporation since its founding in the 1930s. The mountain chart shows how Toyota progressed from a rapidly successful but primarily domestic auto manufacturer in the 1960s and 1970s to a global powerhouse in the decades to follow, with international production eclipsing domestic output. Key Toyota innovations are noted within the plot frame, with car models at the top showing the evolution of Toyota's car design and technology and with a timeline at the bottom of the chart showing key international events. Toyota's story dovetails with the rise of Japanese manufacturing and the multinational company, both of which are embedded in the macro-narrative of global postwar prosperity.

Figure 4.24. Chart from the *Integrated Report 2023* of the Toyota Motor Corporation that narrates its history from its founding (Toyota Motor Corporation 2024, 102). Copyright: Toyota Motor Corporation. Used with permission.

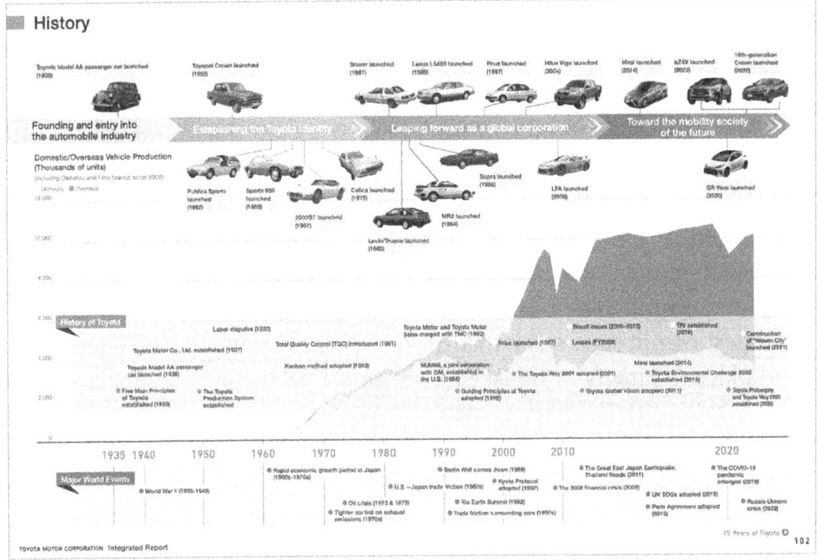

Egalitarian Narratives: Visualizing the Civil Rights of Minorities and Women

These broad national narratives of westward migration and economic prosperity obviously did not encompass everyone's lived experience. The chart in figure 4.25 is embedded in another national macro-narrative—the quest for civil rights and racial equality in the aftermath of the Civil War. One of a series of charts that W. E. B. Du Bois and his students from Atlanta University created for the Paris Exposition Universelle in 1900, this chart appeared on a world stage with a global audience. Collectively, Du Bois's charts visualized key aspects of the lives of African Americans, mostly after the Civil War: their economic status, places they lived, and their occupations and education. The chart in figure 1.11 showed the distribution of African Americans in cities and rural areas in the state of Georgia, with numbers skewing heavily toward rural living, as it did for the U.S. population as a whole, though urban migration for whites was already well underway.

The Du Bois chart in figure 4.25 tells a story about post–Civil War education, with the percentage of young people enrolled in public schools skyrocketing after the Emancipation Proclamation and continuing to rise in the following decades, though more slowly. This chart tells this story from several angles: that African American children are being educated (at a high rate), implying that they value learning and the upward mobility it provides. It also shows a significant gap in education, with over 40 percent in 1890 still not enrolled in school; the contrast between the red bars (enrolled) and black bars (not enrolled) heightens the difference and emphasizes that, while huge progress had been made, many more young people need to be enrolled. Anyone at the time (and anyone still today) would read this chart, as well as the population distribution chart (fig. 1.11), through the national narrative on race—slavery, the Civil War that ended it, and the Reconstruction that followed—though much more astutely then than now because of the temporal proximity to those events.

A quite different macro-narrative of civil rights permeates figure 4.26, a modest comic-style story that makes an argument about women in the workplace. Women's workplace rights were a particularly compelling issue after World War II when women like Rosie the Riveter had poured into factories to build military weapons and equipment. Like figure 4.11 about negotiating union contracts, this 1946 panel of pictures appeared in the *Labor Information Bulletin*. The series of pictures situates the characters

Figure 4.25. Bar chart designed by W. E. B. Du Bois and his Atlanta University students for the Paris Exposition Universelle of 1900 that shows enrollment of African American children in public schools (1900, "Proportion"). Courtesy of the Library of Congress, Prints and Photographs Division, Daniel Murray Collection [LC-DIG-ppmsca-33911].

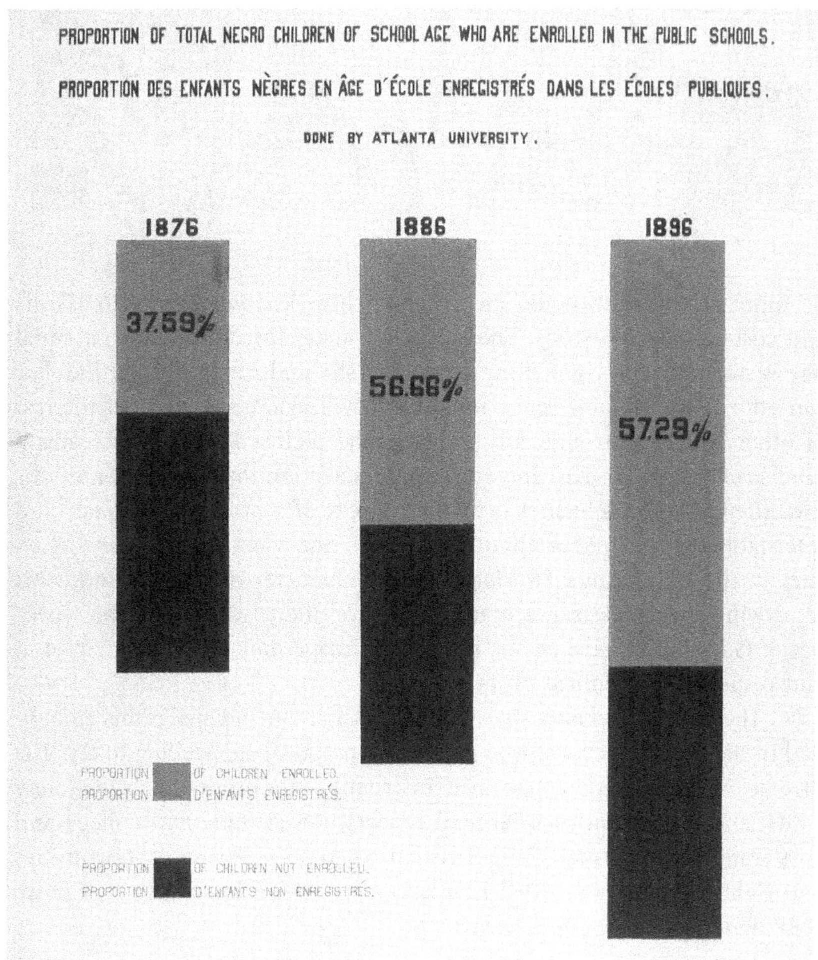

in their workplace, with a female worker in the first frame lamenting that she makes less than male workers, jeopardizing her ability to pay rent and support her children. Subsequent frames visualize a sequence of thoughts inside the mind of a male coworker, who acknowledges the unfairness of the pay difference and realizes that his job would actually

Figure 4.26. Comic-like narrative from the 1946 *Labor Information Bulletin* about equal wages for men and women (U.S. Department of Labor, Bureau of Labor Statistics 1946).

BUREAU STUDIES WHY WOMEN ARE PAID LESS

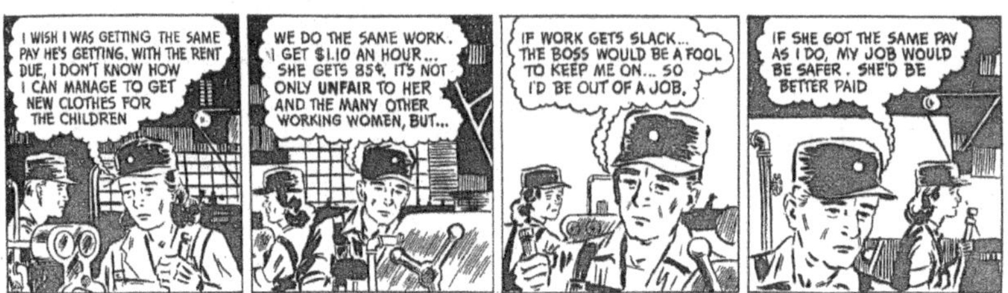

be more secure with equal pay, making him less vulnerable to layoffs and cost-cutting measures. The comic advances the argument that equal pay is not only the right thing to do (so she and other women like her can afford to pay their rent), but that it will also serve the self-interest of other workers through job security. The pictorial sequence creates a kind of enargeia—a vivid and engaging description that stirs the emotions of audiences, who interpret it in the context of a growing national (and international) awareness about women in the workplace. A few years later in the U.S. Army's *PS Magazine*, the character of Connie Rodd was intertwined with the same macro-narrative about women in the workplace. G.I.'s might gaze at her suggestive image, but they also marvel at and respect her technical proficiency.

The macro-level narrative about emancipation and civil rights embedded in these micro-level stories permeates practical storytelling today. The diverse human forms that appear in instructions for products, airline safety cards, and photo stories in annual reports, newsletters, and college and corporate recruiting materials all visualize people with equal opportunities and rights and are embedded in a larger macro-level narrative that more fully democratizes modern society.

Planet Earth Narratives: Visualizing Climate Change

Another macro-narrative, climate change, has emerged in the past several decades that challenges beliefs about the inevitability of progress and prosperity. The famous "hockey stick graph" (based on Mann, Bradley, and Hughes 1999, fig. 3) became a harbinger of future dangers and a scientific and cultural icon about environmental dystopia and public discourse about it. In the years since the hockey stick graph, global warming (or climate change) has achieved the status of an international narrative about its causes and how to mitigate them.

Most of the blame for the climate crisis is directed toward economically prosperous nations because of their elevated carbon emissions. Figure 4.27 shows an animated scatterplot from Gapminder that visualizes the relationship between national per capita wealth and carbon emissions (Rosling, Rosling Rönnlund, and Rosling 2025, bubble chart). Countries from around the world are represented with bubbles proportionate to their populations and are color coded geographically. The slider at the bottom of the chart activates an animated loop that shows the evolution of the data over the past two centuries, revealing that as the per capita income of a given country has increased over time, so have its emissions. Does this chart make an argument? An independent nonprofit, Gapminder

Figure 4.27. Gapminder animated bubble chart showing the relationship between income and carbon emissions over the past two centuries in countries around the world (Rosling, Rosling Rönnlund, and Rosling 2025, "CO2 Emissions"). Free material from www.gapminder.org. CC-BY GAPMINDER.ORG. https://creativecommons.org/licenses/by/4.0/.

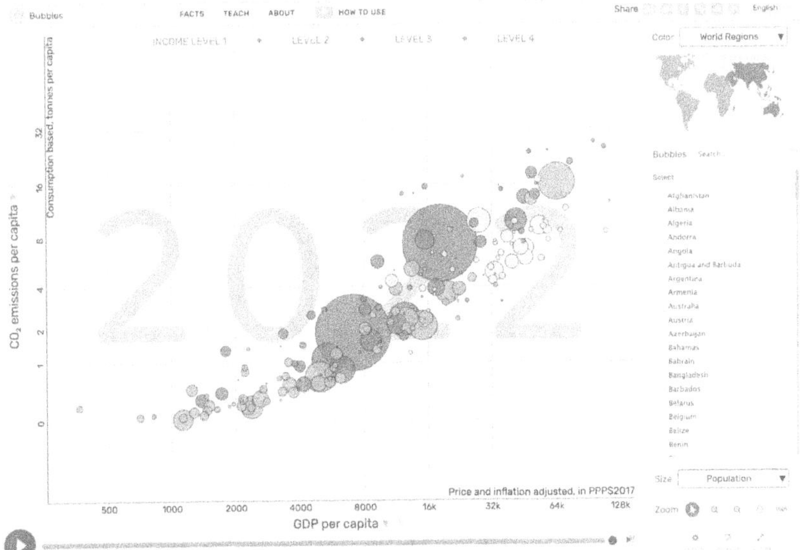

advocates "a fact-based worldview everyone can understand" by creating charts that visualize "the macrotrends that will shape the future" (Rosling, Rosling Rönnlund, and Rosling 2025). Although this approach gives audiences the freedom to interpret this chart as they wish, contemporary perceptions about climate change filter those interpretations, creating an implicit argument. More developed and wealthier countries with larger emissions have the greatest responsibility to reduce them, especially since they have the resources to act. Although this interpretation relies on mere common sense—wealthy countries should fix the problem they've created—it also adheres to the macro-narrative that climate change threatens all life on the planet.

While historical charts enable us to stand atop a mountain and survey the past, other visualizations enable us to pivot around and view the future. Practical visualizations that narrate the future effects of climate change paint a disturbing picture. For example, NOAA (U.S. National Oceanic and Atmospheric Administration) creates maps and pictorial simulations that show the effects of climate change on sea level rise. The map in figure 4.28 shows coastal areas potentially affected by flooding

Figure 4.28. A map of south Florida, along with a picture of a landmark in Key West, that shows the effects of sea level rise (U.S. Department of Commerce, National Oceanic and Atmospheric Administration: NOAA 2025). Image obtained from NOAA. Certain Esri imagery in this work are owned by Esri and its data contributors and are used herein with permission. Copyright © 2025 Esri and its data contributors. All rights reserved.

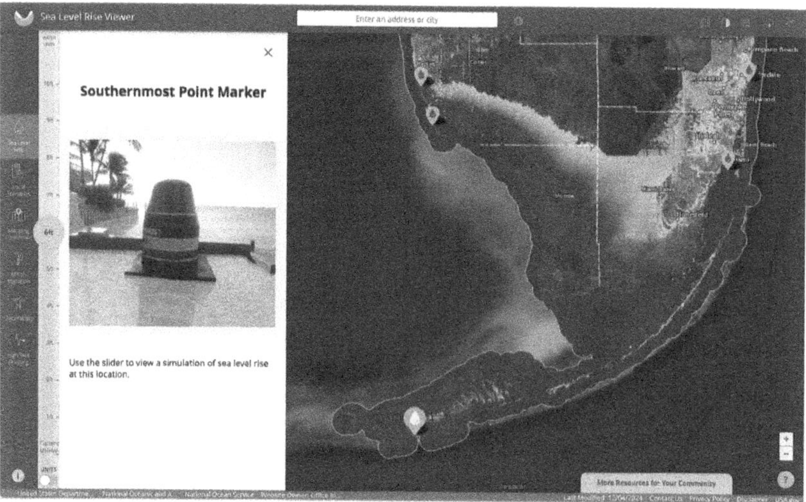

by color-coding the affected areas with shades of blue. Audiences can set the sliding scale from one to ten feet of flooding, with the map here showing the effects of a six-foot sea level rise. Under these conditions, southwest Florida would be largely inundated by seawater, as would areas of the Florida Keys, shutting down the highway in numerous places and making Key West inaccessible by land. The picture on the left shows the effects of the sea level rise on the Southernmost Point Buoy, a concrete marker designating a significant geographical location that's constantly visited and photographed by tourists. In this projected simulation, water has breached the stone wall, inundating the marker and limiting access to it. For people who live in Key West, this scenario would cause emotional distress, given their personal and financial stake in this landmark and their own vulnerability to flooding. And for the thousands of tourists who've visited this place, it would also arouse an emotional response, alarmed by how rising sea level could cause such devastation, how it might affect them wherever they live, and what role they might play in mitigating it.

The Gapminder chart and NOAA's Sea Level Rise Viewer might be considered forms of deliberative rhetoric in that they spur audiences to consider environmental consequences, their causes, and what might be done to address them—all of which benefit audiences by helping them make long-term decisions (Kostelnick and Kostelnick 2016, 175–87).[10] No specific plans are outlined or advanced (for the worse-case scenarios at least), just the motivation to act or accept the possible consequences. By defining problems, these visualizations provoke thinking and planning, which can lead to preemptive action. They're able to do this because their stories are reinforced by a powerful macro-narrative: that the earth is warming (resulting in sea level rise), and steps must be taken to combat it.

How do audiences respond when they encounter a visualization that disrupts a prevailing macro-narrative? The chart in figure 4.29 captures huge swaths of time on an even grander scale, visualizing the glacial ages that have unfolded over hundreds of thousands of years as the planet warmed and cooled in cycles lasting tens of thousands of years. As this super narrative shows, these glacial and interglacial cycles occurred with relative regularly, with one cycle roughly mirroring the next but each having its own unique pattern. If we consider the past as a predicter of the future, another glacial age appears inevitable. However, in our contemporary world, where climate change dominates science and public ethics, how does this chart's story play out rhetorically against this macro-narrative? At the end of the chart (today), will the cycles of cooling and warming be broken, with the down cycle deferred or eliminated? Audiences might

Figure 4.29. Chart showing the ebbs and flows of glaciers over the last 450,000 years (Eldredge and Bieck 2010, 8). Courtesy of the Utah Geological Survey.

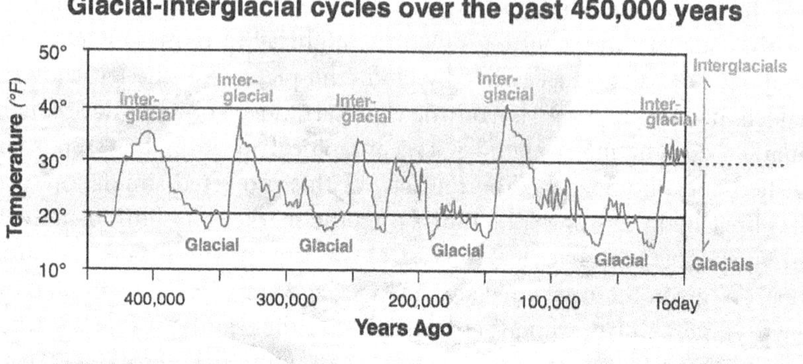

wonder, amid the present period of constant warming, if another ice age is even possible, or they might question whether glacial cycles are even worth discussing when the planet faces a far more immediate climate crisis. Still, super narrative charts visualizing long-term data can at least provide food for thought, even if they seem to swim upstream of the prevailing macro-narrative.

Big Stories Entwined with Small Ones: Macro-Narratives in Our Midst

So how are the micro-narratives I discussed earlier intertwined with these larger macro-narratives? How transparent or opaque are these relationships? Even the tiniest mini-micro narrative embeds bigger stories that, at any given moment, permeate the social and cultural landscape and that regulate our perception of time—that is, how we view the past, present, and future. To explore the relationship between micro- and macro-narratives, let's take another look at a few examples, beginning with the *Encyclopédie* plate in figure 4.10 that illustrates the artificial pearl-making process. This unassuming, matter-of-fact narrative is situated at a convergence of macro-narratives generated by the Enlightenment and its abiding commitment to progress, achievable by escaping the dark ignorance of the past and catapulting humanity into a new age of "light" through rational thinking and a more egalitarian social order. A cornerstone of the Enlightenment, the *Encyclopédie* project aimed to democratize knowledge, a process that began earlier in the eighteenth century with the *Cyclopaedia* of Ephraim Chambers (1728) and that the expansive *Encyclopédie* initiated on a much larger scale. The pearl-making plate was one of thousands that realized that goal of making knowledge fully accessible in print, in this way aligning the little narrative in this plate with the intellectual values of the Enlightenment—that henceforth knowledge should be systematic, transparent, and widely accessible.

This plate also connects with another macro-narrative about the nature of work in the preindustrial age and how it can become more efficient through the division of labor. Like labor narratives in legions of other *Encyclopédie* plates, each woman in this plate performs a task in the pearl-making process. Few narratives probably happened in the exact temporal and spatial proximity envisioned in the *Encyclopédie*'s plates, which tended to idealize these processes by simplifying and streamlining them. However, this visual compression not only economizes the audience's interpretive work but also provides conceptual insights about the division of labor, which was pictured in many of the *Encyclopédie*'s plates and which became a founding principle of Adam Smith's theory of economics in *The Wealth of Nations* (1:5–20; bk. 1, ch. 1–2), first published in 1776 and laying the groundwork for the Industrial Revolution and mass production. The narrative in this little drawing, then, is embedded in larger epistemological, economic, and political narratives that were coalescing across Europe and North America.

Other macro-narratives, especially those about health and disease, have a powerful grip on audiences, exerting a great deal of argumentative force on a national or global level. The flow of population into urban centers in the nineteenth century increased public awareness about health and disease, and charts and graphs fed this narrative by visualizing micro-narratives tracking the course of diseases. Visualizations like John Snow's London cholera map (1855) and Florence Nightingale's charts of monthly mortality rates in military hospitals (1859), which led to reforms in medical practices (Brasseur 2005), can more fully be understood as part of this macro-narrative about understanding and improving public health. If we fast forward to the surgery decision tree in figure 4.8 and COVID-19 test instructions in figure 4.13, these micro-level stories are also entwined with a larger health narrative about the pandemic: its associated risks, its effects on general health care, and the preventative steps (tests, masks, social distancing, vaccines) that everyone needed to take. At the height of the pandemic, when most businesses and schools operated under a lockdown or severe restrictions, graphs and maps of cases became widely available—in the media, online, and in print—with the mantra to "flatten the curve" becoming a national and international imperative. The "curves" on the CDC's weekly 2020 COVID-19 graph, shown in figure 4.30, indicated that positive cases, deaths, and hospitalizations trended up before the holidays, then subsided a bit. Graphs like this one became part of everyday life as the pandemic ebbed and flowed from week to week, with decisions about work, school, church, and social life dependent on

Figure 4.30. A CDC graph showing COVID-19 data for the U.S. for most of 2020, with emphasis on the final two weeks (U.S. Department of Health and Human Services, Centers for Disease Control and Prevention 2021). CDC. Reference to specific commercial products, manufacturers, companies, or trademarks does not constitute its endorsement or recommendation by the U.S. Government, Department of Health and Human Services, or Centers for Disease Control and Prevention. Free material from the CDC: https://archive.cdc.gov/#/details?archive_url=https://archive.cdc.gov/www_cdc_gov/coronavirus/2019-ncov/covid-data/pdf/covidview-01-08-2021.pdf.

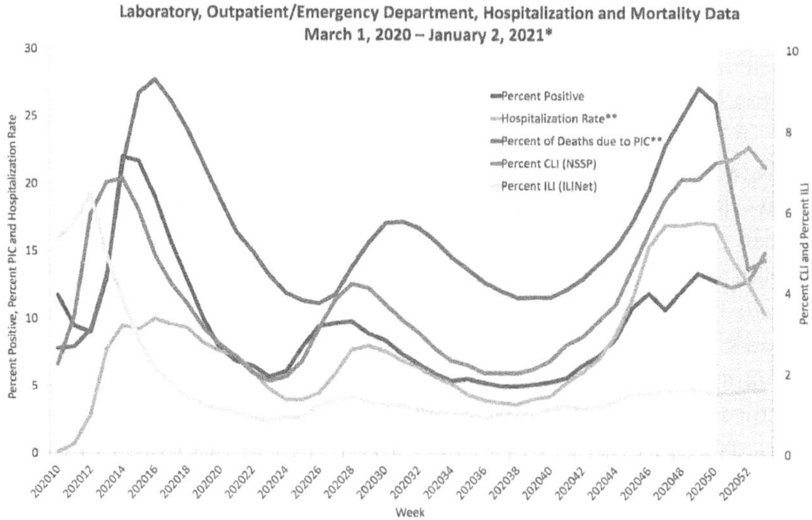

the rise and fall of the data lines. Counter-narratives to the pandemic sprung up, but the dominant narrative that COVID-19 posed a mortal threat that required daily attention and aggressive measures to counter it prevailed globally—a macro-narrative in which the surgery decision tree and COVID-19 test instructions and their rhetorical exigencies were deeply entwined.

In short, many micro-narratives span much broader swaths of time than they actually visualize by integrating larger communal stories shared by their audiences. Even the mini-micro narrative of how to use the car charger (fig. 4.1) intimates a larger narrative: that our personal and professional lives have merged because our electronic devices travel everywhere we do—a global macro-narrative of life in the information age. And the bus and train timetables? Thousands of people use them—and they do so individually, one at a time—private experiences that are enveloped within a macro-narrative that merges public transportation, energy efficiency, and climate change. In short, many micro-narratives connect with a larger temporal phenomenon that articulates them in some way, like a slow-moving current that's visible only on the surface. Some of these embedded macro-narratives—Manifest Destiny, civil rights, economic prosperity, climate change, pandemics—are more transparent than others, but all of them quietly mediate our ability to make meaning from everyday narratives. This interpretive process might explain why someone unfamiliar with a given macro-narrative finds a micro-level story puzzling: They don't have a macro-narrative context in which to situate the interpretation—that schools are vulnerable to multiple threats; that labor unions go on strike over differences with management; that spare parts are crucial to a high-functioning military.

Looking back at macro-narratives through rhetro designs can make these micro/macro connections more transparent. In addition to embodying vestigial visual language that elicits nostalgia, as we saw in chapter 3, rhetro designs invoke macro-narratives from the past, giving us glimpses into worldviews that audiences once embraced. For example, a restaurant menu with a "Wild West" theme situates its patrons in a macro-narrative about the values of frontier life and rugged individualism amid untamed nature, with the rhetro menu guiding their journey. Even though most patrons live in quiet suburban neighborhoods with neatly trimmed yards, they happily embrace the paradox by sidestepping reality during their restaurant visit. Rhetro designs momentarily rekindle macro-narratives that have long dwelled in the shadows, shining a spotlight on them and challenging audiences to confront them or acquiesce to them. Some audiences will, while others will roll their eyes.

Conclusion and Implications for Design

Because practical narratives are so ubiquitous, we can easily overlook their impact on our daily lives: schedules for going places, instructions for getting things done, charts for tracking everything from weather to home energy consumption. These practical narratives represent time with diverse forms of visual language—textual, graphical, and pictorial—that are combined in conventions and genres. Some narrative forms that materialized in the past eventually faded, constrained by technology and cultural values, but many also have continued to thrive today, have evolved into new forms, or been repurposed or reinvented with digital media. With digital technology, practical storytelling has become more dynamic and customizable, individualizing the user's experience while often downplaying the bigger picture.

Practical storytelling has enormous temporal elasticity, ranging from minutes to millennia—from mini-micro-narratives explaining everyday tasks to super narratives offering panoramic views of past and future. For the most part, the lifespans of many forms of practical storytelling we encounter are short lived and fungible: instructions are revised or retired, bus schedules fluctuate, and charts for busy hours evaporate. Whatever their temporal spans, practical narratives perform a variety of vital functions: they inform and persuade, they enable decision-making, and they impel action. Moreover, despite their temporal and material fragility, practical stories often embody powerful national or global macro-narratives that broaden the social and cultural lenses through which we interpret them. Although practical narratives are supple and diverse—visually, rhetorically, and culturally—they all possess the universal allure of storytelling, of immersing us in temporal excursions, however long or brief.

Designing Time with Storytelling Genres

What are the practical implications of these various modes of visual narration? How can past practices inform design today? We have plenty of storytelling conventions to draw from, most of which have their origins in long-standing genres like timelines, flow charts, decision trees, pictorial narratives, and line graphs. Digital technology enables designers to generate and adapt these conventional genres as well as transform them into dynamic interactive narratives. As a result, audiences can personalize time by zooming in on details meaningful to them, like the digital CyRide bus schedule in figure 4.7. Audiences have growing expectations about the temporal scope and immediacy of practical narratives—the more teaspoons they can consume, the better—and designers can meet those expectations with interactive tools. Generative AI will likely expand and diversity the storytelling capabilities of information designers, further enhancing user experience.

Considering the Role of Macro-Narratives

In whatever genre, visualizations of time are often embedded in larger narratives that have cultural, social, and political dimensions and that might span decades or centuries. These larger narratives profoundly affect how we design. If I'm displaying data about weather, crime, or education, some larger social and political narratives are likely to affect my design decisions as well as my audience's interpretations. A chart about long-term warming and cooling, like figure 4.29, will embed the narrative of climate change; a poster showing teachers how to respond to an armed intruder will embed the narrative about gun control. In short, as part of the rhetorical process, we need to consider any macro-narratives linked to visual storytelling and gauge our audience's stake in them.

Infusing Narrative Design with Emotion

Emotion plays a key role in storytelling through heightened suspense and customizing options, which increase audience engagement and motivation. For example, interactive data displays, like the stock market chart in figure 4.14, allow users to control the narrative by choosing the time range and enabling them to explore data in greater detail. Treemaps (fig. 1.7), data maps (fig. 2.27), animated bubble charts (fig. 4.27), and other genres offer these options. Data design programs with accessible tools for creating interactive data displays are increasingly becoming more available, enabling a wider range of designers to integrate them into their data narratives.

Comic-style narratives also provide an especially potent medium for eliciting emotion by picturing characters with personalities and feelings, projecting an inviting and informal tone, and interweaving scenes into a compelling story. Although creating comic-style narratives can be daunting because most information designers lack training in the medium and don't consider themselves "artists," digital technology has made comics design more feasible with grids, word bubbles, clip art, and script typefaces readily available on the desktop.[11] Even if designers can't emulate Will Eisner or Roy Doty, digital technology can enable them and their audiences to realize the rhetorical and psychological benefits of infusing narrative with emotion.

Chapter 5

The Myriad and Mutable Faces of Ethos

Visualizing Trust with Text and Image

Ethos has long been a fundamental building block of the Western rhetorical tradition, one of the three rhetorical pillars along with pathos and logos. From its origins, the concept of ethos—the ability of the rhetor to foster an audience's trust—has been associated with the character of the speaker and the extent to which it can be effectively projected in persuading an audience. Indeed, Aristotle observes that "character is almost, so to speak, the most authoritative form of persuasion" (39; bk. 1, ch. 2, sec. 4). Aristotle and his successors inhabited an oral culture dominated by speech and gestures, where rhetors practiced their art before live, in-person audiences. However, a modern world saturated by print and digital images affords many new *visual* avenues for projecting ethos that lie beyond the scope of the traditional rhetorical canons. Scholars have examined visual ethos-building in several genres and media, including corporate visual identity (Veltsos 2009; Wei 2004), data displays (Hutto 2008; Fanning 2018), industrial design (Buchanan 1985), websites (Wei 2008), and indeed virtually any form of information design (Kostelnick and Roberts 2011).

Both verbally and visually, rhetors generate ethos with strategies that have evolved as culture and the history of ideas have altered the social, moral, and intellectual landscape. Over the past few millennia, rhetoricians have pondered the nature and function of ethos: how it's generated both before and during a rhetorical act, what constitutes a rhetor's character (and self), how audiences respond to ethos appeals, the emotional aspects of ethos, and its social and collective domains (Baumlin 2001; Baumlin and Meyer 2022; Halloran 1982; Jasinski 2001, "Ethos," 229–34). Although all of these aspects of ethos are intertwined, strategies for building ethos visually, like other rhetorical processes we've seen in previous chapters, shift over time as audiences, values, and technology change. So how has visual ethos-building evolved? What vestigial strategies and genres still resonate today? How is digital design transforming them? I'll try to answer these questions by juxtaposing historical and contemporary examples and by comparing and contrasting their visual strategies, the technology used to create them, and social and cultural factors that shaped them.

To explore the visual rhetoric of ethos, I'll focus first on traditional forms of ethos-building: initial (preexisting or extrinsic) and acquired (invented or intrinsic). The former results from the status and character of the individual rhetor, and the latter is established by the rhetor's communication skill and virtuosity (see Crowley and Hawhee 1999, 107–43). I'll then examine ways in which acquired ethos has shifted from the individual ethos of the Aristotelian model, the traditional domain of rhetorical theory, to a collective ethos that socially shapes and absorbs individual ethos as well as redefines the self (Alcorn 1994; Baumlin and Meyer 2022; Halloran 1982). The notion of a collective ethos fundamentally alters our understanding of how contemporary designers construct it, how audiences interpret it, and how cultural, social, and political forces shape it. To illustrate collective ethos, I'll show how it's acquired by deploying visual conventions and genres and by invoking various forms of visual identification. Toward the end of the chapter, I'll also consider how novelty and misdirection factor into ethos-building, and I'll supply some design guidelines for visualizing individual and collective ethos.

Initial Ethos: Touting Character Upfront

As I just mentioned, rhetors build ethos in two long-standing modes: initial and acquired. How do they accomplish this with visual language? Rhetors gain initial ethos by visually brandishing their credentials (name, academic degrees, certifications) and by associating themselves with high-credibility surrogates (pictures of famous people, places, myths, and events). On the other hand, rhetors gain acquired ethos by showing skillful and consistent design habits, deploying visual conventions embodied in genres and other design elements, using an aesthetically pleasing style, and creating identification with the audience. Although the boundaries between initial and acquired ethos can sometimes be a bit nebulous, both methods for projecting ethos visually have the same persuasive goal: to inspire confidence and build trust.

Both methods of ethos-building have their origins in classical rhetoric. Aristotle believed rhetors must project ethos in the speech itself by adapting it to the wants, needs, and disposition (the "character") of the audience (149–56, bk. 2, ch. 12–17)—in other words, through acquired ethos—rather than rely on the rhetor's preexisting merits or reputation. This viewpoint was later contradicted by the Roman rhetorician Quintilian (1959–1963), who believed that the speaker's own character played an essential role in practicing the rhetorical arts, that "the ideal orator . . . should be a good man" (1:315; bk. 2, ch. 15, sec. 33). Only virtuous rhetors can actively (and effectively) engage in oratory, "since no man can speak well who is not good himself" (1:315; bk. 2, ch. 15, sec. 34). The rhetor's character alone won't ensure that audience members will find the message persuasive, but it's the sine qua non for garnering their trust. As such, the rhetor's reputation matters greatly to how much the audience can trust the speaker. In laying out these qualifications, Quintilian set a high moral standard for his students of oratory, though fortunately, most individuals, whatever their stations in life, can build trust through "character" by possessing the qualities of a "good" person.

Designing Good Character

Of course, Quintilian's maxim begs the question of how we define a "good" person and how someone projects this "goodness" visually. "Goodness" has always been associated with integrity, moral rectitude, sincerity, and often power and prestige. We assume that people in leadership positions who are "at the top of their game" in business, politics, academia, and art—and in the contemporary Western world, sports and entertainment—should be accorded some respect, if not our complete trust. In a meritocracy, achievement and fame define "goodness" and lay the foundation for rhetorical success, though other factors play an integral role: religious and political convictions, contributions to the community, social connections, and so on. So, our collective values play a constant role in what constitutes the "goodness" of individuals—or, for that matter, of organizations as well.

How does ethos-building through extrinsic "goodness" play out rhetorically in practical communications? Understandably, rhetors invest in first impressions, and visual language provides the means to seize those opportunities. Perhaps no other genre affords more immediate and visible ethos-building than a signed message on letterhead stationery from a famous individual, especially one who resides in the White House, or the "Executive Mansion" as it was called in the nineteenth century. Figure 5.1 shows an August 1864 letter from Abraham Lincoln to his cabinet about his bleak prospects for reelection and his intent to preserve the Union during the transition to a new administration the following year. The

Figure 5.1. A letter from Abraham Lincoln to his cabinet in August 1864. Courtesy of the Library of Congress, Manuscript Division.

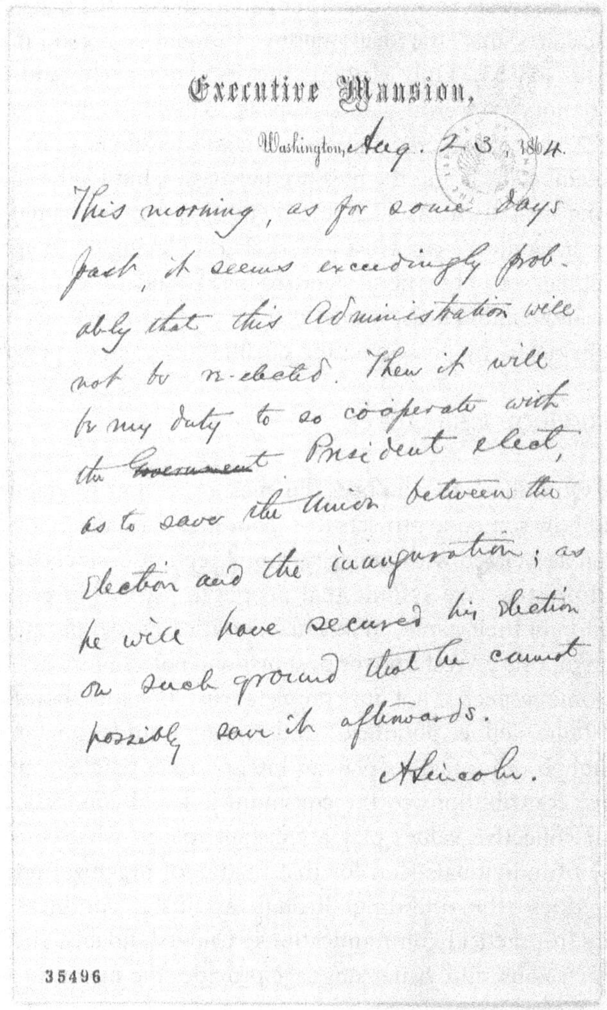

Executive Mansion,

Washington, Aug. 23, 1864.

This morning, as for some days past, it seems exceedingly probable that this Administration will not be re-elected. Then it will be my duty to so co-operate with the ~~Government~~ President elect, as to save the Union between the election and the inauguration; as he will have secured his election on such ground that he cannot possibly save it afterwards.

A. Lincoln.

35496

Executive Mansion letterhead obviously carried a huge amount of rhetorical weight, as did Lincoln's own handwritten message and signature. It's hard to imagine anyone in American life, then or since, with more initial credibility, which powerfully looms over this lean transactional document, with its understated visual and rhetorical presence.

Establishing initial credibility has also been achieved with another long-standing genre—the business card. Over the past three centuries, private individuals, businesspersons, and other professionals have used business cards to foster their ethos, typically with new acquaintances who knew little or nothing about them.[1] Although this petite genre has its own ethos-building capabilities (more on that later), prominent individuals didn't really need the card itself to establish their credibility, as their names and reputations did that instantly. Such was the case with "calling cards" in the nineteenth century, whereby aristocrats and other luminaries used them as a means of announcing their presence, touting their status, and enhancing their social interactions (Green 2016). Good character back then was mostly a matter of pedigree: the higher the rank the greater the reputation, with dukes and duchesses claiming the top tier. The associations with aristocrats who offered their cards, and the initial credibility that accompanied those gestures, were highly prized by their recipients, as readers of Jane Austen novels like *Persuasion* ([1818] 2003) readily understood.

The criteria for measuring the character of card holders differed in the U.S. in the nineteenth century, with initial credibility based largely on merit rather than genealogy and class. Today, public officials, executives, and other high-profile dignitaries who possess upstanding reputations still use business cards—not merely to introduce themselves, given that their audiences already know about them, but to initiate further communication about their agendas, services, and products with their clients or constituents. The reputations of these individuals precede the rhetorical act of placing a card in someone else's hand, instantly mediating the interaction.

This kind of initial trust extends to many newer forms of information design. The newsletters, websites, and social media pages of public figures, financial gurus, or celebrities will engender the trust of the audiences that admire or believe in these individuals, in much the same way that audiences trust advertisements that feature famous people. The mere presence of someone's face or name invokes trust, motivating us to explore the communication. Indeed, up to a point, we'll even forgive its design flaws—illegible typefaces, low-resolution images, gaudy colors—because we *believe in* the person it represents. Initial ethos makes a powerful and tenacious first impression.

Visualizing Character Through Affiliations

Initial trust can also be acquired through an individual's affiliation with an organization, what David Hutto (2008) refers to as "institutional" ethos (112). The collective weight of a prestigious entity engenders confidence in the messages of its representatives, and trust in an individual develops by association. Visually, this preexisting ethos is often invoked by a letterhead at the top of the page—of a proposal, job recommendation, tenure review, or a certification about the authenticity of a product, service, or collectible. Although the "letterhead" is often subjected to glib comments, as if it shouldn't matter to audiences, the visible presence of a highly credible letterhead—of a school, respectable firm, or elected body—can carry a good deal of weight. Presidents, senators, representatives, governors, and other elected or appointed officials all benefit from their letterheads and signatures, a long-standing rhetorical practice in modern governance, as we saw in Lincoln's letter from the Executive Mansion.

The same effect occurs with people holding other high positions: chief executive officers, college presidents, coaches, leaders of nonprofits, and others who represent organizations that remain constantly in the public eye and from which their own individual ethos continually benefits. Without that institutional letterhead, their ethos would likely plummet. Of course, visual language won't likely engender trust among audiences whose viewpoints strongly differ from those of the rhetor; still, the *office* itself holds weight and temporal permanence, well beyond its current occupants.

Building individual ethos by affiliation with an organization occurs in many other print and digital media. Bios or photos of individuals that appear on websites associated with prestigious organizations—universities, research institutes, government agencies, and the like—get an instant ethos boost by their presence among esteemed company. Similarly, certificates or plaques hanging on the office walls of physicians, lawyers, scientists, or engineers certify the degrees they've earned, licensing exams they've passed, or other distinctions they've achieved. In each case, the organization granting the credential gives the recipient instant credibility, bolstering the confidence and respect of visitors. For example, an anxious patient waiting in a doctor's office will see framed certificates from the doctor's alma mater and licensing board, fostering trust and reassurance. These accolades hanging on the wall could be displayed privately in the doctor's home, but their rhetorical force would largely be squandered—except for the few admiring eyes of family and friends.

In all of these ways, individuals build initial ethos with their audiences based on their accomplishments or affiliations *before* an interaction begins with a client, constituent, patient, or total stranger. If none of these ethos-building elements was visible to the audience to bolster the credentials of the rhetor, the interaction with the audience might change direction: public officials would become mere citizens, executives become employees, and coaches mere talking heads. Audiences instinctively know these things, and their trust fluctuates accordingly.

Visual Surrogates: Building Ethos with People, Myths, Places, and Events

Oftentimes trust is built visually not so much with the credentials or reputation of the rhetor, or institutions the rhetor is affiliated with, but by association with other people or things the audience holds in high regard: prominent leaders, famous buildings and monuments, mythological figures, and key historical events. Visualizing these surrogates to build ethos depends on the values of audiences at a given historical moment. Some surrogates have more resilience and longevity than others, with their ability to build trust waxing and waning as they are reinterpreted according to the moral, cultural, and political values of their audiences.

Mutable Faces: Currencies and Commerce

Facial recognition is one of the most primal and powerful perceptual acts, enabling us to connect with people we know, either in person or through pictures. Identification through facial recognition arouses emotions ranging from affection, empathy, and respect to fear and revulsion. When we encounter an image of a famous face, we immediately make a judgment about whether to trust it based on our associations with the person it represents. Benjamin Franklin's image, for example, has appeared on behalf of countless commercial enterprises ranging from food and investments to dime stores, hotels, soaps, and wood stoves. An emblem of practical wisdom, Yankee ingenuity, and diligence, Franklin has long retained his capacity to build ethos, appearing on the hundred-dollar bill for over a century (and earlier on the half dollar coin). Washington, Lincoln, and other national leaders have experienced similar staying power, both on U.S. currency and for commercial purposes.

The reputations of few historical figures remain inviolate forever, as cultural values can lower or elevate their ethos-building capacity over time, in recent years particularly in relation to civil rights, slavery, and the Civil War.[2] Cultural and political literacy also plays an important role because this capacity relies on audience recognition of individuals and their accomplishments. Ethos-building can also be reciprocal process: The appearance of Susan B. Anthony on the one-dollar coin, the first woman to appear on U.S. currency, recognized her achievements and also created greater awareness of them. Internationally, images on coins and bills of monarchs, revolutionaries, and other national figures don't typically resonate with noncitizens, though they might feel an affinity with artists, writers, and scientists visualized on foreign currencies (see McKeever 2021). The future of ethos-building images on physical currencies will likely diminish as digital currencies proliferate, leaving behind a legacy that lasted for millennia, with Caesar, Constantine, and their successors replaced by electronic digits and codes. As a result, the rhetorical presence of facial images as public ethos-building entities—on currencies at least—will likely plummet, while facial recognition technology will continue to skyrocket, authorizing users to enter buildings, cars, and public spaces.

The Mythic Past: Ethos-Building in Agricultural Promotions

Invoking myths from the past offers another method of building ethos through visual surrogates. Images of gods and goddesses like Apollo, Venus, Hercules, Mercury, and Cupid saturate the marketplace, an ethos-building strategy that has its roots in the Victorian period when designers deployed visual surrogates to foster trust for commercial enterprises, especially to promote agricultural products. In 1890 nearly two-thirds of the U.S. population still lived in rural areas, where catalogs for seeds, farm machinery, and other equipment were familiar items in many households. Facing stiff competition, many agricultural suppliers and manufacturers used visual association and surrogates, including classical myths and mythological figures, to boost their ethos with their customers. Appeals based on classical myths resonated broadly during this era, as the Greek Revival in architecture and visual art was still a prominent force in visual aesthetics as well as in literature and education.

This strategy for ethos-building is illustrated in figure 5.2, which shows a promotional piece for a Triumph reaper that was manufactured in the 1880s by D. S. Morgan and Company in Brockport, New York. This reaper is situated in a mythical scene painted by Guido Reni in the early seventeenth

Figure 5.2. Promotional piece for a Triumph reaper manufactured by the D. S. Morgan Company (1887). Courtesy of Iowa State University Library Special Collections and University Archives.

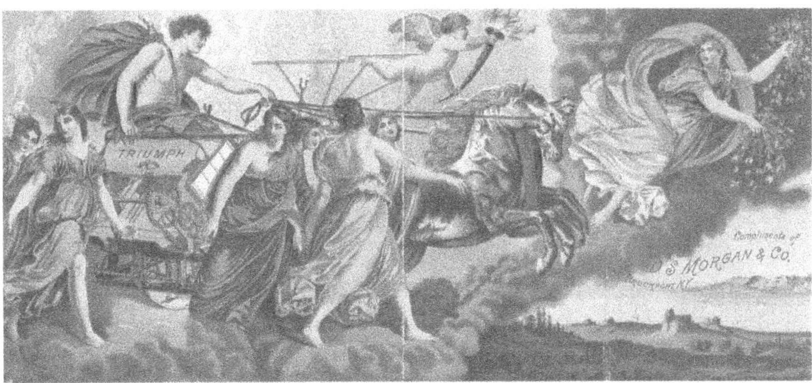

century (1613–1614) as a ceiling fresco in Rome. In Reni's painting, Apollo and his chariot, drawn by surging stallions, are led by Aurora with roses in her hands and are surrounded by jubilant women representing hours of the day, while the cherub Phosphorus, symbolic of the morning star, lights the way through the sky (Pritchard 2015). In D. S. Morgan's version of this celestial scene, Apollo sits atop a Triumph reaper, its mechanical parts clearly visible as it flies through the air. This repurposing of Reni's painting elevates the reaper to mythical status as a marvel of the industrial age. Late Victorian audiences probably wouldn't know that the scene was a re-creation of a picture by Reni centuries earlier, but they would have vaguely understood the classical origins of the allusion—and its power to build credibility. Comparing farm equipment to classical mythology illustrates just how highly these enterprises valued their technology, envisioning its transformation of the landscape as the stuff of myth in order to heighten their ethos.

A similar mythological image appears in figure 5.3, a catalog cover for agricultural equipment manufactured by Adriance, Platt, and Company of Poughkeepsie, New York. The female figure echoes the Roman goddess Libertas and the Statue of Liberty, installed the previous decade in New York Harbor. Wrapped in an American flag and carrying an American shield in her hand, she places a garland around a flaming torch attached to the company's nameplate. Other garlands embellish the dates of the company's operation and, at the very bottom, the countries and continents that have benefited from its products. With Daniel Webster's famous quotation beneath her feet, patriotic Lady Liberty oversees the universal reach of this company's equipment, elevating its reputation and

Figure 5.3. Cover of a catalog for harvesting machinery manufactured by Adriance, Platt and Company (1897). Courtesy of Iowa State University Library Special Collections and University Archives.

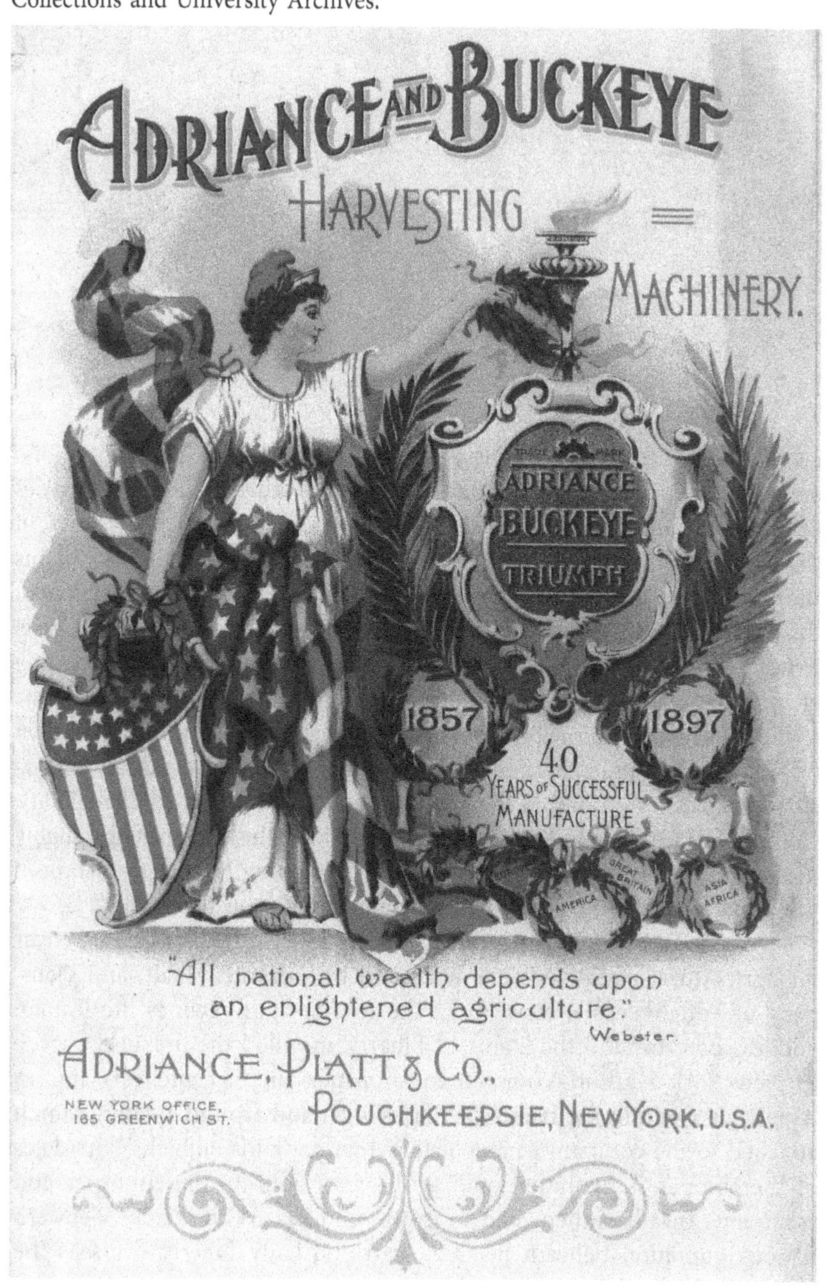

success to mythical status. Another mythical figure graces the cover of the 1897 catalog of the Massey-Harris Company (fig. 5.4), a Canadian farm equipment company. A goddess-like figure in a flowing dress with a garland in her hair stands atop the globe with her bare feet planted on

Figure 5.4. Cover of a catalog for agricultural machinery manufactured by the Massey-Harris Company (1897). Courtesy of Iowa State University Library Special Collections and University Archives.

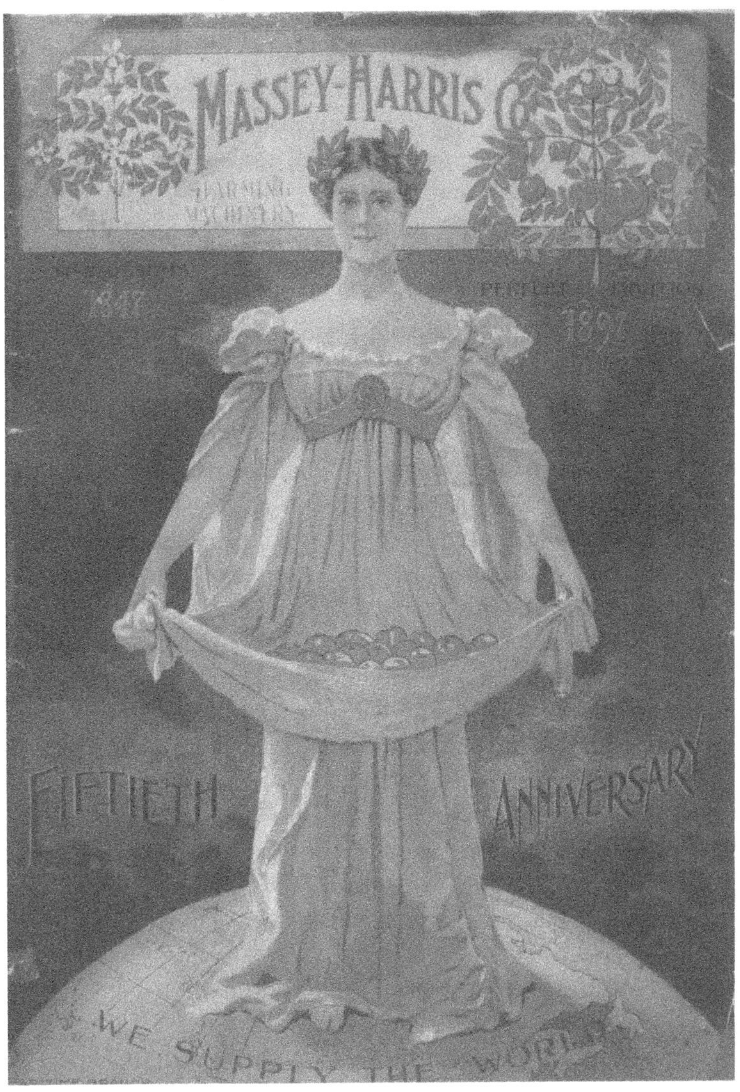

Canada and her garment filled with fruit, with the phrases "It blossoms" and "Perfect Fruition" floating around her. Combining larger-than-life myth with wholesome sensual appeal, she creates a strong, credible presence for Massey-Harris that also extends its reach around the globe. Both of these surrogate figures are deployed to build confidence and trust in their respective companies—*before* the audiences even begin thumbing through their catalogs searching for farm machinery.

In the late nineteenth century, mythic comparisons to historical events also provided a source of initial (extrinsic) ethos through visual surrogates. Figure 5.5 shows the cover of the 1893 catalog for the McCormick Harvesting Machine Company, with Cyrus McCormick's portrait in the middle, ensconced in a "historical egg" and flanked by images of Queen Isabella (left) and Christopher Columbus (right). A drawing at the top envisions Columbus and his crew landing in the New World, with wary Native Americans looking on. At the bottom appears a drawing of McCormick's famous reaper, an invention whose significance seemingly parallels the very discovery of America by Europeans. This explicit comparison attempts to build ethos through association with a world-changing event, and whether this connection appears to the audience as appropriate (or boastful or hyperbolic), the reaper certainly transformed rural America, where many people still lived and worked. For them, at least, the visual association may have been a plausible and persuasive method for reinforcing their trust.

During this era, visual surrogates in closer spatial and temporal proximity were also deployed to boost the ethos of agricultural products. The cover of the 1899 catalog for the Iowa Seed Company, shown in figure 5.6, envisions a circular garden and stone balustrade that set the stage for three prominent edifices in the distance—the Washington Monument, the White House, and the Capitol Building—traditional symbols of national power and pride, associated here with the field of flourishing seeds. The Spanish-American War, which transpired the previous year, fueled patriotic fervor across the nation, and these images, along with the colors and headline type, seized the kairotic moment to bolster ethos.[3] In the late nineteenth century, other monumental objects like Egyptian pyramids and obelisks (fig. 4.21) provided similar ethos-building surrogates based on their status of place and associations with antiquity.

Modern Surrogates: The Past Reimagined

In the following decades, as North America rapidly transitioned to a more industrial and urban culture, ethos-building through mythical surrogates

Figure 5.5. Cover of the 1893 *McCormick Harvesting Machine Company Catalog* that compares Cyrus McCormick to Christopher Columbus. Courtesy of Iowa State University Library Special Collections and University Archives.

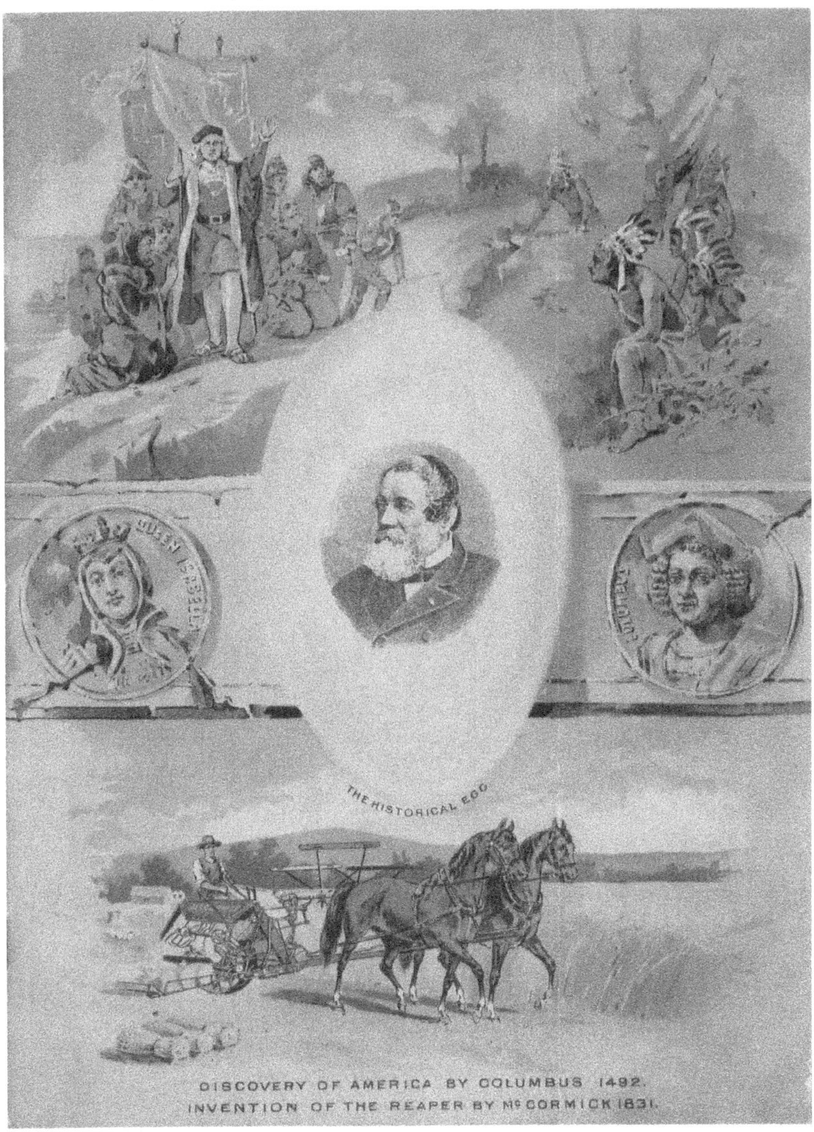

reflected those changes. A worldlier goddess figure appears on the poster for the 1906 National Business Show (fig. 5.7), a one-week event in the Chicago Coliseum that featured "all modern and time saving business

Figure 5.6. Cover of the 1899 *Iowa Seed Company Annual Catalogue* that pictures famous buildings to boost ethos. Courtesy of Iowa State University Library Special Collections and University Archives.

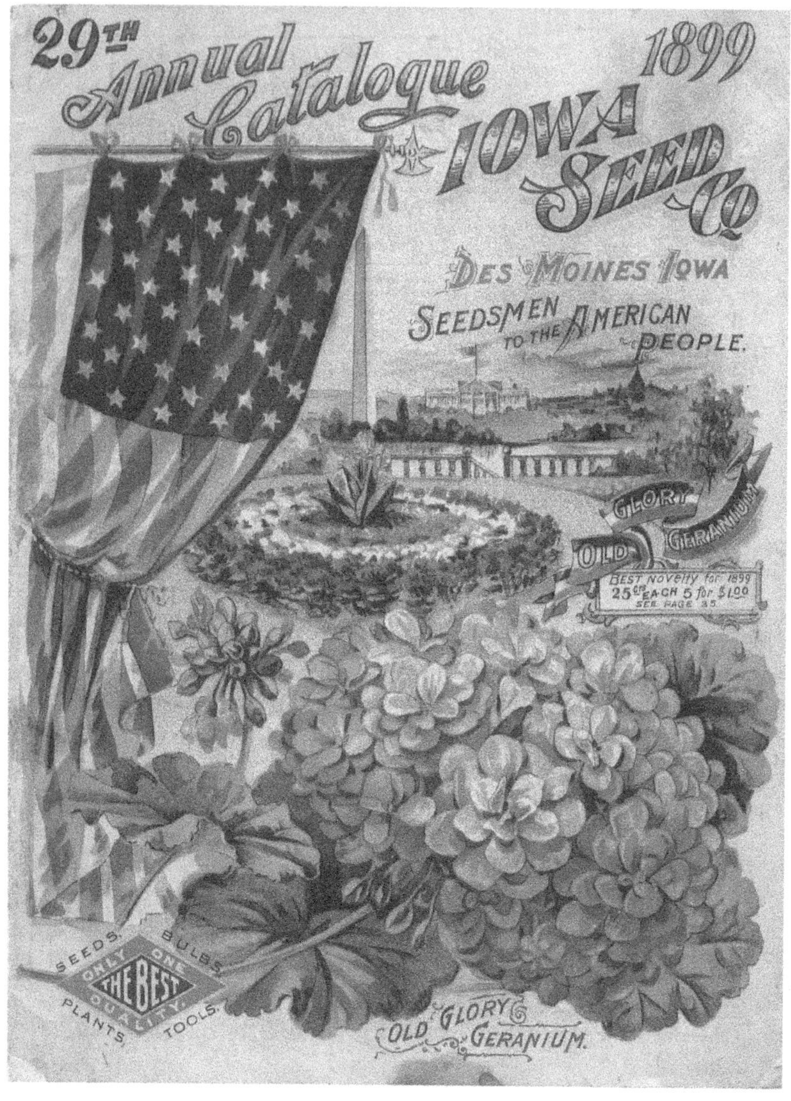

appliances and systems on exhibition." With her heart-shaped garments flowing around her, the goddess with the fiery hair holds a book in which, seemingly from some ethereal realm, she has penned a proclamation about the event. Her feet turn the winged wheel of "progress" as she hovers

Figure 5.7. Poster for the 1906 National Business Show in Chicago (National Business Show Company). Courtesy of the Library of Congress, Prints and Photographs Division.

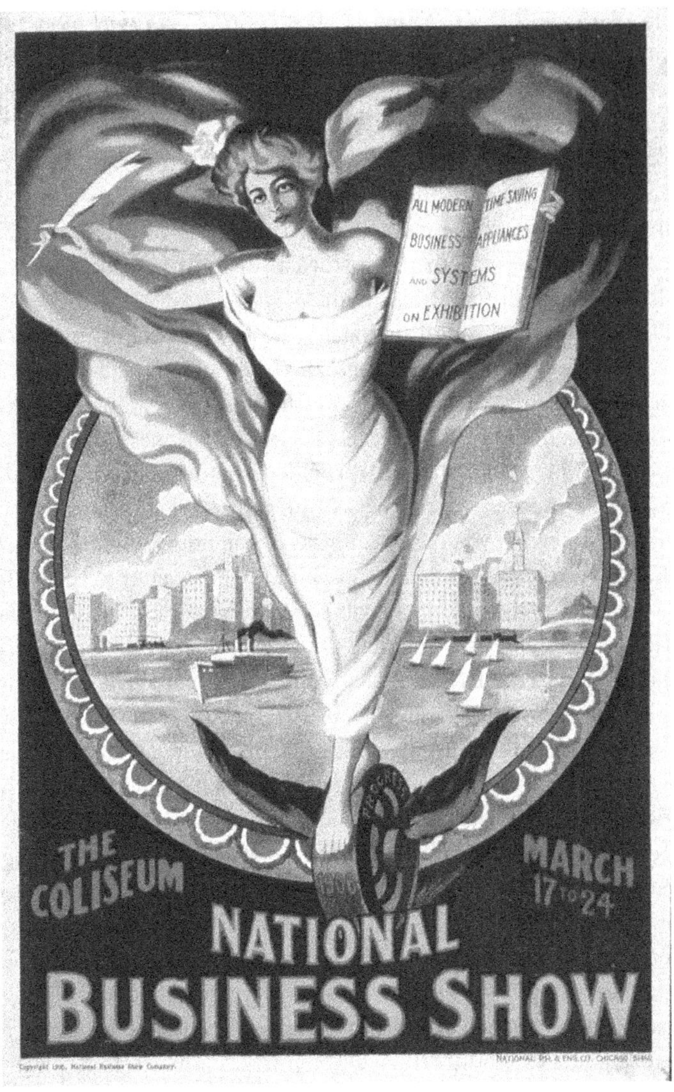

over Chicago's lakefront skyline—the epitome of a modern city with its burgeoning skyscrapers. This cool, seductive female figure elevates the exhibition to mythical heights, the goddess's powerful presence building ethos for the event even if the audience can't quite fathom her identity.

Today businesses and other organizations continue to associate themselves with mythical entities and historical figures, places, and events to boost their ethos. Some businesses visualize mythological figures—for example, Pegasus, Cupid, Minerva, Poseidon, Ceres—in icons or logos to boost the ethos of their brands. Historical figures like Ben Franklin have sustained their ethos-building power in today's world despite controversies surrounding some of their contemporaries. Landmarks like the Statue of Liberty, the Eiffel Tower, and the Egyptian pyramids still appear on promotional materials, their gravitas and integrity seemingly timeless. Invoking entities from the past—myths, people, events, places—assumes that audiences possess both the historical insight to recognize them and the cultural sensibility to value them.

Ethos-building through surrogates, however, remains in constant flux and depends heavily on the rhetorical power of public memory. To the extent that modernism tried to escape the past, we can assume that contemporary audiences find historical associations less compelling than their ancestors did, for whom the past provided rhetorical ballast for the fluctuations of the present. Moreover, as our shifting values constantly adjust our lenses for viewing the past—uplifting some entities while purging others—rhetors have to adapt their ethos-building strategies to these changing circumstances. Creating ethos becomes an iterative enterprise of then and now: The past enables us to validate the present just as the present compels us to honor (or demonize) the past. Will visual surrogates from the past retain their ethos-building capacity in the future? Hard as it might be to imagine a world without Ben Franklin's image, we can only speculate on whether prominent individuals, landmark buildings, and world-changing events will continue to galvanize public memory as a resource to build trust.

Technology has played a key role in shaping and sustaining visual surrogates, though their rhetorical functions have shifted. In the Victorian age, despite the labor-intensive hand work in creating images, inexpensive color printing made visual ethos-building practical and widespread, illustrated by the examples I've shown. Today, creating and circulating visual surrogates has never been easier and cheaper with digital technology, which enables designers to produce, mix, and repurpose images, both for print and for distribution on the internet. Saturating websites and social media with digital surrogates, often in the form of memes laced with levity and irony, broadens their ethos-building power well beyond the commercial purposes they've long served. Ben Franklin's image can still sell products and services, but with a few clicks he also can be repictured in various costumes and poses to critique an idea, add levity, or lambast a leader.

Acquired Ethos: Projecting Good Character with Design Habits

In addition to building ethos by the rhetor's own reputation, or through association with surrogate agents, most rhetors use the communication itself, as Aristotle prescribed, to build trust with their audiences. This "intrinsic" ethos must be acquired (or invented) by the rhetor: orators analyze their audiences carefully, sharpen their persuasive skills, and deliver speeches that showcase their competence. Visually, rhetors do these things (and others) in a variety of ways: projecting their character through good design habits, deploying conventions and genres that meet audience expectations, creating identification with the audience, and enhancing visual decorum through style and tone. Rhetors also occasionally use novelty to engender trust. All of these methods for building ethos visually are constantly evolving as cultural values, available technology, rhetorical exigencies, and audiences' perceptions of themselves change over time.[4] I'll examine these methods by first focusing on the good design habits of visual rhetors, then address the other methods in the sections that follow.

Building trust through good design habits is probably the most visible evidence of "character" because these habits are readily observed on the surface of a communication. Rhetors who design their messages clearly and consistently make positive first impressions on their audiences, showing not only exactitude but also discipline, steadfastness, and moral rectitude. The media and methods for projecting these qualities visually, however, have evolved drastically over time. As I discussed in earlier chapters, the art of penmanship dominated practical communication for centuries, as handwritten text served as the lifeblood of companies, professions, monarchies, and empires as well as high society. The "command of hand" that writers sought to achieve required not only artistic ability (and fine motor skills) but extensive practice and dedication. In the nineteenth century, when handwriting instruction became more vernacular and universal, its affirmation of moral character became more pronounced and explicit (Thornton 1996, 47, 49–50). Those who wrote in a "fair" hand were deemed to be virtuous, partly because they had the discipline, tenacity, and self-control to master the craft for personal or professional purposes. In short, handwriting revealed the rhetor's character, with its bodily movements a form of functional etiquette and rhetorical decorum. Those who wrote in a clear and elegant hand engendered trust, while bad handwriting signaled the opposite—a lack of discipline and integrity (Thornton, 52–53).

The exigency to produce handwritten documents that were both clear and elegant reached a fever pitch in the nineteenth century as commerce, bureaucracy, and literacy rapidly increased. Figure 5.8, for example, shows a letter written in 1864 appointing someone to a committee of the U.S. Sanitary Commission for the Pennsylvania/New Jersey/Delaware region (Queen 1864). The impeccable handwriting exemplifies the Spencerian

Figure 5.8. Letter of appointment to serve on a committee for the U.S. Sanitary Commission during the Civil War (Queen 1864). Courtesy of the Library of Congress, Prints and Photographs Division, Marian S. Carson Collection [LC-DIG-ds-00296].

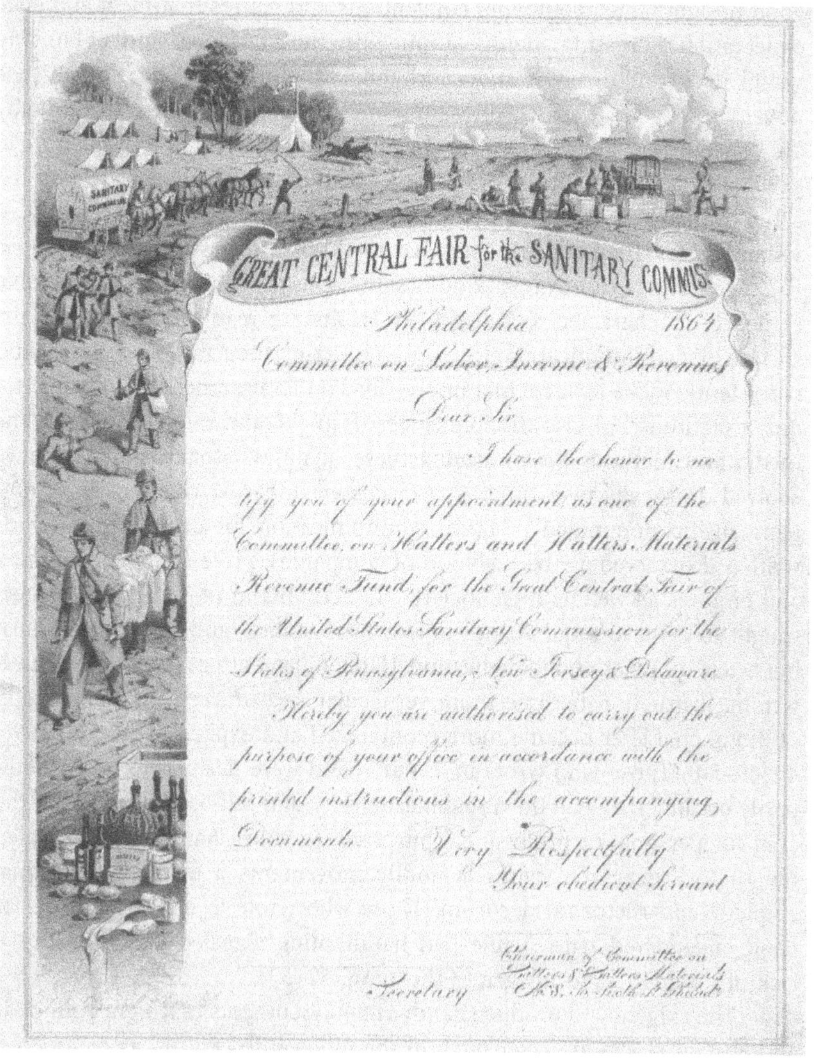

GREAT CENTRAL FAIR for the SANITARY COMMIS.

Philadelphia 1864

Committee on Labor, Income & Revenue

Dear Sir

I have the honor to notify you of your appointment as one of the Committee on Hatters and Hatters Materials Revenue Fund, for the Great Central Fair of the United States Sanitary Commission for the States of Pennsylvania, New Jersey & Delaware.

Hereby you are authorised to carry out the purpose of your office in accordance with the printed instructions in the accompanying Documents.

Very Respectfully
Your obedient Servant

Secretary

Chairman of Committee on
Hatters & Hatters Materials
[illegible]

style popular at the time, eponymously named for handwriting innovator and entrepreneur Platt Rogers Spencer (1866). The Spencerian style enabled writers to compose text legibly, quickly, and elegantly, with curvilinear forms and a high degree of contrast between thick and thin strokes, features strikingly apparent in figure 1.2 and evident here as well in the Sanitary Commission letter. The handwriting is consistently slanted, follows the horizontal lines precisely, and fills all of the available space, showing tremendous dexterity and discipline, a testament to the writer's unwavering moral character (and by extension, the Sanitary Commission's), especially important at a time of national crisis when trust and empathy were desperately needed.

The moral authority of handwritten text, and more generally of hand design, continued throughout the nineteenth century and still carries rhetorical weight today. Moral associations with artifacts created *by hand* were promulgated in the Victorian era by the Arts and Crafts movement and earlier by the writings of John Ruskin about the dehumanizing effects of industrialism.[5] The rural/urban divide also influenced perceptions about the virtues of making things by hand: In the early days of typewritten correspondence, Sears, Roebuck and Company handwrote letters to its farm customers, who disliked those that were mechanically transcribed (Boorstin 1973, 126, 399; Douglas 1985, 130). In the twenty-first century, advocates in favor of teaching handwriting in schools tout its educational and emotional benefits, claiming implicitly that the self-discipline and sensibility required by handwritten text can't be achieved on a keyboard (see Chemin 2014). Although handwriting instruction has mostly disappeared from contemporary education, its moral and emotional associations remain. A CEO who composes a handwritten letter to a disgruntled customer, an elected official who pens letters to campaign volunteers, an athlete who scribbles a note on an autographed picture for a sick child—all of these reveal an empathetic and virtuous character because the rhetor chose to communicate *by hand* rather than delegate the task to a subordinate or an automated system. Although these rhetorical acts are exceptional, time hasn't eroded their ethos-building capacity. If anything, time has only intensified it. In a digital age of email, text messaging, auto responses, and artificial intelligence, a handwritten communication far exceeds most audiences' expectations.

Do we still hold rhetors accountable for their visual dexterity and authenticity, or is that mindset merely a vestige of punctilious Victorians? Indeed, we do still hold rhetors accountable, though technology requires them to demonstrate their character with a much wider array of methods and tools: for example, with font choices, arrangement of text, headings, margins, and graphical cues. Audiences notice when typographical details are awry, and attending to them shows a degree of self-control and respect, both for the content and the audience, akin to editing text for grammar and mechanics. The presence and placement of pictures also reveal character. The pictures accompanying the handwritten text in the Sanitary Commission letter visualize its work on the Civil War battlefield tending the wounded and saving lives, illustrating that they act on what they profess in order to achieve their humanitarian goals.

Today, designers build ethos by picturing inclusivity—human forms that show sensitivity to disabilities, gender, race, and ethnicity; the absence of these has the opposite effect. Character can also be enhanced by complying with ethical codes—for example, by including clear, emphatic warnings in a set of instructions, by avoiding distortions or perceptual challenges in charts and graphs, and by adhering to design guidelines for accessibility. All of these intrinsic aspects of ethos are embedded in the communication itself, visible habits of practice and thinking that reveal the character of the rhetor and consequently, if done well, increase the audience's engagement and confidence.

From Business Cards to Annual Reports: Building Collective Ethos with Conventions and Genres

Because meeting an audience's visual expectations enhances a designer's ethos, visual conventions play a key role in that rhetorical process. Audiences enter interpretive moments informed by their collective experiences with visual conventions, and designers who tap into that collective knowledge will boost an audience's comfort level and confidence. Breaching or flouting visual conventions may have the reverse effect. Although novelty can engender trust when an audience expects creativity (more on this later), the well-worn path of conventions usually offers a safer, more direct route toward establishing trust.

Audience Expectations About Visual Conventions

Conventions solidify over time, with some reifying into a marbled state of durability that can last for centuries, though even the most robust of them eventually degrade, become obsolete, and lose their ethos-generating power.

Handwriting conventions, for example, once adhered to styles designated for specific purposes: blackletter (Gothic) style for manuscripts, secretary style for business, civil for governmental affairs, Italian for personal correspondence, and so on. Although we might still admire the elegance and consistency of these handwriting styles, today these conventions are relegated to the historical record, and their ethos-building capacity has long since dissipated.

Some forms of visual language, however, are resilient and evolve rather than succumb to the ravages of time. Pie charts, for example, became popular in the late nineteenth century, appearing in the first *Statistical Atlas of the United States* (F. Walker and the U.S. Census Office 1874, plates 51–54), which included explanations for readers on how to interpret them (2–3). A quarter of a century later, pie charts had become a fairly conventional genre, with audiences having assimilated their visual language into their repertoire of interpretive skills. Figure 1.8 showed a pie chart from the 1898 *Statistical Atlas* (Gannett and the U.S. Census Office, plate 33) that visualizes the relative size of church denominations, from Baptists to Methodists to Catholics and several others in between. Although contemporary audiences can interpret this chart, its ethos would suffer because it's out of synch with contemporary conventions: the pie chart has seventeen slices, nearly three times the recommended limit today; the slices don't begin at noon, as they typically do today; and they are arranged in a comparative sequence but in inverse order of what contemporary audiences would expect (large to small). Moreover, the labeling inside many of the slices is cramped and illegible, rather than placed outside the chart as call-outs, and the color scheme is vintage Victorian and rather gaudy by today's standards. All of this doesn't mean that a contemporary audience wouldn't trust this chart or be fascinated by it; its visual language just doesn't immediately resonate, and its credibility would have to be earned through interpretive effort, usually not a good thing with data displays. Also, to late nineteenth-century audiences, the sheer skill and labor to create a chart like this *by hand* would generate its own level of respect, unlike a contemporary pie chart, like the one I created for figure 5.9, which virtually anyone can readily produce. Still, the simple pie chart in figure 5.9 looks more normal and credible—or at least more unassuming—because it adheres to conventions that have evolved over the past century and a half. It won't quicken the pulse, but it also won't look odd or exotic.

Some genre conventions have stronger and more explicit ethos-building capacities than others, with annual reports, newsletters, and data

Figure 5.9. Pie chart with typical contemporary conventions of the genre.

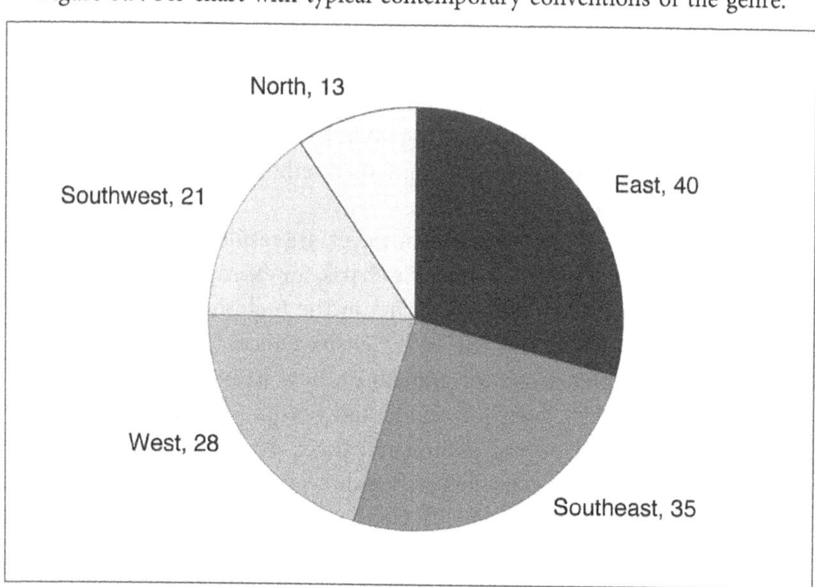

displays embodying well-defined conventional elements that extend over varying stretches of time. The temporal effects on genre conventions have been mixed as they've evolved from print to digital media, with their digital incarnations both preserving and transforming their visual language, as we saw earlier with the transportation timetables.

Ethos-building conventions can also be much more granular than full-blown genres, functioning within and across them. Drop caps, sidebars, page headers and footers, gridlines on tables and charts, pull-down menus—these and many more micro-level conventions will be readily familiar to audiences, certifying the designer's competence and thereby building trust. Although we mostly take them for granted, these design elements provide ballast for our perceptual encounters with practical communications—like familiar, trusted landmarks that guide us through our daily travels on foot, in vehicles, or online.

Business Cards as Rhetorical Gestures

As I observed earlier, perhaps no single genre is more designed to generate ethos than a business card, in the traditional sense of a lone agent attempting to establish trust, often through face-to-face interactions. And perhaps no other genre's visual language is more instantly recognizable, given its distinct diminutive shape (3.5 by 2 inches). Figure 5.10 shows an example of a business card that I created with generic typographical elements: name, position, and contact information arranged in landscape orientation (which is more typical than portrait). The card also includes a small graphical element to establish Karen's identity, optional but also conventional for the genre. So, when an audience member encounters the card—in a face-to-face meeting, as an enclosure with a document, or attached to a bulletin board—it generates immediate ethos because it embodies the visual language of the genre.

Ancestors of contemporary business cards also exploited the visual language of the genre to generate credibility. Figure 5.11 shows two business cards from the 1870s, one for a New Jersey photographer and artist and the other for three Chicago attorneys. Like the visual language of other cards from the era, as well as those today, these conform to the design conventions of the genre: their size (3.5 inches by 2 inches, about the same as the one in fig. 5.11), their landscape orientation, the names in large elegant typefaces that are centered to give the cards their formal and professional tone, and the smaller fonts for details. The restrained visual language of these two cards boosts their ethos: They conform to the conventions of the business card and thereby meet their audiences' expectations, and they don't try to capture their audiences' attention with lavish, eye-catching graphics. Their simplicity and their conformity to the visual conventions of the genre quietly do their ethos-building work, just

Figure 5.10. Business card with typical contemporary conventions of the genre.

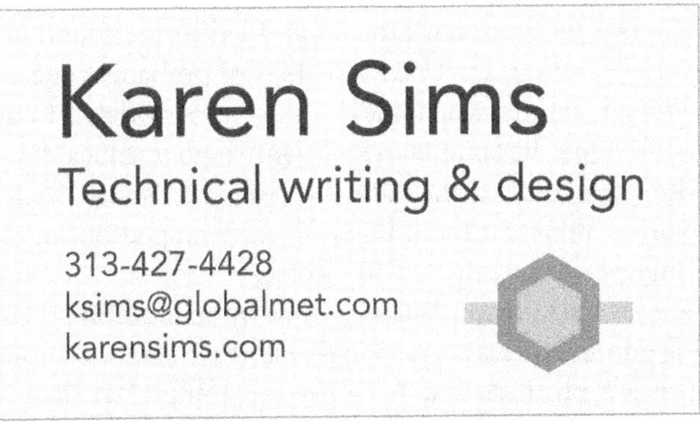

Figure 5.11. Late nineteenth-century business cards, the one on the left for a New Jersey photographer and artist (Stacy 1880) and the one on the right for three Chicago attorneys (Spafford, McDaid, and Wilson ca. 1875). Courtesy of the Library of Congress, Prints and Photographs Division, Larry Gottheim Collection [LC-DIG-ppmsca-51345] and the Manuscript Division, American Colony in Jerusalem Collection.

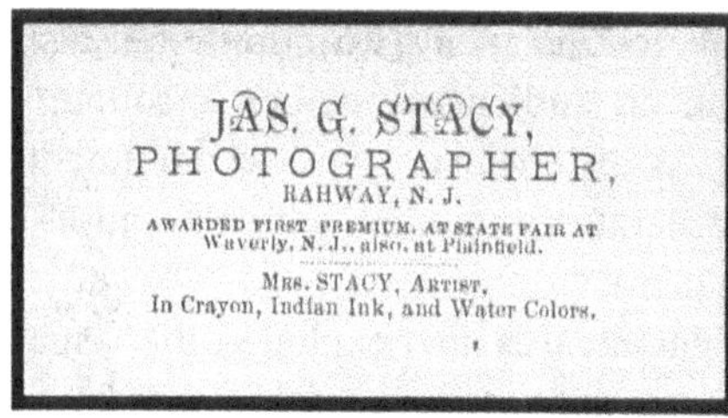

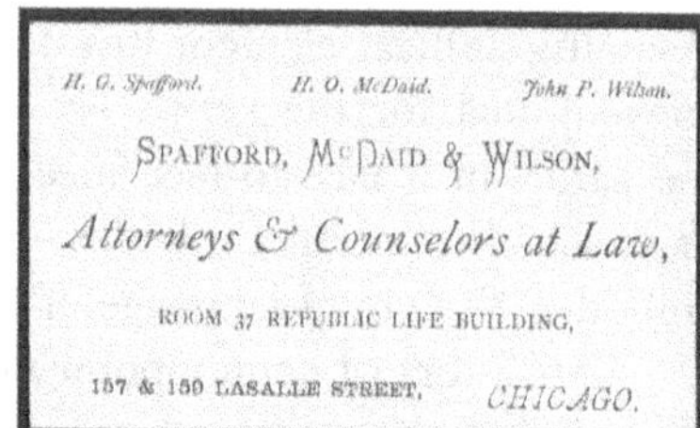

as they have for generations of professionals seeking to build trust with clients and colleagues.

Other variations of the business card also emerged in the nineteenth century, like the immensely popular carte de visite, a calling card that featured a photographic portrait of an individual, often accompanied by a signature. Originating in France in the 1850s (McCauley 1985), these cards featured images of prominent businesspeople, political and military leaders, artists and writers, and anyone who sought instant status through the emerging technology of the photograph. Circulated among friends, family members, and business associates, these cards inexpensively generated a visual presence for their subjects, an incipient form of LinkedIn. Figure 5.12 shows a carte de visite for Justin S. Morrill, the U.S. congressman (and later senator) from Vermont who sponsored the Morrill Act (1862), which was signed by President Lincoln and led to the founding of many public universities across the U.S. Morrill's card probably appeared during the Civil War at a time when these cards were widely circulated and when Morrill's name became entwined with higher education, as it still is today. The popularity of the carte de visite was short lived, however, and this hybrid calling card and business card slipped out of fashion, its ethos-building power having run its course.

Sometimes business cards foster ethos by abandoning their restraint and creating more expressive personas. Earlier I showed an example of highly flourished business and personal cards (fig. 2.3) that, like other

Figure 5.12. Carte de visite for Justin Morrill, U.S. congressman and U.S. senator from Vermont. Courtesy of the Library of Congress, Manuscript Division, James Wadsworth Family Papers [LC-MSS-44297-33-063].

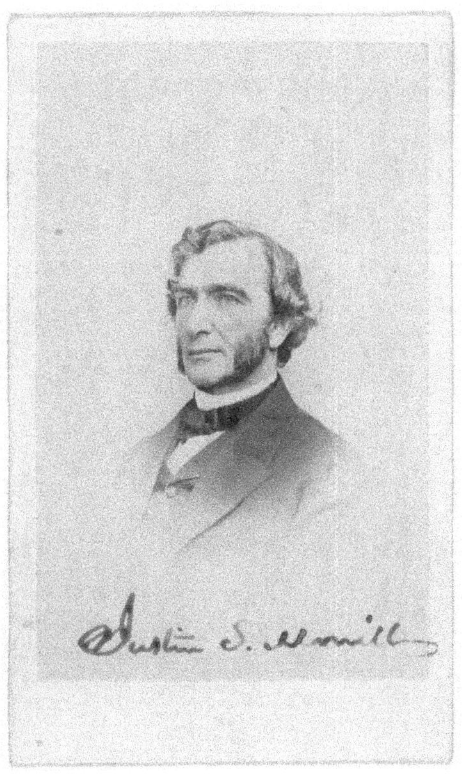

handcrafted messages of the era, created an authentic persona capable of building ethos through manual dexterity and moral character. At the same time, as printing and chromolithography became more widespread and affordable, these cards were able to include more graphical elements, a reflection of both the technology and Victorian aesthetics (Green 2016). Figure 5.13, for example, shows a late nineteenth-century promotional piece for Shearman & Hart, "practical lithographers" that produced printed "show cards" for various trades and professions, among them shipping companies, clothing stores, insurance agencies, livery stables, factories, and manufacturers of pianos, furniture, and sewing machines. These "beautifully illuminated" business cards envision these trades much as business cards

Figure 5.13. Ethos building through enargeia and ornament in business cards and other professional documents (Shearman & Hart). Courtesy of the Library of Congress, Prints and Photographs Division, Popular Graphic Arts Collection [LC-DIG-pga-04144].

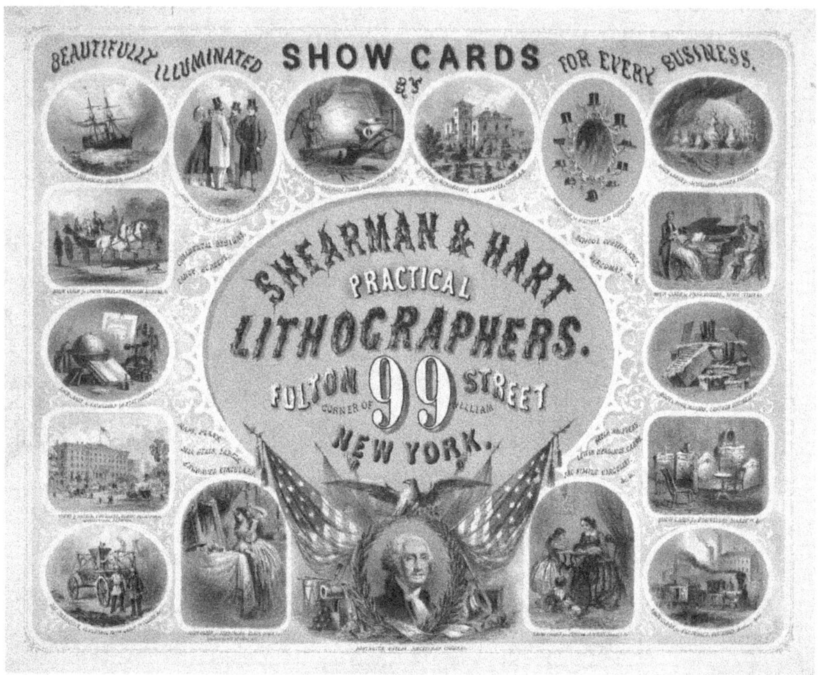

still do today, with a small picture describing the nature of the business, though here in lavish detail. These cards create vivid, compelling pictures of various trades and professions (a form of enargeia), heightening their credibility. Printed cards like these democratized visual rhetoric by picturing virtually any occupation, a display of socioeconomic ethos-building that's amplified by the image of George Washington, flanked by American flags and graphical embellishments. Today business cards continue to feature graphical elements, but like the one on Karen's card (fig. 5.10), with more streamlined images burnishing the credentials of their owners.

An even more daring and inventive approach to constructing professional ethos appears in the business card in figure 5.14, which is nearly twice the size of the standard variety. Representing a Philadelphia lithographer in the late 1860s, the card visualizes the designer (on the

Figure 5.14. An oversized and highly embellished business card that exemplifies the trades it represents (Moras ca. 1867–1869). Courtesy of the Library of Congress, Prints and Photographs Division, Marian S. Carson collection [LC-DIG-ppmsca-24839].

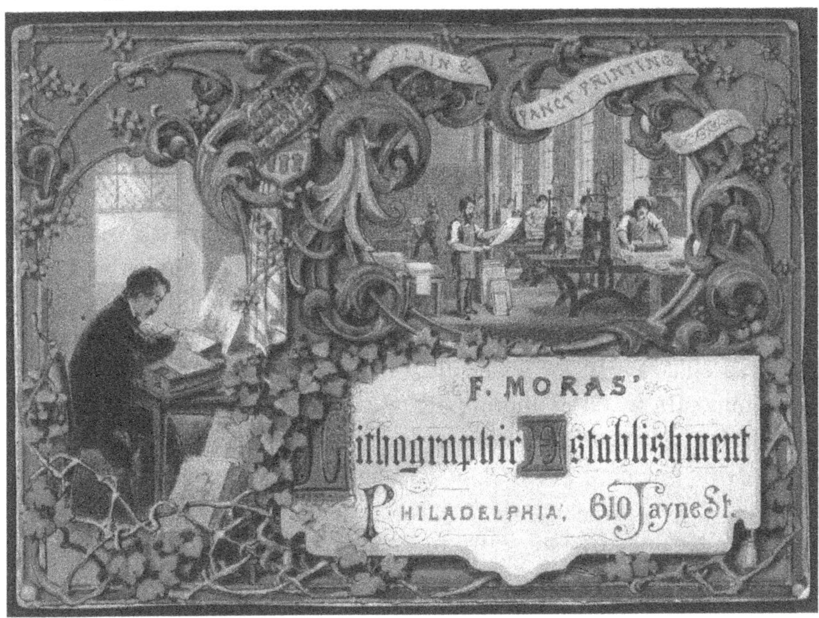

left) at his studio desk and the production team in the background manufacturing lithographic prints, showing both the artistic and functional capabilities of the company. These scenes are framed by elaborate vegetative embellishments that express the intricate, natural aesthetic of the period, part of the legacy of Romanticism and its emphasis on organic forms. This colorful, lavish design is a feast for the eyes, a piece of pure Victoriana. Then as now, expressive designs have found a place in business cards representing entities engaged in visual and other creative arts, generating audience trust by illustrating imagination and craft. In most other situations, this approach would amount to sheer graphic hyperbole, subverting ethos rather than enhancing it. Here, however, the extravagant design aptly serves its purpose.

The business card genre has continued to evolve today as technology and social interactions change. Like many other genres, business cards have begun to navigate from print to digital media, maintaining many of the same conventions of print in terms of size, shape, and design, creating continuity with the past by invoking genre conventions that contemporary audiences (for now at least) still widely understand and expect. Whether digital cards can generate a similar level of ethos, however, remains unclear because print and digital cards diverge in some important ways. Print cards must be touched, received in the hand, and stored in a wallet, purse, or drawer, and this tactile interaction intensifies the audience's emotional attachment. Received in person, the card serves as a visual gesture—an extension of a handshake, smile, or greeting—and the audience can look the rhetor in the eye and judge the rhetor's character. Digital cards forfeit this physical, tactile presence; however, they're extremely portable on mobile devices: no searching through the wallet or desk drawer, and perhaps more importantly in an age of pandemics, *no* direct contact with the audience. Digital cards are typically activated with QR codes, sacrificing immediate personal engagement for wider distribution and networking.

LinkedIn serves as another digital alternative to print business cards by offering a structured, spacious, and interactive platform for building ethos with potential clients and colleagues. Like digital cards, rhetors forfeit the personal contact vital to judging character, some of which might be regained by interactive messaging and, moreover, by the digital gesture itself: that rhetors use this technology says something about their capabilities and willingness to engage with digital natives. This potential for wide circulation benefits lesser-known rhetors—real estate agents, consultants, lawyers, caregivers, engineers—by extending their reach to a broader array of audiences. But that calculus remains the same regardless of the medium: Ethos building with business cards or their digital proxies serves primarily to initiate interactions with audiences.

Letters, Letterheads, and Organizational Brands

The contemporary commonplace that hardly anyone writes personal letters anymore represents a sea change from previous eras when letters served as *the* essential genre, both personally and professionally. Its practice in the eighteenth and nineteenth centuries was so popular and widespread that the epistolary genre dominated early novels.[6] Although personal letter-writing has fallen off the cliff, letter writing still serves an important function in professional communication—in business, marketing, academia, government, and law. What residual ethos-building capacity does letter design have for business and the professions?

By the late nineteenth century, the basic page layout of business letters had become highly prescriptive, consisting of a letterhead, date, inside address, salutation, body, complimentary close, and signature. As technology transitioned from handwritten to typewritten text, these design components remained remarkably stable throughout the twentieth century, even as technology shifted again with advent of laser printing (Kostelnick 1994).[7] Direct descendants of the late Victorian letter, these page design conventions remain largely intact today.

The letterhead embodies the most prominent visual convention of a business or professional letter, and because it immediately captures the audience's attention, it has a heightened ethos-building capacity for the rhetor whose signature appears at the bottom. A mixture of textual, graphical, or pictorial elements that often features a company logo,[8] letterheads carry a good deal of rhetorical weight in generating ethos because of the organizations they represent—universities, corporations, nonprofits, and government agencies. However, for businesses without this initial (extrinsic) ethos, letterheads have long provided a fertile arena for invented ethos, with commercial stationery becoming widely available in the nineteenth century as business, branding, and printing coalesced and flourished. As letterheads proliferated, many of them featured buildings of their respective businesses (like the hotel letterhead in fig. 2.9) as well as their products and values (Wisconsin Historical Society 2025). The letterhead continued to blossom in the twentieth century with the increasing demand for typewritten correspondence in business and the professions. Today, with digital tools readily available to design and reproduce letterheads, and sales and fundraising letters continuing to stuff our physical mailboxes, letterheads won't relinquish their ubiquitous presence anytime soon. However, as digital messaging supplants print letters, and audiences become more removed from these material artifacts, the letterhead as a medium for building ethos may inevitably be repurposed, its vestiges providing rhetorical ballast for other genres not yet imagined.

Annual Reports as Exercises in Visual Ethos-Building

Annual reports have a strong and abiding ethos-building function for collective entities—corporations, nonprofits, government agencies, and universities. From the outset of the twentieth century (and even earlier), annual reports have aimed to impress clients, investors, competitors, and other stakeholders by telling positive and convincing stories about organizations. A kind of collective biography, annual reports embody several visual conventions: a cover page, a letter from the organization's leader, then a mix of text, photos, and charts and graphs on the inside, followed by detailed tables of financial data. Over the last century, this combination of visual elements has typified annual reports of North American and European companies, as well as multinationals from around the globe, collectively building ethos with their audiences through the visual conventions of the genre.

Several of these conventions appear in figure 5.15, which shows the cover and inside page of an annual report of Alliant Energy (2020), a publicly traded company (Nasdaq) that serves electric and natural gas customers in Wisconsin and Iowa. The cover shows Alliant's key values and brand: its dedication to clean energy with the image of the sun and the abstract collage with Alliant's color scheme. The simple, lean design meets audience expectations for an efficient, technologically advanced, and environmentally responsible company of the twenty-first century. The inside page displays genre conventions by combining text and image (charts, map, and picture), as do subsequent pages in grayscale containing financial tables, sans serif text explaining the details, and images of key

Figure 5.15. Cover and inside page of Alliant Energy's *2019 Annual Report* (Alliant Energy 2020). Copyright © 2025 Alliant Energy Corp. Used with permission.

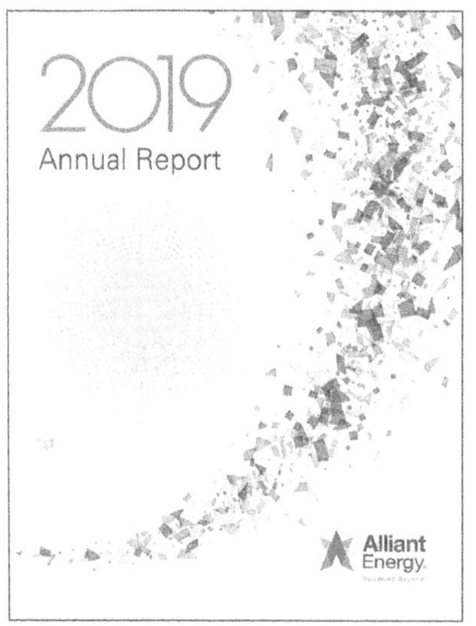

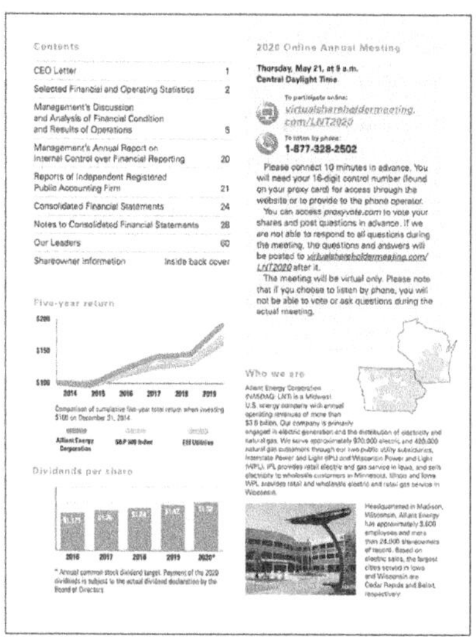

Contents

CEO Letter	1
Selected Financial and Operating Statistics	2
Management's Discussion and Analysis of Financial Condition and Results of Operations	5
Management's Annual Report on Internal Control over Financial Reporting	20
Reports of Independent Registered Public Accounting Firm	21
Consolidated Financial Statements	24
Notes to Consolidated Financial Statements	28
Our Leaders	60
Shareowner Information	Inside back cover

Five-year return

Dividends per share

2020 Online Annual Meeting

Thursday, May 21, at 9 a.m.
Central Daylight Time

To participate online:
virtualshareholdermeeting.com/LNT2020

To listen by phone:
1-877-328-2502

Please connect 10 minutes in advance. You will need your 16-digit control number (found on your proxy card) for access through the website or to provide to the phone operator.

You can access *proxyvote.com* to vote your shares and post questions in advance. If we are not able to respond to all questions during the meeting, the questions and answers will be posted to *virtualshareholdermeeting.com/LNT2020* after it.

The meeting will be virtual only. Please note that if you choose to listen by phone, you will not be able to vote or ask questions during the meeting.

Who we are

leaders. By combining all of these visual elements, the report generates ethos by meeting audience expectations for the genre.

As annual reports gravitate from print to digital form—integrating ubiquitous color, interactivity, live links, animations, and other digital features—they become more accessible to their public audiences as well as raise the stakes for ethos-building. Although some organizations have exploited these tools, many U.S. companies have abandoned graphically rich annual reports, preferring instead to publish only their 10K reports, which are required by the SEC and feature mostly dense text and tables of financial data. Nonetheless, many companies in the U.S. and around the world are now creating graphically rich corporate social responsibility reports that redefine their ethos by visualizing their sustainable practices, diversity initiatives, and contributions to the global community in data displays, diagrams, and pictures. In these ways, this emerging subgenre of the annual report, from which it borrows its ethos-building conventions, visually projects the good character of a company by aligning it with the values and aspirations of clients and investors.

Newsletters and Their Evolving Ethos-Building Capacity

While annual reports generate collective ethos primarily for external audiences, newsletters generate it *within* organizations and local communities by telling their stories mainly to themselves. Initially borrowing key visual elements from newspapers, newsletter design has congealed around a stable set of visual conventions that includes nameplates, multiple columns, spanner headings, pull quotes, narrow margins, and portrait orientation. The two pages of the newsletter in figure 5.16, created by the Center for Industrial Research and Service (CIRAS 2020), show many of these visual conventions: the prominent nameplate on the cover page (left), unjustified text set in three columns, vertical lines between them, and sans serif headings that span them. The inside page (right) features a shaded sidebar with a table of contents, three-column arrangement, a pull quote in the center, and a picture anchoring the upper right corner. Encountering this combination of conventions, most American audiences would immediately identify this genre as a newsletter, fostering an interpretive ease that heightens the document's ethos. By embedding the visual conventions of the genre, the newsletter instills confidence in its audience that this organization takes its communications seriously and has the know-how and resources to implement them.

Figure 5.16. Cover and inside page of a print newsletter for CIRAS (Center for Industrial Research and Service 2020). Reprinted from *CIRAS News*, a publication of Iowa State University Center for Industrial Research and Service. Used with permission.

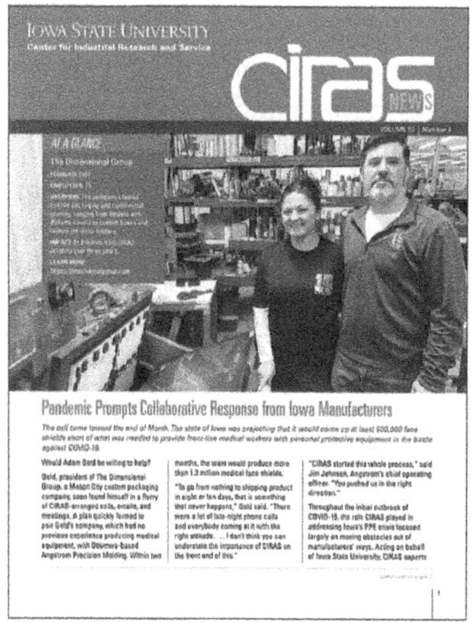

Iowa State University
Center for Industrial Research and Service

ciras NEWS

AT A GLANCE
The Dimensional Group

Pandemic Prompts Collaborative Response from Iowa Manufacturers

The call came toward the end of March. The state of Iowa was projecting that it would come up at least 600,000 face shields short of what was needed to provide front-line medical workers with personal protective equipment in the battle against COVID-19.

Would Adam Gold be willing to help?

Gold, president of The Dimensional Group, a Mason City custom packaging company, soon found himself in a flurry of CIRAS-arranged calls, emails, and meetings. A plan quickly formed to pair Gold's company, which had no previous experience producing medical equipment, with Ottumwa-based Angstrom Precision Molding. Within two months, the team would produce more than 1.3 million medical face shields.

"To go from nothing to shipping product in eight or ten days, that is something that never happens," Gold said. "There were a lot of late-night phone calls and everybody coming at it with the right attitude. . . . I don't think you can understate the importance of CIRAS on the front end of this."

"CIRAS started this whole process," said Jim Johnson, Angstrom's chief operating officer. "You pushed us in the right direction."

Throughout the initial outbreak of COVID-19, the role CIRAS played in addressing Iowa's PPE crisis focused largely on moving obstacles out of manufacturers' ways. Acting on behalf of Iowa State University, CIRAS experts

1

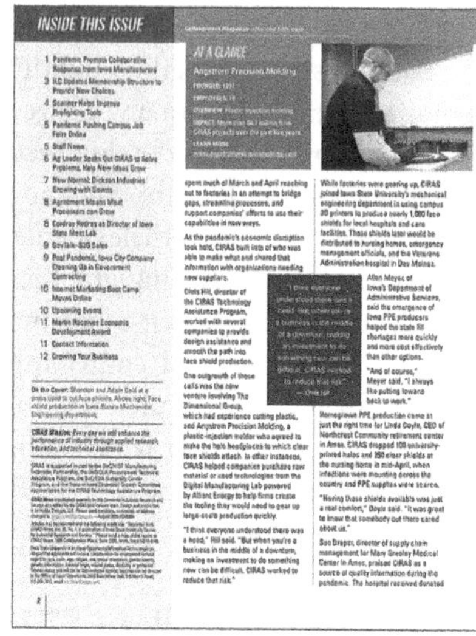

INSIDE THIS ISSUE

AT A GLANCE
Angstrom Precision Molding

spent much of March and April reaching out to factories in an attempt to bridge gaps, streamline processes, and support companies' efforts to use their capabilities in new ways.

As the pandemic's economic disruption took hold, CIRAS built lists of who was able to make what and shared that information with organizations needing new suppliers.

Chris Hill, director of the CIRAS Technology Assistance Program, worked with several companies to provide design assistance and smooth the path into face shield production.

One outgrowth of those calls was the new venture involving The Dimensional Group, which had experience cutting plastic, and Angstrom Precision Molding, a plastic-injection molder who agreed to make the halo headpieces to which clear face shields attach. In other instances, CIRAS helped companies purchase raw material or used technologies from the Digital Manufacturing Lab powered by Alliant Energy to help firms create the tooling they would need to gear up large-scale production quickly.

"I think everyone understood there was a need," Hill said. "But when you're a business in the middle of a downturn, making an investment to do something new can be difficult. CIRAS worked to reduce that risk."

While factories were gearing up, CIRAS joined Iowa State University's mechanical engineering department in using campus 3D printers to produce nearly 1,000 face shields for local hospitals and care facilities. Those shields later would be distributed to nursing homes, emergency management officials, and the Veterans Administration hospital in Des Moines.

Allen Meyer, of Iowa's Department of Administrative Services, said the emergence of Iowa PPE producers helped the state fill shortages more quickly and more cost effectively than other options.

"And of course," Meyer said, "I always like putting Iowans back to work."

Homegrown PPE production came at just the right time for Linda Doyle, CEO of Northcrest Community retirement center in Ames. CIRAS dropped 100 university-printed halos and 250 clear shields at the nursing home in mid-April, when infections were mounting across the country and PPE supplies were scarce.

"Having those shields available was just a real comfort," Doyle said. "It was great to know that somebody out there cared about us."

Sue Draper, director of supply chain management for Mary Greeley Medical Center in Ames, praised CIRAS as a source of quality information during the pandemic. The hospital received donated

On the Cover: Shandon and Adam Gold at a press used to cut face shields. Above right: Face shield production in Iowa State's Mechanical Engineering department.

CIRAS Mission: *Every day we will enhance the performance of industry through applied research, education, and technical assistance.*

2

The CIRAS newsletter is available both in print and online as a PDF, so audiences can adjust their screens to navigate the pages for enhanced accessibility. In addition, CIRAS publishes an online e-news version (fig. 5.17), which contains many of the same stories as the print newsletter, arranged in temporal sequence. How has the transition from a print to an electronic newsletter transformed its visual language, and how has digital design affected its ethos-building capacity? The e-news version displays the CIRAS nameplate, then lists the stories in headline type along with a summary and a link to the full story. This hybrid design mixes print and digital elements, maintaining the headings, text design, and color of the print version but omitting the multiple columns, boxed inserts, and other page design elements. The e-news also gives its audience several options for accessing stories—by year, subject category, and searchable topic—making it interactive and dynamic. Overall, the e-news design clearly echoes the print version, so for audiences accustomed to print, it provides a cohesive transition to an all-digital design. In that way, the past (print) is firmly embedded in the present (digital) design, ensuring some carryover of its ethos.

Figure 5.17. Web page for e-news stories about CIRAS (Center for Industrial Research and Service 2025). Reprinted from www.newswire.ciras.iastate.edu, a digital publication of Iowa State University Center for Industrial Research and Service. Used with permission.

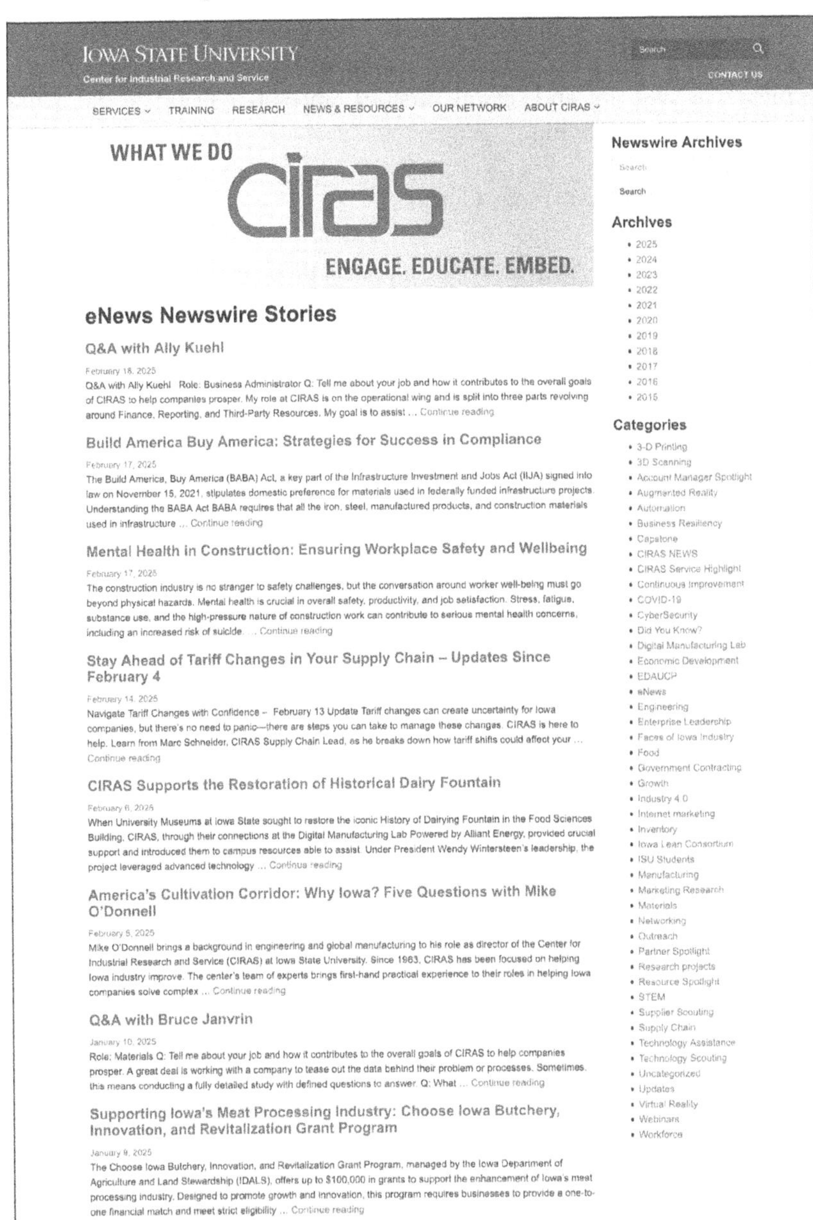

IOWA STATE UNIVERSITY
Center for Industrial Research and Service

Search
CONTACT US

SERVICES ⌄ TRAINING RESEARCH NEWS & RESOURCES ⌄ OUR NETWORK ABOUT CIRAS ⌄

eNews Newswire Stories

Q&A with Ally Kuehl

February 18, 2025
Q&A with Ally Kuehl Role: Business Administrator Q: Tell me about your job and how it contributes to the overall goals of CIRAS to help companies prosper. My role at CIRAS is on the operational wing and is split into three parts revolving around Finance, Reporting, and Third-Party Resources. My goal is to assist … Continue reading

Build America Buy America: Strategies for Success in Compliance

February 17, 2025
The Build America, Buy America (BABA) Act, a key part of the Infrastructure Investment and Jobs Act (IIJA) signed into law on November 15, 2021, stipulates domestic preference for materials used in federally funded infrastructure projects. Understanding the BABA Act BABA requires that all the iron, steel, manufactured products, and construction materials used in infrastructure … Continue reading

Mental Health in Construction: Ensuring Workplace Safety and Wellbeing

February 17, 2025
The construction industry is no stranger to safety challenges, but the conversation around worker well-being must go beyond physical hazards. Mental health is crucial in overall safety, productivity, and job satisfaction. Stress, fatigue, substance use, and the high-pressure nature of construction work can contribute to serious mental health concerns, including an increased risk of suicide. … Continue reading

Stay Ahead of Tariff Changes in Your Supply Chain – Updates Since February 4

February 14, 2025
Navigate Tariff Changes with Confidence – February 13 Update Tariff changes can create uncertainty for Iowa companies, but there's no need to panic—there are steps you can take to manage these changes. CIRAS is here to help. Learn from Marc Schneider, CIRAS Supply Chain Lead, as he breaks down how tariff shifts could affect your … Continue reading

CIRAS Supports the Restoration of Historical Dairy Fountain

February 6, 2025
When University Museums at Iowa State sought to restore the iconic History of Dairying Fountain in the Food Sciences Building, CIRAS, through their connections at the Digital Manufacturing Lab Powered by Alliant Energy, provided crucial support and introduced them to campus resources able to assist. Under President Wendy Wintersteen's leadership, the project leveraged advanced technology … Continue reading

America's Cultivation Corridor: Why Iowa? Five Questions with Mike O'Donnell

February 5, 2025
Mike O'Donnell brings a background in engineering and global manufacturing to his role as director of the Center for Industrial Research and Service (CIRAS) at Iowa State University. Since 1963, CIRAS has been focused on helping Iowa industry improve. The center's team of experts brings first-hand practical experience to their roles in helping Iowa companies solve complex … Continue reading

Q&A with Bruce Janvrin

January 10, 2025
Role: Materials Q: Tell me about your job and how it contributes to the overall goals of CIRAS to help companies prosper. A great deal is working with a company to tease out the data behind their problem or processes. Sometimes, this means conducting a fully detailed study with defined questions to answer. Q: What … Continue reading

Supporting Iowa's Meat Processing Industry: Choose Iowa Butchery, Innovation, and Revitalization Grant Program

January 9, 2025
The Choose Iowa Butchery, Innovation, and Revitalization Grant Program, managed by the Iowa Department of Agriculture and Land Stewardship (IDALS), offers up to $100,000 in grants to support the enhancement of Iowa's meat processing industry. Designed to promote growth and innovation, this program requires businesses to provide a one-to-one financial match and meet strict eligibility … Continue reading

Newswire Archives

Search

Search

Archives

- 2025
- 2024
- 2023
- 2022
- 2021
- 2020
- 2019
- 2018
- 2017
- 2016
- 2015

Categories

- 3-D Printing
- 3D Scanning
- Account Manager Spotlight
- Augmented Reality
- Automation
- Business Resiliency
- Capstone
- CIRAS NEWS
- CIRAS Service Highlight
- Continuous Improvement
- COVID-19
- CyberSecurity
- Did You Know?
- Digital Manufacturing Lab
- Economic Development
- EDAUCP
- eNews
- Engineering
- Enterprise Leadership
- Faces of Iowa Industry
- Food
- Government Contracting
- Growth
- Industry 4.0
- Internet marketing
- Inventory
- Iowa Lean Consortium
- ISU Students
- Manufacturing
- Marketing Research
- Materials
- Networking
- Outreach
- Partner Spotlight
- Research projects
- Resource Spotlight
- STEM
- Supplier Scouting
- Supply Chain
- Technology Assistance
- Technology Scouting
- Uncategorized
- Updates
- Virtual Reality
- Webinars
- Workforce

In the distant future, will the print version of the CIRAS newsletter (and others like it) eventually lose its audience, and if so, what will become of the visual ethos it has accrued, especially for audiences who've never encountered the print newsletter firsthand? Will a new set of conventions emerge that extend, or depart from, the legacy design in figure 5.16? For the time being, the CIRAS newsletter remains immune from these developments, as its print and digital versions amicably cohabitate the genre for mutual ethos-building.

Ethos-Building in Other Genres

Other commonplace genres also generate ethos because their familiar visual conventions meet audience expectations: manuals, home pages, cookbooks, invitations, online shopping sites, brochures, concert programs, posters, PowerPoints, bookmarks, among many others. Deploying the visual language associated with genre conventions will streamline the audience's work, reduce their interpretive stress, and consequently induce some level of trust. Some genres have especially powerful ethos-building capacity. For example, the genre of the table, Lee Brasseur (2003) argues, builds ethos because it organizes facts in a "rational" and transparent system that persuades audiences of its validity (110).

Ethos-building with genre conventions, however, is a fluid process. Visual conventions continually mature as the purposes, audiences, and situational uses of genres evolve and as the affordances of technology increase: instructions that embed animations, graphs that display interactive data, schedules that visualize destinations in real time. As genre conventions mutate into more agile forms, their vestigial traces continue to linger. We may not fully recognize these antecedents, but they continue to animate, visually and rhetorically, the artifacts that embody them.

Us and Them: Generating Ethos with Identification

Many of the intrinsic (invented) ethos-building strategies that I've discussed are generated socially through discourse communities that shape and sustain genre conventions. Collective ethos, moreover, also has deep roots in culture in both the near and distant past. For example, the concept of ethos in "classical Chinese rhetoric," Hong-Kang Wei explains (2008, 2004), differs from traditional Western ethos because Chinese ethos resides within a group rather than in an individual. In contrast, many modern

Western forms of collective ethos based on cultural and social norms are more diverse and mutable and generated by various forms of visual identification. When audiences see themselves in a visual message, they feel an attachment to it, both perceptually and emotionally, and they're inclined to trust it. Collectively identifying with visual language binds them together, in the same sense that Kenneth Burke (1950) defined the unifying ability of rhetoric to make people "consubstantial" with one another (20–22). The collective ethos that renders us "consubstantial" with one another through identification can be generated visually with virtually any form of visual language—perhaps most immediately with pictures but also with colors, typefaces, icons, and logos.

Because forms of visual identification are closely tied to the cultural and social norms of a given historical moment, their ethos-generating power varies enormously over time. Contemporary audiences, for example, won't likely identify with the human figures in the Victorian examples we've seen because their dress and demeanor don't resonate with them. The woman holding the apples in the Massey-Harris catalog cover (fig. 5.4), the farmer on the McCormick reaper (bottom of fig. 5.5), the medics in the Sanitary Commission letter (fig. 5.8)—some audience members might harbor nostalgic sentiment for these figures (as we saw in chapter 3), but they don't see *themselves* in these images, nor do they see people they know.

The trust-building power of identification can be found in everyday practical communications that visualize human forms—instructions, warning signs, promotional materials—though the efficacy of these, too, wax and wane with time. For example, if a brochure for prospective college students contains images of students the audience can identify with—if minority students see other minority students, or international students see other students who look international—they'll more likely trust the message—at least until the corrosive effects of time take their toll. Audiences can tell in a heartbeat whether images of people they identify with are current or passé. As clothing, hair styles, and student demographics change, the aging brochure will relinquish its trust-building capacity until its images are refreshed. The same aging process occurs with images in instructional materials and warnings, even if they're sanitized with a universal Everyperson of modernist aesthetics.

Audience identification is also effected by images of place and space, by pictures of buildings or interior spaces that audience members currently occupy (or imagine themselves in) and that they can identify with. Looking back in time helps us see how identity and space intertwine. As we saw previously, businesses in the nineteenth century pictured themselves architecturally as images of success with impressive hotel façades (the Athearn Hotel in fig. 2.9) and bird's-eye views of manufacturing plants (the match factory in fig. 2.7, McCormick's Chicago plant in fig. 4.20). Commercial success was also visualized with attractive interior spaces like the one in figure 5.18, which shows the inside of a business that

Figure 5.18. Elegant interior as the setting for a promotional piece for a New York magnetic machine company (Major ca. 1848). Courtesy of the Library of Congress, Prints & Photographs Division, Popular Graphic Arts Collection [LC-DIG-pga-02095].

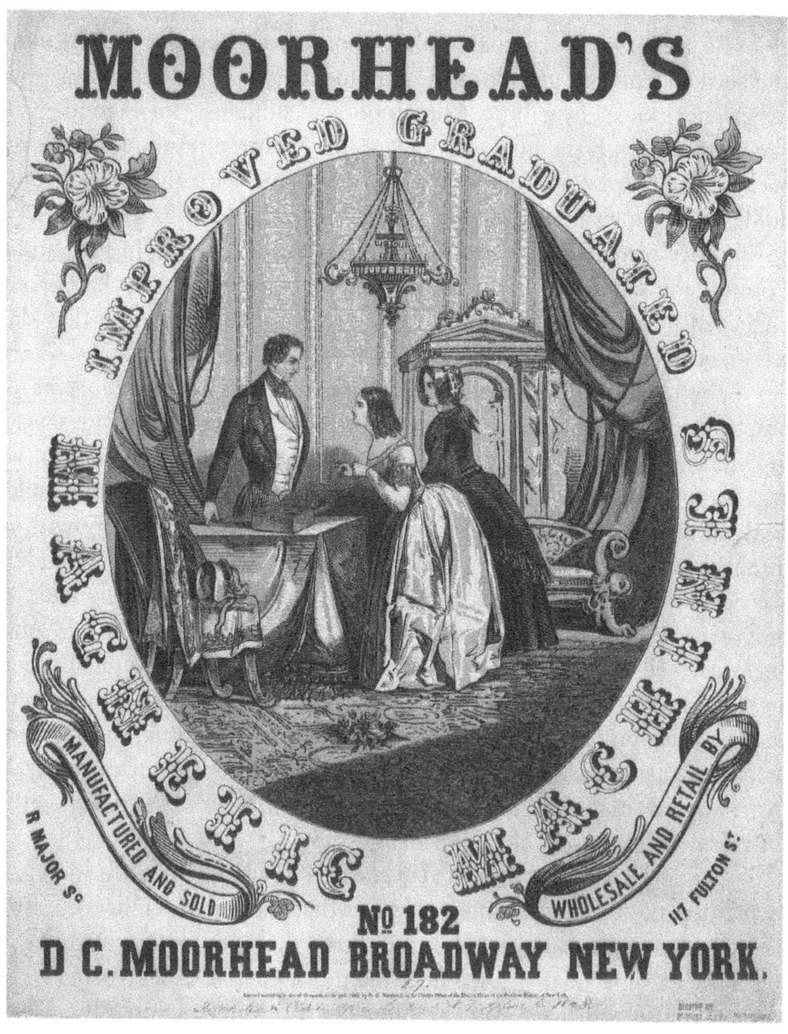

produces and sells magnetic machines to treat diseases. This experimental health technology appears in a drawing room fit for an aristocrat, the space amplified by wallpapered surfaces, long curtains, elegant furniture, chandelier lighting, and architectural detailing. Occupied by a dapper, confident salesman and two women standing in rapt attention, this interior space fosters a high level of ethos, especially for wealthy, status-conscious New Yorkers who expect top-drawer service and professionalism and who identify with—literally see themselves in—such luxurious spaces. For them, the oval picture ensconced in its embellished frame envisions an appealing and trustworthy business worthy of their patronage.

Identification with space and place occurs in virtually any picture that depicts an environmental context. To people dwelling in cities, the rural setting pictured on the bottom of figure 5.5 probably seemed distant and provincial, while the lakefront image with the skyscrapers and boats for the National Business Show poster (fig. 5.7) appeared familiar and engaging. For Civil War soldiers the scenes in figure 5.8 must have struck an immediate chord, visualizing places and events that they recently experienced or that family and friends witnessed vicariously. Moreover, most Americans would have identified with the composite scene of Washington landmarks in the seed catalog cover (fig. 5.6), even if those edifices appeared in fictional space.

Although identification can most easily be generated by pictures, where audiences can immediately recognize themselves (or people or things they value), other forms of visual language also foster it. Colors, typefaces, icons, and logos are often closely allied with the identity of an organization, so deploying these for audiences affiliated with an organization will boost trust as well. Many organizations blend these (and other) visual elements into their own identity systems as a way to project their collective (and often corporate) ethos. For example, the CIRAS newsletter and online e-news (figs. 5.16 and 5.17) feature cardinal red and gold, which are the colors of Iowa State University, with which CIRAS is affiliated. For those who are connected with Iowa State University—faculty, students, staff, alumni, sports fans—the cardinal and gold are extremely familiar, trustworthy colors that appear throughout Iowa State University's print and digital communications. In short, audiences *identify* with these colors. By representing the institution where they collectively work, study, cheer, and share memories, these colors bind audiences together, just as colors do with constituencies at other universities.

However, visual language operates on many different cultural levels—corporate, ethnic, religious, national—each with its own temporal dynamics. Like other forms of visual language, time erodes its capacity to foster communal trust: colors take on new meanings, logotypes gain and lose popularity, brands come and go. The organizations that deploy these forms of visual language change as well, growing and shrinking and often being subsumed into other organizations that redefine their visual identities. Where does that leave the audience? Like a team that changes mascots and colors when two schools merge, audiences have to balance their nostalgia for the old identity with their tepid acceptance of the new one and the community it binds together. Eventually, the ethos of the old identity fades, with few left to defend it or to place their trust in it. Audience attrition and visual atrophy take their toll.

Visual Decorum: Fostering Trust with Style and Tone

Building trust also requires an appropriate style, one of the five traditional rhetorical canons that refers to the rhetor's ability to use language that's appropriate for a given rhetorical situation, as described at length by Aristotle (197–229, bk. 3, ch. 2–12). A form of rhetorical decorum, an effective style generates collective trust by meeting audience expectations, and in the modern world it does so both verbally and visually, especially when it embodies aesthetic elements that immediately resonate with an audience. Sometimes this visual decorum is easier to spot when viewed from the distance of time. In the nineteenth century, for example, large "slab" typefaces had huge public currency, appearing on signs, letterheads, promotional materials, and (perhaps most notoriously) "wanted" posters of infamous criminals. This bold, heavy-handed style, often referred to as "Egyptian," appears in figure 5.19, an 1870s poster for the St. Nicholas Restaurant in Cincinnati (Morris), which specialized in fresh oysters. The typeface has extremely tall letters and thick serifs, further accentuated by red shadows. These blocky, gargantuan letterforms permeated commercial typography in the late nineteenth century, and audiences in that era would have found them as credible as today's audiences find them curious, though no more so than the waiter's conical beard.

Several decades after the flamboyant oyster restaurant poster, changes in aesthetic taste induced by early modernism altered audiences' expectations of what they found credible. Advocates of modernism severed ties with the past and sought an entirely new aesthetic that matched the lean, functional efficiency of the new industrial age. With its high-contrast geometrical forms stripped of the perceived excesses of the Victorian

Figure 5.19. Poster with a large slab typeface promoting a Cincinnati oyster restaurant (Morris ca. 1873). Courtesy of the Library of Congress, Prints and Photographs Division, Popular Graphic Arts Collection [LC-DIG-pga-02243].

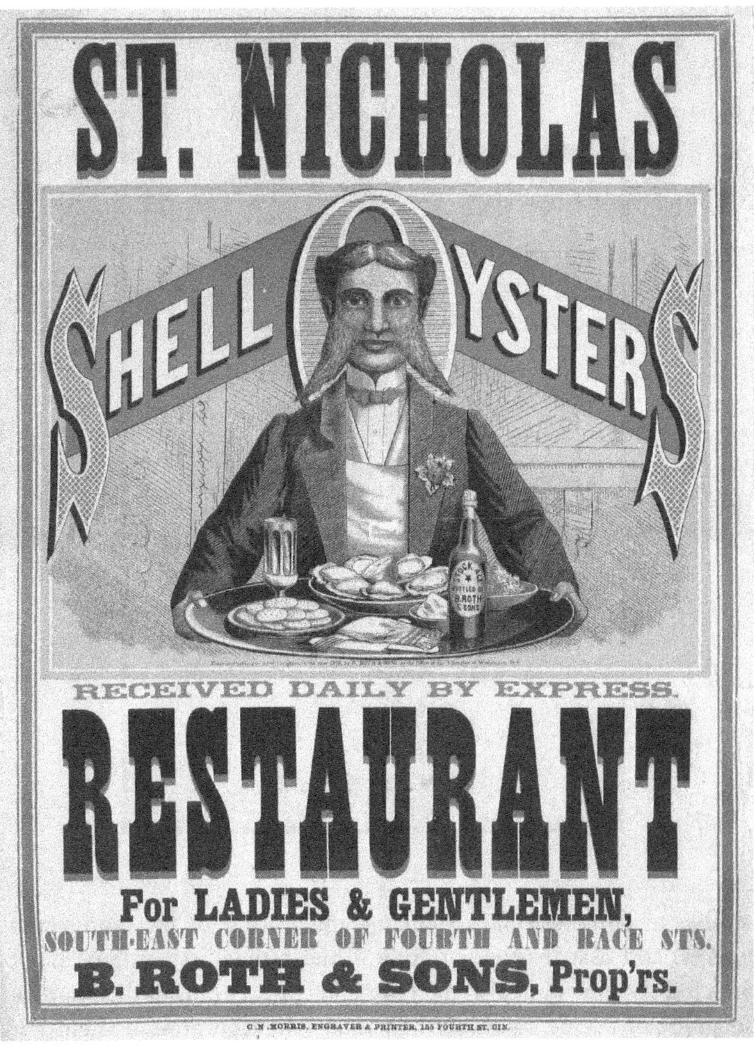

age, modernism transformed all aspects of design, from architecture to typography. Some of the figures in chapter 3 (3.13, 3.14, and 3.15) illustrate this dramatic shift in style, as does the Works Progress Administration educational poster in figure 5.20, which contains several quintessential modernist elements: a sans serif typeface, abstract geometrical forms (that

Figure 5.20. Modernist poster for the WPA educational services during the 1930s (Works Progress Administration 1936/1937). Courtesy of the Library of Congress, Prints and Photographs Division, Work Projects Administration Poster Collection [LC-USZC2-954].

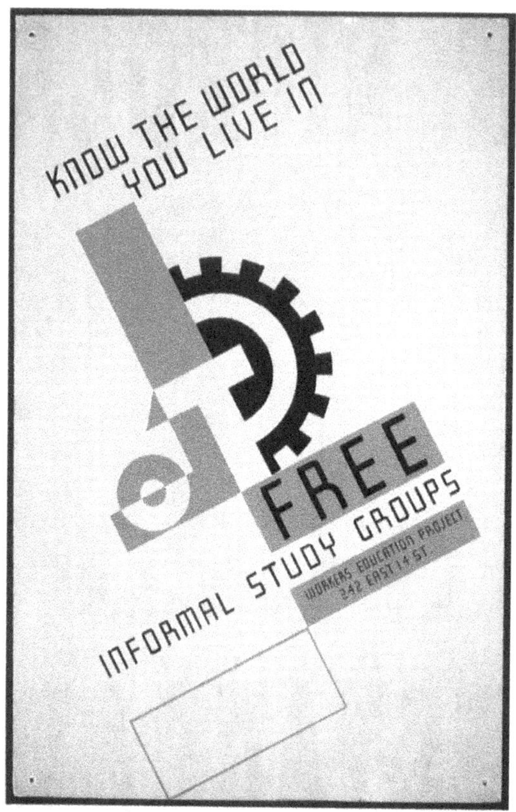

resemble machine parts), a dynamic integration of text and graphical elements, and a sparse use of color. This minimalist aesthetic not only redirected designers but also the taste and expectations of their audiences, who gradually came to resee the visual language around them and to redefine what looked stylish—and credible.

Rhetorical decorum also encompasses tone, both verbal and visual, either of which can engender or squander trust, depending how well tone matches a given situation. Visually, text that's right justified in a formal report will likely match audience expectations for a serious-looking document that speaks in a confident but measured voice. A detailed mechanical drawing intended for a manufacturer will foster trust, but the same drawing placed in a set of instructions will intimidate and frustrate the audience, squandering trust. In short, the visual tone of a design that matches the rhetorical situation will more effectively build trust. The visual tone of the CIRAS newsletter (fig. 5.16) is serious, businesslike, and upbeat, as it should be in telling the success stories of Iowa manufacturers. The tone of the Sanitary Commission letter (fig. 5.8) is formal and somber, matching its content, while the tone of the oyster restaurant poster (fig. 5.19) is playful and brash, projecting confidence and pluck. Visual tone also has very subtle nuances: Just like a rhetor giving a speech, the slightest variation (in voice, gesture, or facial expression) can enhance or erode an audience's trust. Capturing the nuances of visual tone through the eyes of audiences a century (or more) ago can be challenging but also enlightening—perhaps in ways audiences back then couldn't entirely discern themselves.

Novel and Faux Design: Playing with the Audience's Trust

Building trust sometimes calls for visual language that follows a less traveled road, veering off into novelty or into designs that playfully misdirect audiences. These moves can be risky and unorthodox, but some situations warrant them in order to earn the trust of reluctant or demanding audiences.

Rolling the Dice with Novelty

Traveling the familiar, well-worn path of visual conventions makes us feel comfortable and trusting, but what about designs that are unfamiliar and novel, that jar and prod us, that disrupt our familiar interpretive patterns? An invitation shaped like a flower, a newsletter arranged landscape, a PowerPoint with animated icons—how do these unconventional designs affect ethos? Depending on the audience and the situation, a couple of different outcomes might be possible:

- If the rhetorical situation invites (or requires) novelty, audiences might find it highly credible. A business card or résumé with lavish graphics (fig. 5.14) might meet the audiences' expectations if they're hiring people for their creative abilities. In contrast, a plain conventional design in these situations might diminish the rhetor's ethos.

- If the design competes for attention with other communications, the audience will likely be more receptive to novelty, depending on the genre and context. A novel flyer for an amusement park might heighten its ethos among a rack of competing flyers in a hotel lobby, while a novel brochure for an academic major might forfeit ethos compared to more conventional brochures from other colleges.

Given these mixed effects of novelty on ethos, how did novelty play out in the past? W. E. B. Du Bois's population chart in figure 1.11 with its swirling bar must certainly have attracted attention across a spectrum of audiences in Georgia, America, and around the world. This chart's engaging ingenuity no doubt garnered their trust and respect as well as heightened the moral imperative of monitoring the progress of African Americans. Building trust with novelty for an entirely different purpose unfolds in figure 5.21, a late 1800s promotional piece for farm equipment.

Figure 5.21. Heart-shaped cards (front and back) promoting farm products of the Buckeye Agricultural Works (Mast, Foos & Co. n.d.). Courtesy of Iowa State University Library Special Collections and University Archives.

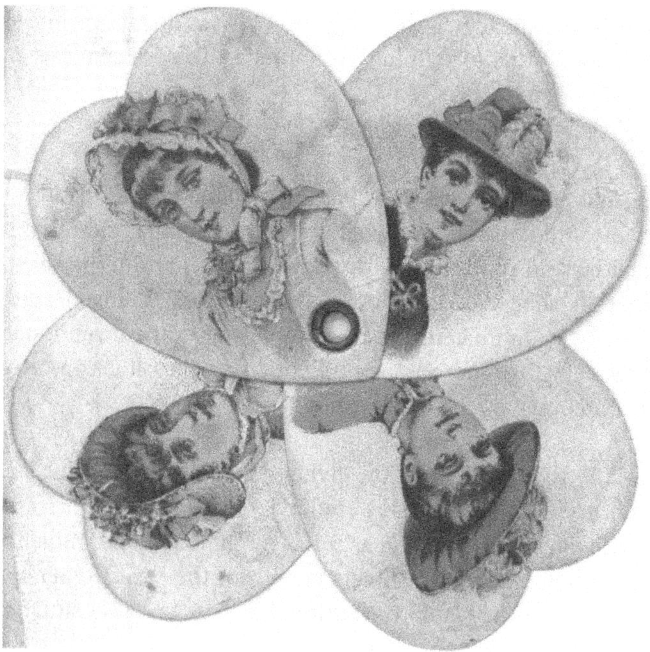

Here four heart-shaped cards are attached on a pinwheel, with women pictured on one side of the cards and farm products (a pump, push mower, wind turbine, wrought iron fence) on the other. As we've already seen in figures 5.2, 5.3, and 5.4, women frequently appeared in designs for agricultural companies in a variety of roles—mythical, maternal, romantic, and sensual—so placing female faces on one side of the cards would not have seemed particularly novel, though their heart-shaped design signals a romantic tone. Moreover, the sheer labor and expense to produce these pinwheel cards must have impressed rural customers, suggesting that the company was innovative and successful. But like most attempts to build ethos with novelty, the design is a gamble, as some audiences might find it engaging and persuasive, while others might find it a costly extravagance that inflates prices.

Pushing the Bounds of Trust with Misdirection and Camouflage

Sometimes visual language that's intended to foster trust can misdirect audiences or arouse their suspicion. For example, genres are sometimes repurposed to achieve persuasive goals—sales materials that are disguised as invitations or newsletters or that mimic "official" documents in the form of legal letters, faux checks, certificates, awards, or express mail envelopes (see Kostelnick and Hassett 2003, 212–16). Rhetro designs often aim to achieve similar ends: handwriting on envelopes, along with scribbles and underlining on the documents inside, to make audiences feel as if they're receiving personal messages. All of these uses of visual language are intended to build trust—it's their raison d'être. Although many readers may not take these attempts seriously, if only a small fraction of them do, a sales promotion might be hugely successful.

In the realm of politics, sometimes the misdirection can be aggressive and hyperbolic, the stakes exceedingly high. In 1864, when Abraham Lincoln was facing bleak prospects for reelection, as he confessed in his letter to his cabinet in figure 5.1, his political opponents created a faux business card (fig. 5.22) pitching his services as a lawyer about to return to Springfield, Illinois, his hometown. The lighthearted quip at the bottom of the card describes his imminent return to legal practice and other activities, driving home the point by citing the date of the next inauguration. Clearly, the creators of this faux business card were expecting, and fairly confident, that Lincoln would soon be packing his bags for Springfield, and

Figure 5.22. A business card with Abraham Lincoln's name, profession, and hometown ("A. Lincoln" 1864). Courtesy of the Library of Congress, Rare Book and Special Collections Division, Printed Ephemera Collection [2020771698].

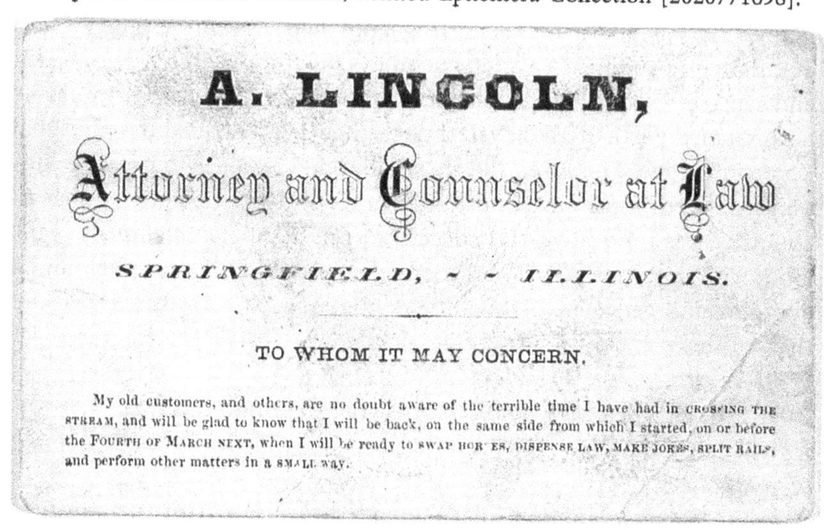

no doubt its recipients realized that the card was a fake and understood its persuasive purpose. As it turned out, Lincoln won in a landslide, and the rhetorical value of the card plummeted.

Today digital design offers far more opportunities to use visual language to foster trust, both ethically and deceptively. Because digital graphics are so portable and reproducible, visual language can be easily blended and manipulated to create ethos-building designs. Templates for awards and certificates with lavish borders, graphics, and typography can easily be purchased and downloaded online—for good or ill, leaving audiences to wonder if they're authentic or self-serving fakes. Online images are equally accessible (and manipulatable) for visualizing client testimonials, dream vacations, and investment options that attempt to build trust. Even if we don't believe these—and even if the businesses circulating them don't really *expect* us to—we might give their purveyors credit for merely being playful and hyperbolic. But like many attempts at humor, the rhetorical outcome can be risky.

AI will proliferate the creation and circulation of faux designs, but like batters in baseball or softball, high failure rates won't necessarily hamper their success. Because their canned ethos repels most audiences, faux designs rely on wide dissemination: bulk mail, spam emails, memes, and in earlier eras (before consumer protection laws) mass media like public notices and newspapers. Are audiences truly duped by these ploys? Many of us are compelled to open a letter that looks like it came from a government agency or a law office, only to discover something completely different inside. In other cases, faux designs provide a kind of rhetorical play, and we might just find ourselves going along with the ruse, crediting the rhetor's ingenuity while rolling our eyes.

Faux design takes a more serious turn with charts and graphs, which build trust because they explicitly visualize numbers, giving audiences the security of seeing them with their own eyes. However, designers can exploit our inherent trust by skewing the data: starting the Y-axis above 0, stretching the plot frame horizontally to flatten variations (or vertically to heighten them), using double scales to show false correlation (or causation), and many other tricks that distort an audience's perception of the data.[9] However, audiences have been coached to distrust, partly due to Darrell Huff's *How to Lie with Statistics* (1954) and Edward Tufte's *Visual Display of Quantitative Information* (1983), in which Tufte explained how certain graphical displays can deceive their audiences (Tufte, 53–77), some of which were identified several decades earlier by Willard Brinton (1914, 20–40), others of which have been revealed by an extensive body of empirical research (Cleveland and McGill 1984; MacDonald-Ross 1977).[10] Still, data designers largely have free rein to create charts and graphs as they wish, lacking universal standards to constrain them, though they walk a precarious line between audiences that overly trust their designs and audiences that habitually question them.

How does the move from static to interactive charts and graphs change their ethos-building capacity? Interactive displays give audiences control over which data they view and in what form (e.g., bar chart, line graph, pie chart), and that enhanced agency might itself heighten trust (see Rawlins and Wilson 2014). The more users believe that they *own* a data display, the more they'll trust it. Moreover, the flow of dynamic data in weather and crime maps (figs. 1.22, 2.27), stock charts (figs. 1.7, 4.14), bus schedules (figs. 4.7), and other data displays enables users to see data in the moment, as a seemingly unmediated reality. Such a detached, automated medium isn't completely possible, of course—someone has to *design* the interface—but the very notion of pure, transparent design will further heighten trust.

Conclusion and Implications for Design

If trust is the key to successful rhetorical acts, as Aristotle claimed, and if visual language plays an increasingly important role in that process, everyone who designs or interprets practical communications has a stake in understanding how rhetors cultivate trust visually. The perspective of time can provide some insights into which aspects of that process are stable and immutable, like boulders in midstream, and which are flexible and fluctuating, like the waters flowing around them. Visual ethos-building in its stable traditional forms, both initial and invented, has evolved over the past few centuries, as conventions, technology, and cultural values have shifted. Initial (extrinsic) ethos generated through preexisting circumstances occurs directly by visualizing the rhetor's credentials (certificates, letterheads) or indirectly by visualizing surrogates (famous people, places, or events). Ethos-building through these strategies shifts along with the changing perceptions of who or what is worthy of association, which every era defines with its own set of values.

Ethos-building through invention (intrinsic) is created through features of the design itself that reveal the competency and character of the rhetor—through the use of visual conventions and stylistic elements that meet audience expectations and through identification with familiar human forms and graphical elements that reinforce community cohesion. Like a rhetorical kaleidoscope, all of these forms of invented ethos continually evolve as rhetorical situations, the cultural forces that shape visual language, and the technology used to articulate it continually evolve.

Still, despite my attempts to categorize and describe various forms of ethos-building through visual language, developing trust can be an idiosyncratic and deeply personal process where audiences might be ambivalent about, or resistant to, rhetors seeking their trust with letterheads, visual surrogates, or familiar conventions. Moreover, visual ethos-building strategies continually evolve and age, undermining their stability and our assumptions about their relevance and utility. Whatever forms visual ethos-building takes, we're obliged to acknowledge its uncertainties and nuances—its myriad and mutable faces—even as we recognize its immense rhetorical power, perceptually and emotionally, to engage and persuade audiences in the near and distant past.

Visualizing Individual and Collective Ethos

How do temporal forces affect visual ethos-building in information design today? What effect has collective ethos had on individual ethos? Of course, individual rhetors still claim ownership over certain genres—résumés, business cards, social media pages, and the like—and those rhetors bear personal (and ethical) responsibility for these information designs. Visualizations of individual ethos have enduring rhetorical value, though they often depend on optimal kairos (as we saw in chapter 1) for their power.

Collective ethos, in contrast, is a shared activity that unfolds over the longer term. Compared to the fast-moving streams of individual ethos, collective visual identities often evolve at a glacial tempo, spanning decades and generations. As a result, logos of organizations typically *evolve* rather than being reinvented from scratch (see Carter 2005), mainly because sudden, sweeping changes from one design to another may confuse or alienate audiences. Like the green patina of a copper roof, visual aging often benefits collective ethos rather than degrades it, leading rhetorical-savvy designers to *tinker* with long-standing designs rather than to replace them.

Counteracting Visual Corrosion

Time also has a corrosive effect that degrades visual ethos. The act of designing the "new" to "look like the old," as Donald Norman puts it in *The Design of Everyday Things* (2013, 159), cuts both ways rhetorically. On the one hand, as Norman points out, the "old" provides emotional reassurance and "comfort" (159), rendering a design more appealing and endearing as we saw with the visual rhetoric of pathos. However, the "old" can also appear corroded and fossilized, damaging ethos. Just as clothing, cars, consumer products, and interiors show their age, so do typefaces, color palettes, pictures, icons, and other graphical elements. Like clothes no longer in style, we relegate aging designs to the back of the closet. Consequently, designers have to walk the median between two temporal lanes—the comfortable, endearing ethos of old designs and the tired, flagging ethos caused by their corrosion—and decide which outweighs the other.

When visual corrosion threatens to diminish ethos, designers can apply several strategies to counteract it:

- Periodic inventories of visual language can identify sources of visual corrosion. Any organization's documents can begin to show their age: dated typography, pictures, clip art, icons, color schemes, graphical elements, and navigational tools. Some usability testing can help spot moribund features that might not be as obvious to legacy users.
- Continually refreshing visual language can sustain (or restore) its ethos. Just like maintaining a car or other equipment, designs need servicing to stay in top form. Documents like procedure manuals, fact sheets, and promotional materials should be periodically updated, especially pictures, color, data designs, and icons. Public-facing digital designs also need constant visual tune-ups of key features: navigational tools, images, animations, and links.
- Deploying modernist forms can slow the aging process. Modernism aimed to create a more universal and durable language free from the corrosive effects of time. Regardless of what we think about their aesthetic or cultural value, modernist forms like geometrical icons, abstract human figures, and high-contrast images can extend a design's useful life by minimizing contextual elements like clothing, background noise, and stylistic features that are more time sensitive.

Implementing any of these ethos-boosting strategies requires rhetorical judgment: The endearing patina of a copper roof in the eyes of one audience might be viewed as visual corrosion to another, increasing or forfeiting trust. The myriad faces of ethos age at different paces, depending on an audience's temporal resilience and appetite for change.

Chapter 6

Making Visual History

Applications for Teaching, Learning, and Discovery

As we've seen in the preceding chapters, time is embedded in virtually every aspect of information design, however transparent or opaque. Throughout the book, I've argued that time constantly shapes visual language—both its design and its interpretation—in the form of visual conventions that have well-worn paths, in rhetorical strategies that have evolved from the classical tradition, and in designs that both exploit and resist the past. Whatever our conscious level of awareness, we can't escape the influence of the past, nor should we *want* to, given its value to the work we do as practitioners, teachers, and scholars. At the conclusion of each of the previous five chapters, I explained some implications for practicing information design. In this chapter I'll explore how time and history can enrich our lives as teachers, learners, and scholars, and I'll provide some guidelines about how to realize these benefits more fully.

Teaching the history of visual communication can occur in a wide variety of forms and offer several benefits—both pragmatic and humanistic as well as a combination of the two. Rhetorically, interrogating the past enables students to understand *why* they make certain design choices and how audiences will interpret them—in short, to see visual language through the eyes of their audiences. In doing so, they can gauge the effects of chronos and kairos, of seeing both the big temporal picture over the long term as well as the more immediate effects of time. Awareness of the past also enables students to integrate conventions into their designs as well as to gauge the effects of novelty on audiences steeped in those conventions. Students can also assess the corrosive effects of time and find ways to counteract them. Furthermore, analyzing sentimental and nostalgic designs and the technology enabling them gives students insights into the values underpinning them (Kurlinkus 2021). Finally, and perhaps most importantly, students can discover past design artifacts, critique them in their historical context, and connect them to contemporary issues in politics, education, immigration, health, and civil rights.

Later in this chapter I'll illustrate opportunities for scholarly inquiry into past design practices, and I'll survey some approaches for analyzing historical artifacts. Opportunities abound for scholarship on the history of visual communication, and I'll point to some avenues, resources, and methods for productively engaging in it.

Rethinking History: Assessing Its Role in the Curriculum

Learning about visual design occurs in many different ways—both in formal education and on our own as we experience or study visual language in the disciplines we inhabit. Students in engineering, science, business, agriculture, and medicine learn the visual languages of their disciplines, both in academic programs and through immersion in the field. Moreover, students in degree programs in business, technical, and professional communication learn about design primarily through their academic coursework (see Brumberger and Northcut 2013; David and Richards 2008), though many also have design experience in the workplace or degrees in cognate areas. My undergraduate degree in architecture, for example, provided a solid foundation in design practice, theory, and history, but like others, I've also learned many things about design and its development from scholars and practitioners in the field.

So how should we interweave design history into academic curricula? There isn't an easy one-size-fits-all answer because teaching and learning design occur in different forums—from general education across disciplines to undergraduate majors and graduate programs. Moreover, design covers a wide spectrum of media, periods, and locations—from art and architectural history in the ancient and modern world, both in the West and around the globe, to visual and pop culture, including film and advertising. In most academic design programs—architecture, interior design, graphic design, and the visual arts—history has remained a staple of the curriculum, with one or more courses devoted to its study.

What role should history play in business, technical, and professional communication programs where only one or two courses are devoted to *all* aspects of visual design—document, data, web, and multimedia? Given such constraints, how can students in these (and related) programs benefit from design history in their coursework as they prepare for their future careers? In programs where a single course covers a gamut of visual modes (typography, illustrations, data design, digital design), history understandably becomes a lower priority. In our textbook *Designing Visual Language: Strategies for Professional Communicators* (2011) Dave Roberts and I include little history, partly because the book covers so much territory already: typography and text design, tables, pictures, charts and graphs, and whole documents, both print and digital. Moreover, undergraduate students preparing to enter the job market want to learn marketable skills and therefore consider historical content less beneficial to their immediate goals.

Nonetheless, in the classroom I show historical examples to deepen students' understanding of design—where it originated, how it evolved, and what rhetorical work it did (and still does) for its audiences. Charles Joseph Minard's chart of Napolean's Russian campaign (Marey 1878, 72), charts from the *Statistical Atlases of the United States* (F. Walker and the U.S. Census Office 1874; Gannett and the U.S. Census Office 1898), and pictures from early engineering books (Ramelli 1588; Besson 1578) or from the *Recueil de Planches* of the French *Encyclopédie* (Diderot and d'Alembert 1762–1772)—the creativity and rhetorical impact of these designs still resonate today. History can also give students insights into practical everyday design elements like leading, paragraphing, serifs, script typefaces, and drop caps, all of which originated in older practices and technology. Students can enhance their understanding of these, and many others, by learning about their past incarnations, the functions they served, and how these relate to their contemporary roles.

Design courses that have a narrower scope offer instructors additional freedom to include historical visuals. For example, when I team taught an undergraduate course on data visualization with two colleagues in statistics, we briefly covered the evolution of data design, including the "golden age" in the second half of the nineteenth century (Funkhouser 1937, 330).[1] When I taught a graduate seminar in data visualization, students read historical scholarship from several fields on the evolution of charts and graphs from the eighteenth century to the present. In a graduate course in visual communication, students read scholarship on the history of graphic design, data design, and rhetoric and professional communication to gain an understanding of the key intellectual, social, and cultural forces that steered design in new directions, particularly the shift to modernism in the early twentieth century.

In short, virtually any course that has a significant design component can integrate historical elements into the curriculum. In the next several sections, I'll provide some specific methods on how students can benefit from doing so, both through analysis of historical artifacts and through reckoning with past design practices and their ongoing influence on the present.

Gauging Change: Observing the Effects of Time on Information Design

Studying historical examples shows students how design constantly changes and evolves. This fluid process allows some design innovations to thrive and achieve long, prosperous lives, while others flounder and never gain much traction. Design is constantly in flux, an iterative process between past, present, and future: Designs we create today look backward and forward, drawing from the past to satisfy audiences in the present while anticipating fresh eyes in the future. As such, designs are never frozen in time. What looks appropriate, engaging, and effective to an audience today might not look that way at all in a decade or a year, or even a month or a week. Looking back, then, helps students look forward.

However stable visual language may appear to students at a given moment, then, it's vulnerable to time and the forces that push and pull on it. To become thoughtful, self-aware designers, students can benefit greatly from understanding these temporal forces and the factors that drive them—both long term (chronos) and short term (kairos). Long term, if students deploy highly decorative elements in their designs—typefaces with slab serifs or page borders with hand-drawn vegetation—they might evoke nostalgia for Victorian culture. Short term, if they deploy a certain color—green, purple, or orange—it might resonate at some times of year better than others, depending on the cultural orientation of their audiences. Students come to understand that design is a moving target, some aspects of which move glacially, others as swiftly as a mountain stream.

Some design stories on the glacial end of the temporal scale take decades or generations to unfold. For example, I frequently show students the mosaic area chart in figure 1.6 from the first *Statistical Atlas* (F. Walker and the U.S. Census Office 1874) as an example of an emerging genre in the U.S. and Europe in the late nineteenth century. Here the chart displays the population of each state using rectangular areas divided into subcategories based on residency (born in or out of the state), race, and immigration status. As I explained in chapter 1, mosaics like this one faced dwindling popularity in the early twentieth century but were recently revived with digital technology, primarily as interactive displays like the stock market treemap in figure 1.7. In this case (and others like it), an incipient form of data design that emerged over a century and a half ago found even greater traction in contemporary digital design.

Observing transformations in information design over time prompts several compelling questions. For example, why have pie charts remained so popular, despite their perceptual deficiencies? How did Helvetica become such a universal typeface? Why do sans serifs dominate the internet? How has popular culture influenced the development of social media icons and emojis? Some of these answers can be complicated but asking them helps students gain a larger perspective on how information design developed—and why—and how that presages future developments.

Students can explore the changing effects of time from many different angles, beginning with tracing the genealogy of designs they encounter every day and how they evolved amid changing conditions. For example, earlier I showed a newsletter (fig. 5.16) that began nearly sixty years ago, with many of its issues archived online. How have the designs of newsletters like this one evolved—their typography, organization, color, pictures, and graphical elements? What effects have technology, aesthetics, and shifts in newsletter conventions had on their designs? What traces of earlier designs still appear?

Students can trace the genealogy of any information design they encounter in their daily lives or that they discover online.[2] Here are a few options for this activity:

- Trace the genealogy of a document design over many years or decades—a company's annual report, a restaurant's menu, a bus/train/subway schedule, a government publication, or an organization's website. How do previous versions differ from the current version? How do they reflect changes in aesthetic taste, genre conventions, or an organization's collective ethos?

- Trace the evolution of a logo for a corporation, university, local business, sports team, campus club, or other organization. What do the design changes reveal about the identity and collective ethos of the organization? About changes in aesthetics, culture, and design conventions? David Carter's *Logos Redesigned: How 200 Companies Successfully Changed Their Image* (2005) can provide an excellent resource for this activity.

Learning Visual Conventions: Applying the Past Today

How does exploring the past benefit someone behind the scenes learning how to design a website, an annual report, or a newsletter? To answer that question, let's return to chapter 1 in which I explored the long-term effects of time (chronos) on information design, especially related to visual conventions. Time nurtures conventions like soil and water, but like most living entities, the amount of time to maturation varies. Initial letters have evolved for centuries, web conventions for a mere few decades. However long conventions need to develop and thrive, they are time capsules that embody sustained practices, and when we deploy them, the residual effects of time cling to them, along with the interpretive habits of their audiences.

Although many students may not be innately drawn to history, they actively participate in it when they learn visual conventions. The choices they make about typefaces, page and screen design, illustrations, and data displays grow out of past practices or reactions to them. Echoes of the past lurk all around in the genres students interpret and deploy, from flyers posted on bulletin boards to invitations, résumés, and PowerPoint presentations. All of these trace their visual language back to prior design practices that discourse communities of designers and audiences understand, share, and come to expect, however formal or loosely connected those communities might be. Some conventions like initial letters, 3D perspective, and Cartesian grids for charting data have long and illustrious lives, while other conventions like home pages, emojis, and online shopping interfaces have much shorter genealogies. In whatever form, conventions provide a stable visual language that students can learn and apply immediately.

The resources for learning visual conventions are readily accessible. Students constantly encounter conventions in a wide variety of genres: brochures, schedules, newsletters, social media, websites, interactive maps—literally any print or digital communication within their purview, either on campus or from the local community. For example, figure 6.1 shows a brochure for a series of concerts sponsored by the Ames Town & Gown Chamber Music Association (2021–2022). The brochure has had a consistent design—cover (shown in fig. 6.1), size and orientation, with

Figure 6.1. Brochure for a chamber music concert series that has used a consistent style and arrangement from year to year (Ames Town & Gown Chamber Music Association 2021–2022). Reprinted with the permission of the Ames Town & Gown Chamber Music Association.

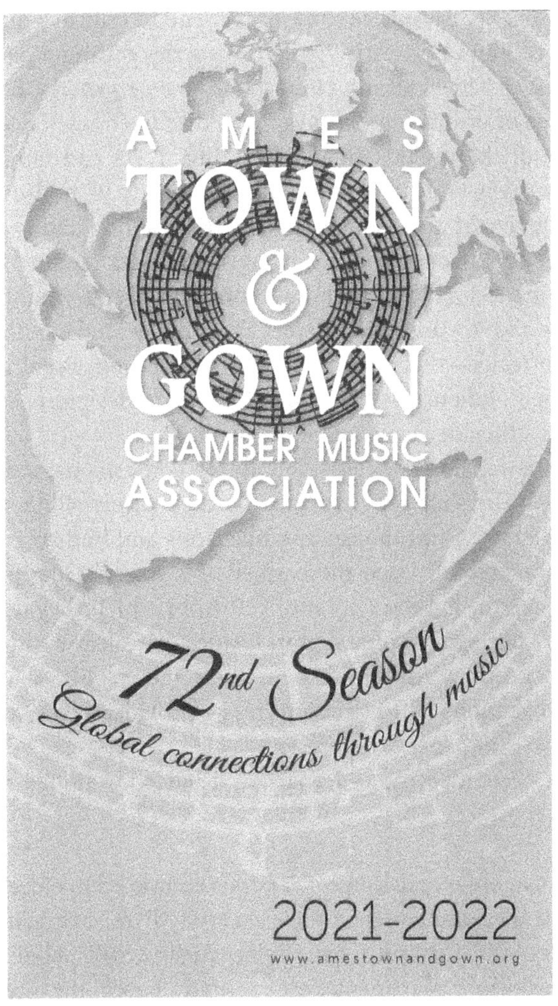

three internal panels, one of which audiences can use to order their tickets. With minor graphic modifications and changes in color, the basic design remained the same from year to year, creating a mini-conventional genre that members of the Ames campus and community came to recognize, creating continuity from one season to the next.

The Town & Gown brochure embodies typical design elements that students encounter as audiences. Of course, students can't be expected to trace the precise visual genealogy of a given convention or genre; still, knowing how they originate can help students decipher its structural features and how they relate to cognate genres. For example, the decision trees in figures 4.8 and 4.9 tell audiences how to respond to critical situations—one for doctors needing to prioritize patient surgeries during a pandemic, the other for school teachers needing to respond to emergencies. Both decision trees are forms of branching displays that I discussed in chapter 4 and are close cousins of the organizational chart (fig. 1.21) and family tree (fig. 2.5). Decision trees typically use an "if this, then that" network of binary decisions to arrive at an outcome, efficiently enabling audiences to work their way through a set of contingencies on subjects ranging from health and job training to finances and how-to projects. Identifying the diagrams in figures 4.8 and 4.9 as decision trees will enable students to situate the designs rhetorically—that they deploy long-standing visual conventions familiar to their audiences—and expand students' design repertoire when they visualize other decision-making situations.

These two decision trees also embody the conventional visual language of modernism. The sans serif typefaces and generous white space in the surgical decision tree and the Isotype-like icons and human figures in the school decision tree all wave the banner of modernist design and echo the WPA posters in figures 3.13 and 5.20 from which they're descended. Students don't need to become design historians to learn and value these insights. Just keeping conventions on their radar—like the genre and modernist conventions of the decision trees—provides a constant reminder of how the past influences the present.

Students can identify and analyze visual conventions in a variety of ways:

- They can observe design conventions embodied in objects and spaces they encounter in their everyday lives: cars, clothing, jewelry, and furniture; hair styles, cuisine, cell phones, and films; dorms, apartments, offices, and classrooms. Students can also identify design conventions that have lost their popularity (Kostelnick and Roberts 2011, 44).

- Before beginning an assignment, they can take an inventory of genre-specific conventions. When students design a newsletter in my visual communication class, I bring several examples to class so students can identify genre conventions: nameplates, narrow margins, multiple columns, headings, boxed inserts, drop caps, images—conventions that have accrued over time and that collectively carry rhetorical weight.
- Students can compare digital and print conventions—for example, conventions of an online shopping website and those in print catalogs, brochures, and flyers. They can list design features that distinguish digital conventions (product images, pricing, ordering, reviews, and other interactive features) from the print ones as well as overlap between the two domains.

Unlike veteran practitioners, who deploy visual conventions as part of their tool belt of reliable resources, students have to *learn* these skills, which offer several benefits as motivation:

- Meet audience expectations. Conventions simplify the audience's work and make them feel confident and comfortable as they enter a document. At the same time, audiences easily spot deviations from conventional practices: hairlines that are too thick, leading too tight or loose, or brand colors off by a shade.
- Streamline the design process. Designs don't have to be created from scratch, which can be exhausting and time-consuming. During the invention process, conventions provide pathways to get started, they build guardrails to stay on track, and they maximize efficiency—benefits that students appreciate.
- Take ownership over conventions. By creating an inventory of conventional elements for a given genre, students can take ownership over conventions rather than rely on software templates. Students can, for example, create a style sheet of conventions for a brochure or website for a campus group or local small business.

Although most students won't consider these activities forays into the past, that's exactly what they're doing—exploiting the power of the past to adapt their communications to the present.

Designing Time: Creating Rhetros and Narratives

In addition to deploying conventions that embody past practices, students can explicitly visualize time. They can do so with rhetro design and narrative design, both of which draw heavily on the past to achieve rhetorical ends in the here and now.

Designing Nostalgia: Rhetro Designs, Technology, and Emotional Appeals

As I explored in chapter 4, design forms have the power to evoke strong emotional appeals, especially ones that trigger memory, nostalgia, and sentimentality. With rhetro design, students deploy older forms to create emotional appeals—ranging from serious and reflective to playful and affectionate. Where can students find the resources for rhetro design? Many are readily available right on their desktops: clip art, fonts, graphical elements, and emojis; public domain images hosted by libraries; and software filters that blur, color, distress, and age images. Because visual language bears the imprint of technology, designers can also use it to create nostalgia.

Here are some activities that can immerse students in *analyzing* rhetro designs:

- Identify designs of clothing, consumer products, and colors that have experienced revivals: tie-dye shirts, industrial style décor, black-and-white films (see Kostelnick and Roberts 2011, 45). Why have these experienced revivals? Consider how culture, age, or gender affect audiences' responses to them.
- Identify contemporary information designs that reveal traces of old technology—script and typewriter typefaces, hand-drawn images, pixelated fonts, or clip art—or that contain other design elements evoking the past—distressed typefaces or images, old photos, or antique or sepia colors. Evaluate

their rhetorical effects, ranging from nostalgia, sentimentality, and curiosity to variations in clarity and ethos.

Here are some suggestions for how students can *create* rhetro designs:

- Create a rhetro design for a new or existing document: a promotional piece (brochure, poster, flyer), an invitation to an event, a restaurant menu, or a program for a concert, play, or dance recital. Choose a historical era as the basis for the design (Victorian, Art Deco, '50s, '60s, '70s, '80s).
- Create a rhetro design that simulates older technologies by using, for example, typewriter fonts, stencil or pixelated images, or photos with sepia tints or scrapbook frames. These design elements evoke nostalgia for a simpler, low-tech era even if students create them with digital technology.
- Create a rhetro design that uses (or simulates) handmade elements such as script text, decorative borders, freehand drawings, or antique or woven watermarks or paper stock. All of these can generate emotional appeals and amplify the epideictic force—for example, in an invitation, concert program, brochure, or award.

Designing Stories: The Universal Allure of Visual Narratives

Students are fascinated with stories and instinctively understand the power of narrative, both verbal and visual. As I explained in chapter 4 with Scott McCloud's survey of pictorial narratives from the past (1993, 10–17), historical examples give us multiple perspectives on visual storytelling, whether it entails making artificial pearls (fig. 4.10), constructing a building, or learning how to dance. Historical examples also humanize visual storytelling: The *Bayeux Tapestry* narrates a conflict (the Norman Conquest) that transpired nearly a thousand years ago, but the gestures, facial expressions, and emotions of its pictures still resonate today. Sequences of pictures enable us to travel through time—both the past (a military conflict) and the present (assembling a table). How we get from one moment in the story to the next requires both technical and rhetorical skill, and historical examples provide insights into how to do this.

Students can *analyze* visual storytelling, both past and present, in many different media and genres. Here are some avenues for doing so:

- Analyze the narrative techniques used in instructions for assembling something (a bike, a tent, a child's toy) or doing a task (cooking a meal, making coffee, planting a tree). What graphical elements—pictures, arrows, lines, color—are used to construct the narrative? What conventions are used (e.g., to see inside of objects, show motion, or reveal details)? How easily can users move from one step to the next?
- Analyze a historical timeline. Search the internet for a timeline that visualizes world history, political events, scientific discoveries, or milestones in literature, art, film, or music. What visual elements (text, graphics, pictures, icons) does the timeline use to create a cohesive narrative? Which people, events, or milestones are chosen for representation, and what does this say about the designer's (and audience's) values?
- Analyze a chart or graph that displays data over time about a contemporary issue. How does the design of the data display—graphical coding, aspect ratio, colors, labeling, title, pictorial elements—influence the audience's interpretation of the narrative?

Here are some suggestions for how students can *design* practical narratives:

- Draft wordless instructions for a familiar task—serving a tennis ball, doing a craft, or operating a piece of equipment. Draw thumbnail sketches that show the temporal movement from step to step, including graphical and textual elements (arrows, borders, numbers). As an alternative to thumbnails, take pictures with a cell phone and organize them into a narrative.
- Design a comic that explains a process or an event. Include at least one character who narrates the story through pictures and words in a series of frames. Use graphic cues to guide the audience from one frame to the next.

- Using historical data from a university source, tell a graphical story of changes in student enrollments (or degrees granted) in various majors or based on gender, ethnicity, race, or nationality. Which data display genre narrates the story most effectively: a line graph, bar chart, radar chart, pie chart, or some other display?

Humanizing Time: Visualizing Social and Cultural Issues

As I've tried to show in previous chapters, visual language provides a panoramic view of the social and cultural factors that shape information design. By studying the humanities, many students have acquired some knowledge of history, literature, rhetoric, philosophy, and art, so they already have some grounding in cultural and intellectual developments that fostered shifts in taste, both ancient and modern. These previous experiences provide a framework for learning about the history of information design, which follows along similar temporal patterns in virtually all of its forms, including typography, illustrations, data design, and color. As a result, historical insights that students gain from studying the humanities enable them to evaluate visual language in a much broader context, well beyond the immediate page or screen.

Studying visual design through the lens of the past also invites exploration of the social, political, and economic forces—empire, immigration, human rights, democracy, industrialism—that are embedded in historical artifacts. The mosaic chart in figure 1.6, for example, prompts discussion about several issues that remain highly relevant in contemporary life: immigration, race, and geographical shifts in population.[3] Abraham Lincoln's slave map (Hergesheimer 1861) served as a moral and geographical compass for the Civil War, illustrating the power of visual language to inspire and motivate leaders in the pursuit of a just and compelling cause. Charles Joseph Menard's *Carte Figurative* of Napoleon's Russian campaign (fig. 2.26) can stimulate discussion about the foolishness of foreign incursions and invasions, a topic poignantly relevant today.

Students can discover a particularly evocative artifact in Charles Booth's poverty maps of London ([1902] 1969, 2025), which during the late Victorian era fostered public awareness of financial disparities among city dwellers. The portion of Booth's map in figure 6.2 visualizes a section of northern London, where the predominantly red buildings show middle-class dwellings, with a few upper class streets in the center and

Figure 6.2. Charles Booth's poverty map of a section of London, 1898–99, from the London School of Economics Library (Booth 1887, 1902–1903, Sheet 3, Northern District). Courtesy of LSE Library.

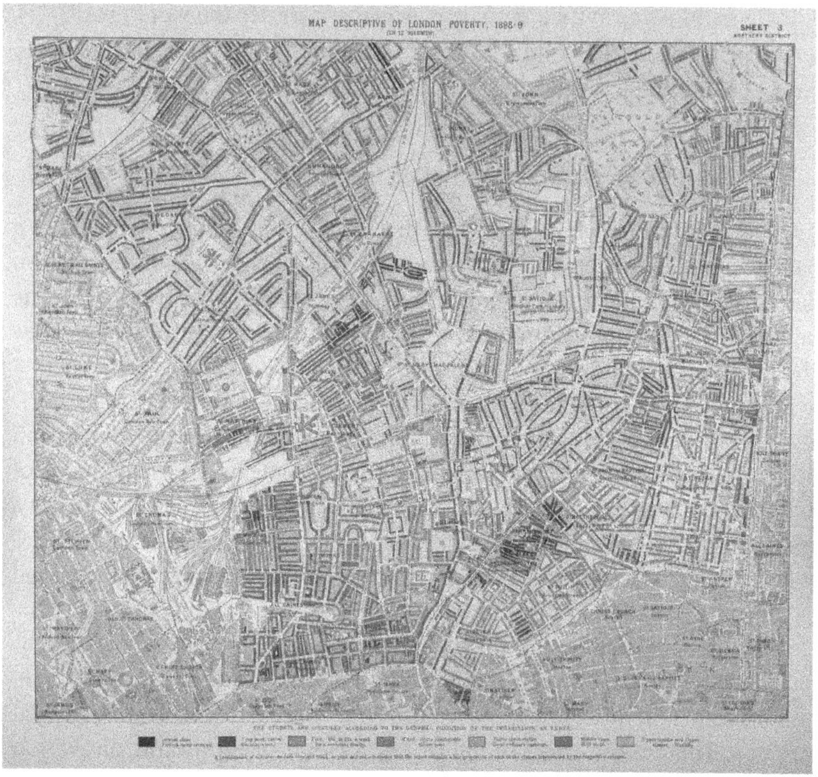

several enclaves of "Poor" or "Very poor" sections in shades of blue. Other maps in the series show an even darker shade representing the "Lowest class," which is also described as "Vicious, semi-criminal." Visualizing poverty as a troubling but "manageable" situation, Booth's charts fostered public engagement with social issues (Kimball 2006, 357–60), eventually paving the way for the contemporary welfare state. Today Booth's maps prompt discussion about the power of data design to stimulate social change, as well as the ethical implications of mapping demographic information about economic status, moral behavior, or crime.

Similar kinds of social and cultural inquiries are prompted by figure 6.3, an area chart designed by W. E. B. Du Bois and his Atlanta University students for the Paris Exposition Universelle of 1900. The rectangles show the business occupations of African American men in the late nineteenth

Figure 6.3. Area chart designed by W. E. B. Du Bois and his Atlanta University students for the Paris Exposition Universelle of 1900 that shows the types of businesses that African American men engaged in (Du Bois 1900, "Negro Business Men"). Courtesy of the Library of Congress, Prints and Photographs Division, Daniel Murray Collection [LC-DIG-ppmsca-33919].

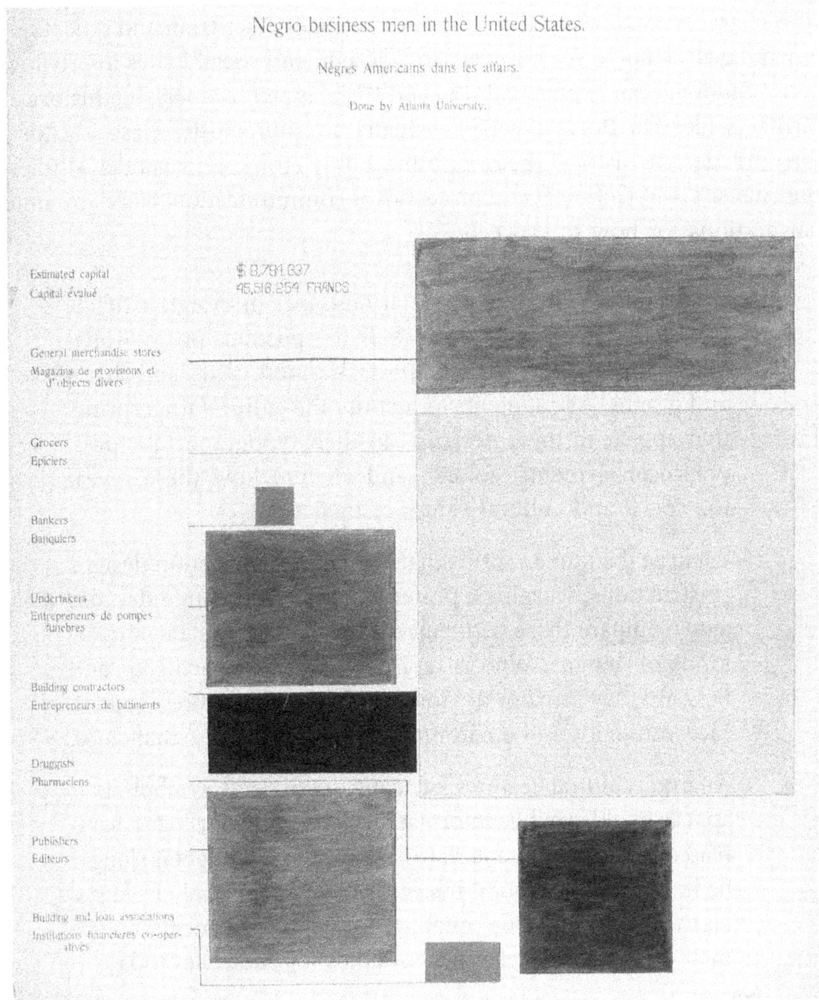

century, with grocers, merchants, and publishers leading the way. From a design standpoint, we can see echoes of the mosaic in figure 1.6, a genre that still had a good deal of currency at the turn of the century, though Du Bois's chart uses rectilinear areas of various shapes and has a wider spectrum of colors. Like its mosaic predecessors, Du Bois's chart prompts exploration of social issues, which might begin with these questions: What can this chart, designed by African Americans, tell us about their lives, then and now? Why were so many African American men engaged in these particular businesses? How might audiences in the U.S., France, and around the world (who attended the Paris Exposition) have interpreted this chart? A single chart like this one stimulates discussion and reflection, enabling students to see how rhetoric, design, and social issues intertwine.

Students can explore cultural and social issues by analyzing historical artifacts like Du Bois's chart in virtually any humanities class—history, art and design history, rhetoric, philosophy, ethnic and gender studies, journalism, and technical and professional communication. Here are some suggestions for how to proceed:

- Analyze the cultural and social messages in design artifacts from the past century and a half—posters, promotional materials, catalogs, annual reports, business cards, and charts and graphs. Ask students to identify the cultural fingerprints that appear in these designs—in their typography, pictures, graphical elements, colors—and analyze how these reveal the social and cultural values of their time.
- Analyze the human forms that appear in information designs (instructions, warnings, posters) from two or three decades ago. Compare these with current images of people in similar kinds of designs. Who is represented in these artifacts, and how are they portrayed? Consider clothing, gestures, gender, race, ethnicity, and environmental or other contextual clues.
- Analyze cultural features of icons, logos, and symbols that appear in older public information signs, company logos, team mascots, or computer interfaces, many of which might no longer be in use. What cultural traces do these aging symbols, logos, and icons reveal? How might these cultural elements reduce their effectiveness, relevance, or ethos for audiences today?

Unearthing Hidden Gems: Telling Bygone Stories with Fresh Eyes

Historical artifacts also offer plentiful opportunities for scholarly inquiry into practices that have previously remained hidden or overlooked.

Innumerable stories remain to be told about how business, technology, health, and science were visualized by people with the skill and tenacity to solve practical problems and communicate their insights and expertise to others with similar needs and interests: stories about pumping and transporting water, mining and processing metals, spinning yarn and weaving textiles, manufacturing furniture and domestic goods, laying out gardens, handwriting documents, curing and preventing diseases, and engaging in leisure activities like dancing, swimming, fishing, and horseback riding. How these activities were communicated visually through text and images, how contemporary theories can inform them, and what bearing they have on contemporary design open up a broad, nearly limitless expanse of inquiry for scholars willing to explore this largely uncharted terrain.

Some of those stories that have already captured the public imagination, and some have had long-term consequences: Leonardo's drawings of technical innovations (collected in Zöllner and Nathan 2003, 570–671); plates from the eighteenth-century French *Encyclopédie* illustrating trades, engineering, and science (fig. 4.10); and documents like the Declaration of Independence and Orville Wright's telegram from Kitty Hawk (fig. 1.5). Historical data displays also tell compelling stories with long-term repercussions—for example, Florence Nightingale's circular charts for hospital reform (1859), John Snow's cholera map (1855), Charles Booth's London poverty maps (fig. 6.2), and W. E. B. Du Bois's charts of the African American experience (figs. 1.11, 4.25, and 6.3). Other stories of consequence like these still remain untold.

Discovering Historical Artifacts

Finding historical artifacts for analysis can be an adventure, full of surprises, challenges, and unexpected discoveries. Artifacts that provide the raw materials (the "data") for scholarly inquiry take many different forms: books, drawings, instructions, letters, posters and promotional materials, charts and graphs, diagrams, maps, photos, and timetables, among others. Many of these items are often labeled as "ephemera" because of their short lives and their obscurity and fragility. Their edges become frayed, coffee gets spilled on them, and they get written on, folded, creased, and stained (see, for example, McCormick's pyramid chart in fig. 4.21). And they get thrown away or lost in folders, languishing for decades between visits by curious hands and eyes. These functional, workaday items weren't created just to be shelved in libraries, displayed under glass, or hung in museums—though many are worthy of these distinctions.

So how can students discover these disparate materials? As archeologists of information designs, students should first look in public places such as libraries, local museums, and digital collections (more on these below). But they can also search in personal spaces: family records, attics, antique malls, and garage sales where artifacts might be readily accessible. In particular, students should look for artifacts that reside on the social, cultural, and economic margins that embed stories, for example, about immigrant experiences, political and religious groups, and communities and companies that once thrived in a given location. Artifacts handled by U.S. presidents, corporate CEOs, and monarchs tell important stories, but those left behind by less famous individuals can be just as compelling.

Locating Print and Digital Resources

Most historical artifacts are printed in books or on paper documents, and many appear in handwritten form. These materials are often confined to limited access libraries (special collections, rare book rooms), museums, or archives. In some cases, only *one* (or a handful) of the original documents might even exist. Overcoming the challenges of hunting for artifacts under these conditions requires planning, patience, tenacity, and luck. Here are a few guidelines for students undertaking these adventures:

- Allow enough time to explore, and be patient. Students might have to search awhile to find what they're looking for in collections that have limited hours and staff. Serendipity plays a major role in hunting for artifacts. I've discovered many historical examples looking for one thing, then finding something even more interesting along the way.
- Overcome the challenges of handling fragile and often rare materials. Many artifacts show their wear because of their age, extensive use, and the fragility of the materials used to produce them. Books on engineering, science, health, gardening, and other practical topics that were printed centuries years ago have greater longevity than "ephemeral" artifacts, but some are so fragile or bound so tightly that they are difficult to scan or photograph. However, none of this should deter students: artifacts stored in libraries, even those in secure vaults, are there to be used.

- Search the internet for artifacts in digital archives. Although print artifacts might be hard to locate, digital archives give students immediate access. Websites like Archive.com and the Hathi Trust, along with the websites of the Library of Congress, Getty Research Institute, and Newberry Library, contain voluminous collections of historical materials, including complete digital copies of early engineering, landscape gardening, and handwriting books.

Choosing Frameworks for Analyses

After identifying potential sources and sifting through print and digital artifacts, what criteria can students use to analyze them? As I've discussed in earlier chapters, most artifacts fit into some larger picture, embodying visual conventions that they share with other designs of their genre and era. In this way, they function like living organisms, revealing a stage of development in which their conventions have evolved. Visual artifacts also embody the values of their era—aesthetic features (relating to a particular style or movement), social preferences or biases, religious or moral codes, and political influences. Below are some guidelines for doing these kinds of analyses:

- Identify visual conventions—typography, picturing styles, graphical elements, page design, and color—that are visible in the artifact. What values do the conventions reflect—disciplinary, aesthetic, social, or ethical—and how do they differ from contemporary values? What technologies were used to deploy the conventions, and what role did they play in designing or printing the artifact? Which of the conventions still appear in information designs today?
- Analyze the rhetorical strategies and impact of artifacts. At a given historical moment artifacts do rhetorical work by persuading audiences to interpret reality a certain way, motivating or inspiring them emotionally, or directing their thinking and actions toward some practical end. However, gauging audience responses to historical artifacts can be challenging. Although literary works, public speeches, paintings, and buildings can often be measured by contemporaneous articles, reviews, or

private letters, practical artifacts leave few interpretive traces. Nonetheless, as I've done in previous chapters, students can scrutinize rhetorical strategies, both classical and modern, which provide multiple lenses for analysis.

- Apply theoretical or ideological frameworks. Semiotics, for example, provides one of many lenses for analysis, treating visual language as systems of signs that engage audiences both perceptually and socially.[4] Scholars have also used feminist, multicultural, and intercultural lenses to interpret artifacts, especially those that envision human forms and narratives. These are just a few among many theories and approaches that students can apply to visual analysis, although depending on the artifact, some yield richer, more insightful analyses than others.

In doing these kinds of analyses, reconstructing the rhetorical and social context can help situate an artifact. For example, the early *Statistical Atlases of the United States* were disseminated in public libraries and universities at a time when most audiences knew little (if anything) about data displays and emerging genres like pie charts, data maps, and mosaic area charts. In the first *Statistical Atlas* (F. Walker and the U.S. Census Office 1874) the data displays were accompanied by explanatory notes, as were the plates (the *Recueil de Planches*, Diderot and d'Alembert 1762–1772) for the *Encyclopédie*, in both cases yielding insights about their audiences. Forward-looking visualizations like those in the *Statistical Atlases* and the *Encyclopédie* had the dual task of *educating* their audiences and *selling* them on their designs—to convince them that seeing the world in new and revealing ways was worth the interpretive effort.

Connecting Artifacts to Narratives Small and Large

Artifacts operate in historical context at both the micro and macro levels. Locally, an artifact is generated (and often interpreted) within an organization or local community, and the story surrounding that artifact can help us interpret it. However, if we step back and widen the lens, we might be able to connect that artifact to larger social, economic, or political narratives in which it was immersed at the time. Here are some suggestions for pursuing these micro- and macro-level stories:

- Connect the artifacts to their local stories. Artifacts generated by an organization (a company, nonprofit, or government entity) reflect its values and the social and economic context at a given historical moment. For example, a menu for a hotel restaurant in the 1920s was immersed in the social, cultural, and financial circumstances of the business and the neighborhood in which it operated. A deep dive into local history can shed light on this artifact.
- Connect the artifacts to larger historical narratives. Artifacts often play a role in (or reflect the influence of) regional, national, political, or disciplinary narratives: for example, the narrative nexus of Abraham Lincoln's slave map (Hergesheimer 1861) and the Civil War, the Emancipation Proclamation, and Lincoln's struggle to retain the Union; Minard's chart of Napoleon's Russian invasion (Marey 1878, 72), nineteenth-century nationalism, and colonial empire building; Florence Nightingale's charts (1859), the Crimean War, women's rights, and the emerging field of modern medicine. Artifacts might also play a role in (or reflect) larger narratives in the history of ideas: for example, how the eighteenth-century *Encyclopédie*'s plates democratized knowledge during the Enlightenment or how the epideictic pictures and charts (discussed in chapter 2) comparing buildings, monuments, mountains, and rivers modeled Darwin's (1859) theories about competition and natural selection.

These are just a few of the many approaches and frameworks that researchers can adopt to analyze historical artifacts, revealing untold stories long hidden from sight. The forums for sharing historical scholarship encompass a wide array of professional organizations, conferences, and journals that span disciplines from business, technical, and professional communication to design studies, statistical graphics, and journalism along with related fields in science, agriculture, education, medicine, and engineering. All of these opportunities await curious and creative scholars willing to excavate artifacts and scrutinize them with lenses that illuminate both past and present.

Conclusion

As I've tried to show, instructors and learners can connect past, present, and future through a variety of media, rhetorical modes, and activities, many of which are summarized in table 6.1. Some of these activities, like

Table 6.1. Pedagogy of Time and Visual Design

Focus	Rhetoric	Artifacts	Activities
Genealogy	ethos	annual report, newsletter, schedule, logo, website, restaurant menu	trace the evolution of a design over several versions
Conventions	genre, audience expectations	typography, arrangement, images, color typical for a given genre	identify and analyze design conventions in a document; compare print and digital conventions
	genre	brochure, newsletter, schedule, résumé, invitation, website, social media, data display	take an inventory of genre-specific conventions before designing a document
Rhetro	pathos, sentimentality, nostalgia, epideictic	any design artifact, from a car to a T-shirt; contemporary documents with rhetro or distressed elements	identify design revivals; analyze rhetro elements and traces of old technology
		images and icons from a previous period; typewriter, script, stencil, and distressed typefaces; sepia colors	create a rhetro design based on a historical era, old technology, or handmade forms

Focus	Rhetoric	Artifacts	Activities
Narrative	storytelling for engagement, clarity, pathos	timelines, instructions, decision trees, flow diagrams, data displays	Analyze narrative techniques in pictorial instructions, a historical timeline, or a data display
	narration of a task, process, or data	instructions, comics, and data displays showing time	design a set of wordless instructions, a comic-style story, or a data display
Humanities	visual language as a touchstone for cultural and social issues, both past and present	Victorian and early modern images, text design, and data displays (Booth, Nightingale, Minard, Du Bois, *Statistical Atlases of the United States*)	analyze cultural and social values embedded in historical artifacts; connect them to contemporary issues
	designs as touchstones for cultural & social issues, both past and present	instructions, warnings, public information signs from previous decades	analyze cultural features of human forms, icons, or logos, and compare with recent versions
Discovery	rhetorical, theoretical, and ideological frameworks for analysis	artifacts from physical and digital archives; artifacts from companies, nonprofits, or government entities	analyze an artifact for conventions and rhetorical strategies; apply a cultural or ideological framework
	micro- and macro-level narratives	artifacts from family records, attics, garage sales, or local museums	interpret an artifact using local community stories and larger historical narratives

analyzing and deploying visual conventions and narratives, are hidden in plain sight; others have to be sought out. Students, of course, have different learning styles, and some of these activities will be more attractive and relatable to some students than others. For example, undergraduate technical communication students want a quicker return on their learning and prefer activities they can apply. Graduate students, especially doctoral students, gravitate toward theory, analysis, and research, with the goal of producing scholarship. Either way, the goal is to nurture a greater awareness of how the past shapes the present and how it can be redirected in the future—an ambition all learners aim to fulfill!

Afterward

Whither Time, Visual Language, and Practical Communication?

If examining the effects of time on visual language can enhance our understanding and practice of information design, as I've tried to show throughout the book, where do we go from here? To establish its identity, a discipline needs to chronicle its past, tell the stories of its origins, trace how its methods and concepts developed, and create a canon of exemplary forms. Exploring the visual history of business, technical, and professional communication can be both exhilarating and daunting, partly because it intersects with the histories of so many other disciplines—rhetoric, art, graphic design, engineering, business, and statistics, among others—which challenges our ability to distinguish it as a discrete entity with its own genealogy. On the other hand, this interdisciplinary synergy provides a breathtaking, panoramic view of information design as well as temporal and intellectual ballast.

Developing a coherent and comprehensive visual history of practical communication poses other challenges and opportunities. Although we can imagine a tentative route toward achieving that goal, the path forward is hindered by the myriad uncertainties of an explorer's map, leaving swaths of blank spaces to speculation and extrapolation. Many more expeditions, however near or far, need to be launched to chart these unexplored regions—more artifacts need to be discovered and more stories about them told and interpreted, one artifact and story at a time. Rather than a linear narrative history, then, we might aim to create a collage of fragmentary maps that are pieced together, microcosms of the whole that together create a coherent and compelling landscape.

To embark on this journey, we have plenty of rich and varied terrain yet to explore. Every day at work, at home, and in public spaces we're deluged with visual language, both print and digital. However, most print and digital communications have fleeting, transitory lives. Time quickly rushes these artifacts downstream, beyond eyesight, before we have a chance to sort through or reflect on them. They're here—and then they vanish. Although the temporal effects on these artifacts may not be as explicit as those that infuse buildings, monuments, or canvases, they are nonetheless just as tangible and ubiquitous and *always* within our purview—if we just know where and how to look.

The goal of this book has been to find ways of doing that looking, of seeing the visual language all around us as both extensions of the past and arbiters of it, each mutually shaping the other while affording glimpses of the future. This temporal reciprocity applies not only to visual artifacts but to other facets of our professional and personal lives in an ongoing interplay that both resonates and beguiles.

Notes

Introduction

1. Our lives constantly bridge the past and the future, with a single life often spanning hundreds of years. For example, members of the Baby Boom generation interacted with people who lived in the late 1800s and who, in turn, interacted with people who lived a half century or more before that, passing on those insights, however dimly recollected. And descendants and students of Baby Boomers (and Gen Xers, Millennials, and Gen Zers) will live into the twenty-second century, rendering Boomers a living link between the distant past and distant future. Most lives hinge past, present, and future in these ways.

2. For a definition of hauntology and its intellectual roots, see Colin Davis (2005) and Mark Fisher (2012). Like other contemporary theoretical constructs, hauntology has been appropriated for many different purposes and by many different fields.

3. Digital "hauntings" also include mimicking the handwriting styles of famous people from the past, as Harald Geisler (2014) did with Sigmund Freud's handwriting.

Chapter 1

1. I wish to thank a former graduate student, Krista Klocke, for first introducing me to the chronos/kairos dichotomy in her own scholarship.

2. Buildings constructed with traditional forms like the *Basilica de la Sagrada Família* in Barcelona, Spain, are visible examples of vestigial design. Designed by Antoni Gaudi (1883–1926), the church has risen from its foundations for over a century, a testimonial to the values and visual conventions of the Gothic style (in the language of Art Nouveau), still living and present long after its origins and revivals.

3. Finding the headwaters of design languages once occupied a good deal of historical scholarship, both in the fine and applied arts; today most design histories still focus on beginnings.

4. Exhausted from copying text morning and afternoon, day after day, week after week, it was no wonder that Bartleby finally declared, "I would prefer not to" (Melville 1853).

5. In Japan, however, a personalized stamp rather than a handwritten signature is often used for official documents ("A Foreigner's Quick Guide" 2025).

6. I've found that handwriting comments on student papers is far more efficient than typing them online, though I can't claim that it's more effective.

7. Richard Polt in *The Typewriter Revolution* (2015) explains how in the twenty-first century the typewriter has become a form of resistance against technology and an icon in popular culture.

8. In the late eighteenth century, pioneer data designer William Playfair had himself experimented with this technique (Tufte 1990, 107), though on a more modest scale and with a slightly different twist.

9. Du Bois also used circular bars for visualizing household possessions (1900, "Assessed"). He and his Atlanta University students prepared numerous charts for the Paris Exposition Universelle (1900) that documented the lives of African Americans after the Civil War. Kevin Van Winkle (2022) analyzes the design, social context, and rhetorical impact of this wide-ranging assemblage of charts, which have been published in a complete collection (*W. E. B. Du Bois's Data Portraits*) by Whitney Battle-Baptiste and Britt Rusert (2018).

10. The technique of showing parts strewn on the ground was used, for example, a few decades earlier in Georgius Agricola's *De Re Metallica* (1556).

11. The emphasis on novelty became one of the hallmarks of modern design, from its inception in the early twentieth century through the design methods movement of the 1960s and 1970s. Christopher Alexander (1964), one of the founders of the design methods movement, claimed that the conventional/traditional ways of generating design forms weren't adequate for solving complex contemporary problems, which demanded more systematic methods for creating effective solutions.

Chapter 2

1. Ironically, a few decades after this chart appeared in the *Recueil de Planches* of the *Encyclopédie*, the French Revolution began eradicating the monarchy, and the epideictic force of this chart switched from praise to blame.

2. Quintilian includes cities and buildings as places and things worthy of praise (1: 477–79; bk. 3, ch. 7, secs. 26–27). In the fine arts in the modern world, place also includes nature—visually sublime and spiritually transcendent—the

epideictic celebration of which appears in Thomas Cole's landscape paintings (Clark, Halloran, and Woodford 1996) and those of Albert Bierstadt of the American West (see also Halloran 1993).

3. These visits often generate other epideictic artifacts for businesses—for example, autographed pictures hung on the walls of hotels, restaurants, and night clubs.

4. Many websites have this epideictic purpose—for example, the Data Visualization Society's (2025) Information Is Beautiful Awards, which recognize excellence in data design.

5. Miles Kimball (2016) insightfully analyzes these natural comparatives in the context of the burgeoning British Empire and the pictorial statistics of Michael Mulhall.

6. For example, Ben Fry (2005–14) uses an interactive slope chart to rank professional baseball teams based on their wins and losses in relation to player salaries.

Chapter 3

1. As Donald Norman (2005) argues in *Emotional Design*, emotion plays a key role in our interactions with all forms of design. However, contemporary assumptions about the benefits of using visual design to stir the emotions weren't always so self-evident. In his introduction to *Visual Thinking*, Rudolf Arnheim (1969) traces the belief in Western classical philosophy, especially Plato, that images can be deceptive, partly based on their ability to arouse the emotions through sensory experience (2–8). See, for example, book 10 of Plato's *Republic* (1901) for his criticism of art.

2. Related to this point, time can be reshaped rhetorically by leaders (and movements) in pivotal moments, as they both look back and look forward to secure the present and to legitimize their power (Lazar 2019, 5, 60–66). Megalomaniac rulers, moreover, often create their own monuments while still in power as way to assert their authority.

3. Some especially painful and important memories that are absent from the public sphere require interventions in order to be recognized (O'Brien and Walwema 2022).

4. Two such examples are William Wordsworth's "Simon Lee" ([1798] 1965, 53–55) and "Resolution and Independence" ([1807] 1965, 165–69).

5. This element of surprise parallels the Chinese approach to gardening, which according to A. O. Lovejoy (1948) had a strong influence on English landscape gardening in the eighteenth century and sowed the seeds for Romantic aesthetics.

6. The rough, irregular shapes of ruins gave them an attractive picturesque character, as William Gilpin ([1748] 1976) observes in the gardens at Stowe

(*Dialogue* 5–6). In *Three Essays: On Picturesque Beauty; On Picturesque Travel; and On Sketching Landscape* (1792) he later formulates a coherent theory of the picturesque, contending "that *roughness* forms the most essential point of difference between the *beautiful*, and the *picturesque*" (6). Other major contributors to picturesque theory included William Chambers (1773), Uvedale Price (1796–1798), and Henry Repton (1806), whose influence on gardening extended well into the nineteenth century (see Hussey 1967), along with René Girardin in France, whose *Essay on Landscape* was published in English in 1783.

7. The obsession with picturing ruins also extended to the future. Joseph Gandy's *Architectural Ruins–A Vision* (1798) illustrates how the Bank of England, newly designed by Sir John Soane, would look as a ruin, presumably after the collapse of the British Empire. A similar theme is visualized by the American artist Thomas Cole in *The Course of Empire* (1833–1836), a series of paintings that visualizes an empire's rise and fall. Gandy's and Cole's images echo Hubert Robert's pictures of ruins (Dubin 2010), including one of the Louvre (1796).

8. Charles Dickens (1854) satirized these dehumanizing effects in his novel *Hard Times*.

9. As S. Michael Halloran (1993) argues, wilderness and picturesque scenery figured prominently in the works of nineteenth-century American writers and had the rhetorical effect of shaping and celebrating a national identity.

10. Children's voices appear in, for example, William Blake's *Songs of Innocence and of Experience* ([1794] 2008), William Wordsworth's "We Are Seven" ([1798] 1965, 49–51), and Elizabeth Barrett Browning's "The Cry of the Children" ([1843] 2009) in which she argues against child labor in British mines and factories.

11. For example, the cover of the McCormick Harvesting Machine Company's 1885 *Annual Catalogue* features a rural scene with children situated amid sheaves of grain.

12. In *Pioneers of Modern Design: From William Morris to Walter Gropius*, Nikolaus Pevsner (2004) makes a compelling and erudite argument that some of the formative ideas and methods of early modernism grew out of the Arts and Crafts movement (14–27). However, modernism eventually took information design in a direction that doesn't look or function like the Victorian designs that preceded it.

13. Researchers like Eva Brumberger (2003, "Awareness," "Persona") have used empirical methods to define the subjective qualities of typefaces. See also Christopher Wyatt and Dànielle DeVoss's collection *Type Matters: The Rhetoricity of Letterforms* (2018), which examines the rhetorical impact of contemporary typography.

Chapter 4

1. Oral storytelling, however, which depended on its transmission from one teller to the next in an unbroken chain, was perpetually vulnerable to its weakest link.

2. Priestley's chart was arranged geographically by continent and region, from 1000 B.C. to the present, with major empires shaded in color and key leaders listed chronologically. An inventive approach to telling the stories of (mostly Western) nations was James Ludlow's *Concentric Chart of History* (1885), a series of cards fastened together to create a fan-like display, an interactive version of which appears on Daniel Rosenberg's website Time OnLine (2016). Other imaginative designs from the nineteenth century include Emma Willard's diagram of the world's empires (1839), Willard's projection of ancient history in architectural space (Sarony 1851), and Elizabeth Peabody's color-coded grids of American history (Peabody 1856; Foster et al. 2017).

3. Some contemporary data designs explicitly serve aesthetic purposes, using digital tools and other media to create two- and three-dimensional works of art (Yau 2013, 74–81).

4. These modernist features are also readily apparent in the elegant, perceptually efficient designs of the current Boston area train schedules for the Massachusetts Bay Transit Authority (2025).

5. Sequential stories can also be created with data displays—for example, DuPont's use of a "chart room" in the early twentieth century to view graphs about the company's performance (Yates 1985).

6. Other techniques for showing short bursts of motion include energy lines in drawings, motion lines in comics, and ghosting and blurring in photography (Jirsa 2024), which show the paths of cars and other moving objects. Before the invention of film, micro-narratives were also created with series of photographs, a technique pioneered by Eadweard Muybridge ([1887] 1955) and Etienne-Jules Marey (1878).

7. Audiences use instructions in different ways, with some reading through them entirely before implementing them. I did that with the COVID-19 instructions because I was concerned about the specific time limits for performing the test.

8. In her expansive study of American pioneer cartography, Susan Schulten (2012) illustrates how the 1874 *Statistical Atlas of the United States* played a key role in mapping the growth of the nation, especially its westward expansion (173–95), and provided inspiration for Frederick Jackson Turner's ideas about the American frontier (185).

9. Hyperbolic data stories still appear today with charts that use area and volume, especially in mass media. These exaggerations also appear in line graphs where the aspect ratio (the plot frame's height vs. width) magnifies the hyperbole, distorting the narrative.

10. Another future projection of a natural disaster appears in a map by the U.S. Geological Survey's Earthquake Hazards Program (2022) showing the likely number of "damaging" earthquakes in regions across the U.S. over the next 10,000 years. The map visualizes the risks of living in certain areas so people can plan accordingly, though some will just shrug their shoulders, the super narrative beyond their comprehension or concern.

11. In *Creating Comics: A Writer's and Artist's Guide and Anthology*, Chris Gavaler and Leigh Ann Beavers (2021) provide an excellent introduction and guide for creating comics, especially for students and novice practitioners.

Chapter 5

1. During the COVID-19 pandemic, the business card lost its currency because of rare in-person contact, temporarily disrupting a pervasive genre that spanned two centuries.

2. Many have advocated replacing Andrew Jackson with abolitionist Harriet Tubman on the twenty-dollar bill, a proposal that's wending its way through the political process. Who appears on currencies reveals a nation's history, values, culture, and conflicts (McKeever 2021).

3. Similar to the patriotic themes pictured on the catalog's cover, in the aftermath of 9/11 many U.S. businesses included flags and patriotic colors on their home pages and promotional materials to show their support for 9/11 victims and to rally American support to defeat terrorism. I wish to thank Amanda Arp, a former graduate student, for drawing my attention to this cover of the *Iowa Seed Company Annual Catalogue* and the online archive that provides access to it.

4. Derek Ross (2008), for example, shows how visitor brochures for the Glen Canyon Dam fostered ethos in response to evolving social, political, and environmental conditions.

5. In his essay "The Nature of Gothic" (*The Stones of Venice,* 1867), John Ruskin argues against the dehumanizing effects of industrialism and celebrates hand labor, which spurs creativity and allows for individual expression.

6. Eighteenth-century epistolary novels like Samuel Richardson's *Clarissa* (1748), Jean-Jacques Rousseau's *Julie, ou La Nouvelle Hélöise* (1761), and Goethe's *Die Leiden des Jungen Werthers* (*The Sorrows of Young Werther*, 1774) were among the most widely read and influential literary productions of their time.

7. As typewritten text proliferated and extended to other genres, several new design conventions emerged (see S. Walker 2003, 2018).

8. An integral part of most letterheads, company logos are also a form of invented ethos that have both conventional and singular features, seen in figure 3.21 (Grand Hotel logo).

9. Annual reports often chart aspirational data about future revenues, production, and carbon emissions. Because the data are plotted graphically, and the numbers for previous years are firm and precise, these graphical extrapolations can engender unwarranted trust.

10. More recently, Claire Lauer and Christopher Sanchez (2023) examined deceptive practices of chart titles and scales with their empirical study using eye tracking.

Chapter 6

1. Michael Friendly and Daniel Denis's website Milestones in the History of Thematic Cartography, Statistical Graphics, and Data Visualization (2001) is an excellent resource for teaching data design that maps its history in an accessible, interactive format.

2. This activity was inspired by Rebecca Burnett, who used an assignment in a graduate professional communication course in which students traced the genealogy of a chosen document.

3. See, for example, Li Li's (2020) insightful analysis of the visualization of Chinese immigrants in the *Statistical Atlases of the United States*.

4. Gunther Kress and Theo van Leeuwen's landmark study *Reading Images: The Grammar of Visual Design* (1996) can provide an excellent guide for this approach.

Works Cited

Adams, Sebastian C. 1881. *Adams' Synchronological Chart of Universal History: Through the Eye to the Mind, A Chronological Chart of Ancient, Modern and Biblical History, Synchronized by Sebastian C. Adams*. 3rd ed. New York, Colby & Co., David Rumsey Map Collection, David Rumsey Map Center, Stanford Libraries, p. 6. www.davidrumsey.com/luna/servlet/detail/RUMSEY~8~1~226106~5505928:Page-6--Adams--Synchronological-Cha?sort=pub_list_no_initialsort%2Cpub_date%2Cpub_list_no%2Cseries_no&qvq=q:Sebastian.

Adriance, Platt and Company. 1897. *Adriance and Buckeye Harvesting Machinery Catalogue*. Poughkeepsie, NY, Adriance, Platt, Iowa State University Library Digital Collections, Lawrence H. Skromme Agricultural Machinery Literature Collection. https://n2t.net/ark:/87292/w9tb0z02r.

Agricola, Georgius. 1556. *De Re Metallica*. Basil, Froben.

Alcorn, Marshall. 1994. "Self-Structure as a Rhetorical Device: Modern Ethos and the Divisiveness of the Self." In *Ethos: New Essays in Rhetorical and Critical Theory*, edited by James S. Baumlin and Tita French Baumlin, 29–62. Southern Methodist University Press.

Alexander, Christopher. 1964. *Notes on the Synthesis of Form*. Harvard University Press.

"A. Lincoln, Attorney and Counselor at Law, Springfield, Illinois." 1864. Business card. Springfield, Illinois, Library of Congress, Rare Book and Special Collections Division. http://hdl.loc.gov/loc.rbc/rbpe.0180070a.

Alliant Energy Corporation. 2020. *2019 Annual Report*. Madison, WI, Alliant Energy Corporation. https://investors.alliantenergy.com/financials/annual-reports/default.aspx. https://s201.q4cdn.com/991130938/files/doc_financials/2019/ar/2019-Annual-Report.pdf.

Alpers, Svetlana. 1983. *The Art of Describing: Dutch Art in the Seventeenth Century*. University of Chicago Press.

Ames Town & Gown Chamber Music Association. 2021–2022. "Ames Town & Gown Chamber Music Association 72nd Season: Global Connections Through Music." Brochure. Ames, IA.

Anker Innovations. 2023. *User Manual: Power Drive+ III Duo.* Model Number A2725. Anker Innovations.

Architects' Small House Service Bureau of Minnesota. 1921. *How to Plan, Finance, and Build Your Own Home.* New Orleans, LA, Southern Pine Association.

Aristotle. 2007. *On Rhetoric: A Theory of Civic Discourse.* Translated by George. A. Kennedy. 2nd ed. Oxford University Press.

Arnheim, Rudolf. 1969. *Visual Thinking.* University of California Press.

Austen, Jane. 1818. *Persuasion.* Rev. ed. Penguin Classics, 2003.

A. W. Coates & Company. N.d. "A. W. Coates and Company Spring Seat Rake Advertisement." Coates Lock Lever, Self Dump & Spring Seat Hay & Grain Raker. Illustrated advertisement. Alliance, OH, A. W. Coates & Company; Buffalo, NY, Gies & Co. Iowa State University Library Digital Collections, Lawrence H. Skromme Agricultural Machinery Literature Collection. https://n2t.net/ark:/87292/w9fj29h5g.

Baggott, Julianna. 2013. "Status Updates: Long Dead Authors on Facebook and Twitter." *New York Times*, January 11. https://www.nytimes.com/2013/01/13/books/review/long-dead-authors-on-facebook-and-twitter.html.

Ballif, Michelle. 2013. "Historiography as Hauntology: Paranormal Investigations into the History of Rhetoric." In *Theorizing Histories of Rhetoric*, edited by Michelle Ballif, 139–53. Southern Illinois University Press.

Banham, Reyner. 1960. *Theory and Design in the First Machine Age.* Architectural Press.

Barrett Browning, Elizabeth. 1843. "The Cry of the Children." In *Elizabeth Barrett Browning: Selected Poems*, edited by Marjorie Stone and Beverly Taylor, 148–55. Broadview Editions, 2009.

Barry, Mike, and Brian Card. 2014. "Visualizing MBTA Data: An Interactive Exploration of Boston's Subway System." mbtaviz.github.io/.

Barton, Ben F., and Marthalee S. Barton. 1988. "Narration and Technical Communication." *Iowa State Journal of Business and Technical Communication* 2, no. 1 (January): 36–48. https://doi.org/10.1177/105065198800200103.

Barton, Ben F., and Marthalee S. Barton. 1993. "Ideology and the Map: Toward a Postmodern Visual Design Practice." In *Professional Communication: The Social Perspective*, edited by Nancy Roundy Blyler and Charlotte Thralls, 49–78. Sage.

Barton, Ben F., and Marthalee S. Barton. 1993. "Modes of Power in Technical and Professional Visuals." *Journal of Business and Technical Communication* 7, no. 1 (January): 138–62. https://doi.org/10.1177/1050651993007001007.

Battle-Baptiste, Whitney, and Britt Rusert, eds. 2018. *W. E. B. Du Bois's Data Portraits: Visualizing Black America: The Color Line at the Turn of the Twen-*

tieth Century. W. E. B. Du Bois Center at the University of Massachusetts at Amherst; Princeton Architectural Press.

Baumlin, James S. 2001. "Ēthos." In *Encyclopedia of Rhetoric*, edited by Thomas O. Sloane, 263–77. Oxford University Press. https://doi.org/10.1093/acref/9780195125955.001.0001.

Baumlin, James S., and Craig A. Meyer. 2022. "Positioning Ethos in/for the Twenty-First Century: An Introduction to Histories of Ethos." In *Histories of Ethos: World Perspectives on Rhetoric*, edited by James S. Baumlin and Craig A. Meyer, 1–26. MDPI. https://www.mdpi.com/2076-0787/7/3/78.

Bayeux Tapestry. Eleventh century. Bayeux Museum, Bayeux, France.

Bell, Alexander Graham. 1896. "Letter from Alexander Graham Bell to Mabel Hubbard Bell." March 17. Library of Congress, Manuscript Division. http://hdl.loc.gov/loc.mss/magbell.03910104.

Bell, Michael. 2000. *Sentimentalism, Ethics and the Culture of Feeling*. Palgrave.

Bertin, Jacques. 1981. *Graphics and Graphic Information-Processing*. Translated by William J. Berg and Paul Scott. Walter De Gruyter.

Besson, Jacques, and François Béroald. 1578. *Theatre des Instrumens Mathematiques et Mechaniques*. Lyons, Vincent.

Best, Will. 2020. "Social Media and Modernist Authority: The Hauntology of Facebook." *Persona Studies* 6, no. 1 (December): 83–99. https://doi.org/10.21153/psj2020vol6no1art872.

Black, Alison, Paul Luna, Ole Lund, and Sue Walker, eds. 2017. *Information Design: Research and Practice*. Routledge.

Blake, William. 1794. *Songs of Innocence and of Experience*, edited by Robert N. Essick. Henry E. Huntington Library and Art Gallery, 2008.

Böckler, Georg Andreas. 1673. *Theatrum Machinarum Novum*. Nuremberg, Fursten.

Booker, Peter Jeffrey. 1979. *A History of Engineering Drawing*. 1963. Northgate.

Boorstin, Daniel J. 1973. *The Americans: The Democratic Experience*. Random House.

Booth, Charles. 1902. *Life and Labour of the People in London*. First Series: Poverty. 5 vols. A. M. Kelly, 1969. Rept., London School of Economics & Political Science. booth.lse.ac.uk/.

Booth, Charles. 1887, 1902–1903. *Maps Descriptive of London Poverty: Printed 1887, 1902–1903*. London School of Economics and Political Science; London School of Economics Library, 2025. https://archives.lse.ac.uk/records/BOOTH/E/1.

Boston and Worcester Railroad. 1849. "*Time Table and General Rules to Take Effect October 8*." October 4. Library of Congress, Rare Book and Special Collections Division. http://hdl.loc.gov/loc.rbc/rbpe.0590220d.

Botticelli, Sandro. ca. 1485. *The Birth of Venus*. Tempera on canvas. Uffizi Galleries, Florence, Italy.

Boym, Svetlana. 2001. *The Future of Nostalgia*. Basic Books.

Brasseur, Lee E. 2003. *Visualizing Technical Information: A Cultural Critique*. Baywood Publishing.

Brasseur, Lee. 2005. "Florence Nightingale's Visual Rhetoric in the Rose Diagrams." *Technical Communication Quarterly* 14, no. 2 (Spring): 161–82. https://www.tandfonline.com/doi/abs/10.1207/s15427625tcq1402_3.

Braun, Georg, Franz Hogenberg, et al. 1612–1618. *Civitates Orbis Terrarum*. Vols. 1–2. Cologne. 6 vols. Library of Congress, Geography and Map Division. https://www.loc.gov/item/2008627031/.

Brinton, Willard C. 1914. *Graphic Methods for Presenting Facts*. Engineering Magazine Company.

Brockmann, R. John. 2002. *Exploding Steamboats, Senate Debates, and Technical Reports: The Convergence of Technology, Politics, and Rhetoric in the Steamboat Bill of 1838*. Baywood Publishing.

Brumberger, Eva R. 2003. "The Rhetoric of Typography: The Awareness and Impact of Typeface Appropriateness." *Technical Communication* 50, no. 2 (May): 224–31. http://www.jstor.org/stable/43089123.

Brumberger, Eva R. 2003. "The Rhetoric of Typography: The Persona of Typeface and Text." *Technical Communication* 50, no. 2 (May): 206–23. http://www.jstor.org/stable/43089122.

Brumberger, Eva R., and Kathryn M. Northcut, eds. 2013. *Designing Texts: Teaching Visual Communication*. Baywood Publishing.

Bruner, Jerome. 1990. *Acts of Meaning*. Harvard University Press.

Buchanan, Richard. 1985. "Declaration by Design: Rhetoric, Argument, and Demonstration in Design Practice." *Design Issues* 2, no. 1 (Spring): 4–22. https://doi.org/10.2307/1511524.

Burke, Edmund. 1759. *A Philosophical Enquiry into the Origin of Our Ideas of the Sublime and Beautiful*, edited by James T. Boulton. 2nd ed. Routledge and Kegan Paul, 1967.

Burke, Kenneth. 1950. *A Rhetoric of Motives*. Prentice-Hall.

Campbell, George. 1776. *The Philosophy of Rhetoric*. 2 vols. London, Strahan and Cadell. HathiTrust. https://hdl.handle.net/2027/njp.32101061812242; https://hdl.handle.net/2027/njp.32101061812259.

Carter, David E. 2005. *Logos Redesigned: How 200 Companies Successfully Changed Their Image*. HarperCollins.

Caxton, William. 1490. *The Mirrour of the World*. London.

Center for Industrial Research and Service (CIRAS). 2020. *CIRAS News* 55, no. 4 (Summer). www.ciras.iastate.edu/ciras-newsletters/ciras-news-issue-list/. https://www.ciras.iastate.edu/files/CIRASNews/2020Summer.pdf.

Center for Industrial Research and Service (CIRAS). 2025. *CIRAS e-news*. CIRAS. https://newswire.ciras.iastate.edu/category/enews/.

Chambers, Ephraim. 1728. *Cyclopaedia, Or, An Universal Dictionary of Arts and Sciences*. 2 vols. London.

Chambers, William. 1773. *Dissertation on Oriental Gardening*. 2nd ed. W. Griffin.

Chandler, Daniel, and Rod Munday. 2020. "Grand Narratives (Metanarratives, Master Narratives)." In *A Dictionary of Media and Communication*, 3rd ed., edited by Daniel Chandler and Rod Munday. Oxford University Press. https://doi.org/10.1093/acref/9780198841838.001.0001.

Chase, J. Richard. 1961. "The Classical Conception of Epideictic." *Quarterly Journal of Speech* 47, no. 3 (October): 293–300. https://doi.org/10.1080/00335636109382490.

Chemin, Anne. 2014. "Handwriting vs. Typing: Is the Pen Still Mightier Than the Keyboard?" *Guardian Weekly*, December 16. www.theguardian.com/science/2014/dec/16/cognitive-benefits-handwriting-decline-typing.

Clarion Safety Systems. 2025. "Hand Entanglement Rotating Gears Label." Warning sign. Milford, PA, Clarion Safety Systems. www.clarionsafety.com.

Clark, Gregory. 2004. *Rhetorical Landscapes in America: Variations on a Theme from Kenneth Burke*. University of South Carolina Press.

Clark, Gregory, S. Michael Halloran, and Allison Woodford. 1996. "Thomas Cole's Vision of 'Nature' and the Conquest Theme in American Culture." In *Green Culture: Environmental Rhetoric in Contemporary America*, edited by Carl G. Herndl and Stuart C. Brown, 261–80. University of Wisconsin Press.

Cleveland, William S., and Robert McGill. 1984. "Graphic Perception: Theory, Experimentation, and Application to the Development of Graphical Methods." *Journal of the American Statistical Association* 79, no. 387 (September): 531–54. https://doi.org/10.2307/2288400.

Cocker, Edward. 1659. *The Pen's Triumph, Being a Copy-Book, Containing Variety of Examples of All Hands Practiced in This Nation According to the Present Mode; Adorned with Incomparable Knots and Flourishes*. London. Newberry Library Digital Collections. https://collections.newberry.org/asset-management/2KXJ8Z812GPMK?&WS=SearchResults&Flat=FP.

Cocker, Edward. 1660. *The Pen's Transcendency: or Fair Writings Store-house: Furnished with Examples of All the Curious Hands Practiced in England, and the Nations Adjacent: Adorned with Variety of Admirable Knots and Flourishes, Invented for the Recreation of Practitioners*. London. Newberry Library Digital Collections. https://collections.newberry.org/asset-management/2KXJ8Z812G50O?&WS=SearchResults&Flat=FP.

Cocker, Edward. 1685. *Arts Glory: or, The Pen-Mans Treasury: Containing Various Examples of Secretary, Text, Roman, and Italian Hands: Adorned with Many Curious Knots and Flourishes, to Render Them Pleasant as well as Profitable*. London. Rare Book & Manuscript Library, University of Illinois at Urbana-Champaign. HathiTrust. https://hdl.handle.net/2027/uiuc.99525619412205899.

Cole, Thomas. 1833–1836. *The Course of Empire*. New-York Historical Society.

Colton, G. W. 1874. *Mountains and Rivers of the World*. Map and drawing. In *Colton's General Atlas, Containing One Hundred and Eighty Steel Plate Maps and Plans, On One Hundred and Nineteen Imperial Folio Sheets, Drawn By G. Woolworth Colton*, nos. 4–5. New York, G. W. and C. B. Colton. David Rumsey Map Collection, David Rumsey Map Center,

Stanford Libraries. https://www.davidrumsey.com/luna/servlet/detail/RUMSEY~8~1~209976~5003812:Mountains-and-Rivers-of-the-World-?sort=pub_list_no_initialsort%2Cpub_date%2Cpub_date&mi=96&trs=101&qvq=q:comparative%20mountain%20map;sort:pub_list_no_initialsort%2Cpub_date%2Cpub_date;lc:RUMSEY~8~1.

Consigny, Scott. 1992. "Georgias's Use of the Epideictic." *Philosophy & Rhetoric* 25, no. 3: 281–97. https://www.jstor.org/stable/40237726.

Cook, Robert, and Howard Wainer. 2016. "Joseph Fletcher, Thematic Maps, Slavery, and the Worst Places to Live in the U.K. and the U.S." In *Visible Numbers: Essays on the History of Statistical Graphics*, edited by Miles A. Kimball and Charles Kostelnick, 83–105. Ashgate.

CrimeMapping.com. 2024. CrimeMapping.com: Helping You Build a Safer Community. Accessed December 23, 2024. https://www.crimemapping.com/.

Crofutt, George A. 1873. *American Progress* or *Westward the Course of Destiny*. After a John Gast painting. New York, George A. Crofutt. Library of Congress, Prints and Photographs Division. http://hdl.loc.gov/loc.pnp/cph.3b49232.

Crowley, Sharon. 1993. "Modern Rhetoric and Memory." In *Rhetorical Memory and Delivery: Classical Concepts for Contemporary Composition and Communication*, edited by John Frederick Reynolds, 31–44. Lawrence Erlbaum.

Crowley, Sharon, and Debra Hawhee. 1999. *Ancient Rhetorics for Contemporary Students*. 2nd ed. Allyn and Bacon.

Currier, Nathaniel, and James Merritt Ives. N.d. *Harvesting: The Last Load*. Lithograph. New York, Currier & Ives. Library of Congress, Prints and Photographs Division. http://hdl.loc.gov/loc.pnp/pga.09065.

Currier, Nathaniel, and James Merritt Ives. 1874. *Home Sweet Home*. Lithograph. New York, Currier & Ives. Library of Congress, Prints and Photographs Division. http://hdl.loc.gov/loc.pnp/pga.09100.

CyRide. 2018–2025. "Estimated Arrival Times." CyRide. Accessed March 17, 2025. https://www.mycyride.com/arrivals.

Darwin, Charles. 1859. *On the Origins of Species by Means of Natural Selection*. London, John Murray.

DataRoyals. 2025. Rank charts. https://www.youtube.com/@DataRoyals_.

Data Visualization Society. 2025. "Information Is Beautiful Awards." www.informationisbeautifulawards.com/.

David, Carol. 2001. "Investitures of Power: Portraits of Professional Women." *Technical Communication Quarterly* 10, no. 1: 5–29. https://doi.org/10.1207/s15427625tcq1001_1.

David, Carol, and Ann R. Richards, eds. 2008. *Writing the Visual: A Practical Guide for Teachers of Composition and Communication*. Parlor Press.

Da Vinci, Leonardo. 1503–1519. *Mona Lisa*. Oil on wood. Musée du Louvre, Paris, France.

Davis, Colin. 2005. "Hauntology, Spectres and Phantoms." *French Studies* 59, no. 3 (July): 373–79. https://doi.org/10.1093/fs/kni143.

Davis Platform Binder Company. 1889. *Davis Platform Binder Company Catalogue.* Cleveland, OH, Davis Platform Binder Company; Buffalo, NY, Geis & Co. Iowa State University Library Digital Collections, Lawrence H. Skromme Agricultural Machinery Literature Collection. https://n2t.net/ark:/87292/w95x25h54.

Davis, Stephen Boyd. 2017. "Early Visualizations of Historical Time." In *Information Design: Research and Practice*, edited by Alison Black, Paul Luna, Ole Lund, and Sue Walker, 3–22. Routledge.

De Jong, Mary G. 2013. Introduction to *Sentimentalism in Nineteenth-Century America: Literary and Cultural Practices*, edited by Mary G. De Jong and Paula Bernat Bennett, 1–12. Farleigh Dickinson University Press.

De Jong, Mary G., and Paula Bernat Bennett, eds. 2013. *Sentimentalism in Nineteenth-Century America: Literary and Cultural Practices.* Farleigh Dickinson University Press.

Den Boer, Pim. 2008. "Loci memoriae—Lieux de Mémoire." In *Cultural Memory Studies: An International and Interdisciplinary Handbook*, edited by Astrid Erll and Ansgar Nünning, 19–25. Walter De Gruyter.

Dickens, Charles. 1854. *Hard Times: For These Times.* London, Bradbury & Evans.

Dickinson, Greg, Carole Blair, and Brian L. Ott, eds. 2010. *Places of Public Memory: The Rhetoric of Museums and Memorials.* University of Alabama Press.

Diderot, Denis and Jean Le Rond d'Alembert, eds. 1751–1765. *Encyclopédie, ou Dictionnaire Raisonné des Sciences, des Arts et des Métiers.* 17 vols. Paris, Briasson.

Diderot, Denis, and Jean Le Rond d'Alembert, eds. 1762–1772. *Recueil de Planches, sur les Sciences, les Arts Libéraux, et les Arts Méchaniques.* 11 vols. Paris, Briasson.

Donne, John. 1663. "The Sun Rising." In *John Donne's Poetry: Authoritative Texts, Criticism*, edited by Donald R. Dickson, 74–75. W. W. Norton, 2007.

Douglas, George H. 1985. "Business Writing in America in the Nineteenth Century." In *Studies in the History of Business Writing*, edited by George H. Douglas and Herbert W. Hildebrandt, 125–33. Association for Business Communication.

Doty, Roy. 1996. *The Family Handyman: Wordless Workshop.* Reader's Digest.

Dragga, Sam, and Dan Voss. 2001. "Cruel Pies: The Inhumanity of Technical Illustrations." *Technical Communication* 48, no. 3 (August): 265–74. https://www.jstor.org/stable/43090430.

D. S. Morgan and Company. 1887. *D. S. Morgan and Company 1887 Catalog.* Brockport, NY, D. S. Morgan and Company. Iowa State University Library Digital Collections, Lawrence H. Skromme Agricultural Machinery Literature Collection. https://n2t.net/ark:/87292/w9mw28j9f.

Dubin, Nina L. 2010. *Futures & Ruins: Eighteenth-Century Paris and the Art of Hubert Robert.* Getty Research Institute.

Du Bois, W. E. B. ca. 1900. "Assessed Value of Household and Kitchen Furniture Owned by Georgia Negroes." In *The Georgia Negro.* Library of Congress. http://hdl.loc.gov/loc.pnp/ppmsca.33887.

Du Bois, W. E. B. ca. 1900. "City and Rural Population, 1890." In *The Georgia Negro*. Library of Congress, Prints and Photographs Division. http://hdl.loc.gov/loc.pnp/ppmsca.33873.

Du Bois, W. E. B. ca. 1900. "Number of Negro Students Taking the Various Courses of Study Offered in Georgia Schools." In *The Georgia Negro*. Library of Congress. http://hdl.loc.gov/loc.pnp/ppmsca.33879.

Du Bois, W. E. B., and Atlanta University. ca. 1900. "Negro Business Men in the United States." In *A Series of Statistical Charts Illustrating the Condition of the Descendants of Former African Slaves Now in Residence in the United States of America*. Library of Congress, Prints and Photographs Division. http://hdl.loc.gov/loc.pnp/ppmsca.33919.

Du Bois, W. E. B., and Atlanta University. ca. 1900. "Proportion of Total Negro Children of School Age Who are Enrolled in the Public Schools." In *A Series of Statistical Charts Illustrating the Condition of the Descendants of Former African Slaves Now in Residence in the United States of America*. Library of Congress, Prints and Photographs Division. http://hdl.loc.gov/loc.pnp/ppmsca.33911.

Du Bois, W. E. B., and Atlanta University. ca. 1900. "Religion of American Negroes." In *A Series of Statistical Charts Illustrating the Condition of the Descendants of Former African Slaves Now in Residence in the United States of America*, ca. 1900. Library of Congress, Prints and Photographs Division. http://hdl.loc.gov/loc.pnp/ppmsca.33924.

Ehses, Hanno H. J. 1984. "Representing Macbeth: A Case Study in Visual Rhetoric." *Design Issues* 1, no. 1 (Spring): 53–63. https://doi.org/10.2307/1511543.

Eisner, Will. 2008. *Comics and Sequential Art: Principles and Practices from the Legendary Cartoonist*. 1985. Rev. ed. W. W. Norton.

Eldredge, Sandy, and Bob Bieck. 2010. "Glad You Asked: Ice Ages—What Are They and What Causes Them?" *Utah Geological Survey Notes* 42, no. 3 (September): 7–8. Utah Geological Survey. https://ugspub.nr.utah.gov/publications/survey_notes/snt42-3.pdf.

Empire Sewing Machine Company. 1870. "Empire Sewing Machine Co., New York." Lithograph. New York, Henry Seibert & Bros. Library of Congress, Prints and Photographs Division. hdl.loc.gov/loc.pnp/pga.02736.

Erll, Astrid. 2008. "Cultural Memory Studies: An Introduction." In *Cultural Memory Studies: An International and Interdisciplinary Handbook*, edited by Astrid Erll and Ansgar Nünning, 1–15. Walter De Gruyter.

Erll, Astrid, and Ansgar Nünning, eds. 2008. *Cultural Memory Studies: An International and Interdisciplinary Handbook*. Walter De Gruyter.

Esser, Hermann. 1877. *Draughtsman's Alphabets: A Series of Plain and Ornamental Alphabets Designed Especially for Engineers, Architects, Draughtsmen, Engravers, Painters, Etc.* 20th ed. New York, Keuffel and Esser.

Fairbank, Alfred J. 1968. *A Book of Scripts*. Rev. ed. Penguin.

Fanning, Shannon. 2018. "Transformations in Genre: An Examination of Visualizations of the Zika Virus." PhD diss., Iowa State University.

Ferguson, Eugene S. 1993. *Engineering and the Mind's Eye*. MIT Press.

Fisher, Mark. 2012. "What Is Hauntology?" *Film Quarterly* 66, no. 1 (Fall): 16–24. https://doi.org/10.1525/fq.2012.66.1.16.

"A Foreigner's Quick Guide to Japanese Hanko Stamp." 2025. LearnJapanese123.com, edited by Kazue. learnjapanese123.com/a-foreigners-quick-guide-to-japanese-hanko-stamp/.

Forman, Janis. 2013. *Storytelling in Business: The Authentic and Fluent Organization*. Stanford University Press.

Foss, Sonja K. 2004. "Framing the Study of Visual Rhetoric: Toward a Transformation of Rhetorical Theory." In *Defining Visual Rhetorics*, edited by Charles A. Hill and Marguerite Helmers, 303–13. Lawrence Erlbaum.

Foster, Caroline, Adam Hayward, Svyatoslav Kucheryavykh, Angela Vujic, Maninder Japra, Shivani Negi, and Lauren Klein. 2017. "The Shape of History: Reimagining Nineteenth-Century Data Visualization." Digital Humanities Conference 2017. https://dh2017.adho.org/abstracts/146/146.pdf.

Fox, Celina. 2009. *The Arts of Industry in the Age of Enlightenment*. Yale University Press.

Friendly, Michael. 2002. "A Brief History of the Mosaic Display." *Journal of Computational and Graphical Statistics* 11, no. 1: 89–107. https://doi.org/10.1198/106186002317375631.

Friendly, Michael. 2008. "The Golden Age of Statistical Graphics." *Statistical Science* 23, no. 4 (November): 502–35. https://doi.org/10.1214/08-STS268.

Friendly, Michael, and Daniel J. Denis. 2001. Milestones in the History of Thematic Cartography, Statistical Graphics, and Data Visualization. www.datavis.ca/milestones/.

Friendly, Michael, and Howard Wainer. 2021. *A History of Data Visualization and Graphic Communication*. Harvard University Press.

Fry, Ben. 2005–2014. "Salary vs. Performance: What Baseball Teams Are Spending Their Money Well, and How Does It Change over the Course of the Season?" Interactive chart. https://benfry.com/salaryper/.

Funkhouser, H. Gray. 1937. "Historical Development of the Graphical Representation of Statistical Data." *Osiris* 3: 269–404. http://www.jstor.org/stable/301591.

Gandy, Joseph Michael. 1798. *Architectural Ruins, A Vision*. Pen and watercolor on paper. Sir John Soane's Museum, London.

Gannett, Henry. 1903. *Statistical Atlas: Prepared under the Supervision of Henry Gannett, Geographer of the Twelfth Census*. U.S. Census Office.

Gannett, Henry, and the U.S. Census Office. 1898. *Statistical Atlas of the United States, Based upon Results of the Eleventh Census*. U.S. Government Printing Office. Library of Congress, Geography and Map Division. http://hdl.loc.gov/loc.gmd/g3701gm.gct00010.

Gaudi, Antoni. 1883–1926 (unfinished). *Basilíca de la Sagrada Família*. Barcelona, Spain.

Gavaler, Chris, and Leigh Ann Beavers. 2021. *Creating Comics: A Writer's and Artist's Guide and Anthology*. Bloomsbury Academic.

Geisler, Harald. 2014. "Hands on the Sigmund Freud Typeface: Making a Font for Your Shrink." *Smashing Magazine*, June 2. www.smashingmagazine.com/2014/06/hands-on-sigmund-freud-typeface-making-fonts/.

General Motors Corporation. 1939. *Thirtieth Annual Report of General Motors Corporation, Year Ended December 31, 1938*. Wilmington, DE, General Motors Corporation. https://digital.library.mcgill.ca/images/hrcorpreports/pdfs/G/General_Motors_Corporation_1938.pdf.

General Surgeons Australia. 2020. "Impact of COVID-19: General Surgery: Decision Tree for Surgeons." https://generalsurgeons.com.au/wp-content/uploads/2025/03/Covid-19_Response-for-General-Surgeons_23-APRIL-2020.pdf.

Gigante, Maria E. 2018. *Introducing Science Through Images: Cases of Visual Popularization*. University of South Carolina Press.

Gilpin, William. 1748. *A Dialogue upon the Gardens of the Right Honourable the Lord Viscount Cobham, at Stow in Buckinghamshire*. Introduction by John Dixon Hunt. Augustan Reprint Society no. 176, William Andrews Clark Memorial Library. University of California Press, 1976.

Gilpin, William. 1792. *Three Essays: On Picturesque Beauty; on Picturesque Travel; and on Sketching Landscape*. London, R. Blamire.

Girardin, René Louis. 1783. *An Essay on Landscape; Or, on the Means of Improving and Embellishing the Country Round Our Habitations*. London, J. Dodsley.

Goethe, Johann Wolfgang. 1774. *Die Leiden des Jungen Werthers*. Leipzig.

Goldenberg, Sally. 2012. "New York City Agencies Still Using Typewriters." *New York Post*, January 30. nypost.com/2012/01/30/new-york-city-agencies-still-using-typewriters/.

Goodchild, H. 1885. *Elevations of Seventy-Six of the Most Remarkable Buildings of Different Countries, Drawn to a Uniform Scale of 100 ft. to 1 and ½ Inches, Collected from the Most Authentic Sources*. Cincinnati, OH, Strobridge Lith. Co. Library of Congress, Prints and Photographs Division. www.loc.gov/item/2003656976/.

Grand Hotel. 2025. "Grand Hotel Logo." Mackinac Island, MI, Grand Hotel. www.grandhotel.com.

Green, Claire. 2016. "Calling Cards and Visiting Cards: A Brief History." Hoban Press, September. https://hobancards.com/blogs/thoughts-and-curiosities/calling-cards-and-visiting-cards-brief-history.

Greenbaum, Julie. 2019. "The Top 10 Typewriter Users by Industry." Typewriters.com, October 10. typewriters.com/blogs/blogthe-top-10-industries-that-still-use-typewriters/.

Gross, Alan G., and Joseph E. Harmon. 2014. *Science from Sight to Insight: How Scientists Illustrate Meaning*. University of Chicago Press.

Hale, Robert. 1865. "Time Table for the Special Train, Conveying the Funeral Cortege with the Remains of the Late President from Chicago to Springfield." Chicago Alton Railroad Company, Illinois, May 2. Library of Congress, Rare Book and Special Collections Division. https://www.loc.gov/resource/lprbscsm.scsm0258/.

Halloran, S. Michael. 1982. "Aristotle's Concept of Ethos, or if Not His Somebody Else's." *Rhetoric Review* 1, no. 1 (September): 58–63. https://doi.org/10.1080/07350198209359037.

Halloran, S. Michael. 1993. "The Rhetoric of Picturesque Scenery: A Nineteenth-Century Epideictic." In *Oratorical Culture in Nineteenth-Century America: Transformations in the Theory and Practice of Rhetoric*, edited by Gregory Clark and S. Michael Halloran, 226–50. Southern Illinois University Press.

Handwrytten. 2025. "Your Words. In Pen + Ink." Handwrytten. www.handwrytten.com/.

Hardwig, Florian. 2025. "Quaint/Desdemona." Fonts in Use. fontsinuse.com/typefaces/12494/quaint-desdemona.

Harris, Robert L. 1996. *Information Graphics: A Comprehensive Illustrated Reference*. Management Graphics.

Hawhee, Debra. 2002. "Kairotic Encounters." In *Perspectives on Rhetorical Invention*, edited by Janet M. Atwill and Janice M. Lauer, 16–35. University of Tennessee Press.

Heal, Ambrose. 1962. *The English Writing-Masters and Their Copy-Books 1570–1800: A Biographical Dictionary & a Bibliography*. Introduction by Stanley Morison. Georg Olms.

Hegel, Georg Wilhelm Friedrich. 1837. *Philosophy of History*. Translated by J. Sibree. P. F. Collier and Son, 1901.

Helmers, Marguerite. 2004. "Framing the Arts Through Rhetoric." In *Defining Visual Rhetorics*, edited by Charles A. Hill and Marguerite Helmers, 63–86. Lawrence Erlbaum.

Hergesheimer, Edwin. 1861. "Map Showing the Distribution of the Slave Population of the Southern States of the United States. Compiled from the Census of 1860." Washington, DC, Henry S. Graham. Library of Congress, Geography and Map Division. http://hdl.loc.gov/loc.gmd/g3861e.cw0013200.

Hewes, Fletcher W., and Henry Gannett. 1883. *Scribner's Statistical Atlas of the United States, Showing by Graphic Methods Their Present Condition and Their Political, Social and Industrial Development*. New York, Charles Scribner's Sons. Library of Congress, Geography and Map Division. http://hdl.loc.gov/loc.gmd/g3701gm.gct00009.

Hickory Park Restaurant Co. 2025. Menu. Hickory Park Restaurant Co., Ames, IA. https://hickoryparkbbq.com/wp-content/uploads/2024/03/Hickory-Park-Menu.pdf.

Hill, Charles A., and Marguerite Helmers, eds. 2004. *Defining Visual Rhetorics.* Lawrence Erlbaum.

Hill, Thomas E. 1888. *Hill's Manual of Social and Business Forms: A Guide to Correct Writing.* Chicago, Hill Standard Book.

Hogarth, William. 1732. *A Harlot's Progress.* Etching and engraving, 6 plates. National Gallery of Art, Washington, DC. https://www.nga.gov/search?keywords=A%20Harlot%27s%20Progress.

Hogarth, William. 1735. *A Rake's Progress.* Etching and engraving, 8 plates. National Gallery of Art, Washington, DC. https://www.nga.gov/search?keywords=A%20Rake%27s%20Progress.

Hogarth, William. 1751. *Gin Lane.* Etching and engraving. National Gallery of Art, Washington, DC. https://www.nga.gov/collection/art-object-page.30433.html.

Hopkins, Albert A., and A. Russell Bond. 1913. *Scientific American Reference Book: Edition of 1913.* New York: Munn & Co.

Howard, June. 1999. "What Is Sentimentality?" *American Literary History* 11, no. 1 (Spring): 63–81. https://doi.org/10.1093/alh/11.1.63.

Huff, Darrell. 1954. *How to Lie with Statistics.* W. W. Norton.

Hunt, John Dixon. 2002. *The Picturesque Garden in Europe.* Thames & Hudson.

Hussey, Christopher. 1967. *The Picturesque: Studies in a Point of View.* Frank Cass.

Hutto, David. 2008. "Graphics and Ethos in Biomedical Journals." *Journal of Technical Writing and Communication* 38, no. 2 (April): 111–31. https://doi.org/10.2190/TW.38.2.b.

IKEA, Inc. 2010/2024. "Malm 2-Drawer Chest Instructions." AA-505147-11. Deft, Netherlands, Inter IKEA Systems B.V. https://www.ikea.com/us/en/assembly_instructions/malm-2-drawer-chest-gray-stained__AA-505147-11-100.pdf.

International Federation of Red Cross and Red Crescent Societies and Save the Children. 2018. "Part C: Standard Operating Procedures for Disasters and Emergencies in Schools." In *Public Awareness and Public Education for Disaster Risk Reduction: Action-Oriented Key Messages for Households and Schools*, 130–56. 2nd ed. Geneva, Switzerland. https://ifrc.soutron.net/public/DownloadImageFile.ashx?objectId=6427&ownerType=0&ownerId=7550.

Iowa Seed Company. 1899. *Iowa Seed Company Annual Catalogue, 1899.* Des Moines, IA, Iowa Seed Company. Iowa State University Library Digital Collections, Iowa Seed and Nursery Pamphlets Collection. https://n2t.net/ark:/87292/w9gd0s.

Iowa State University. 2008. "Biorenewables Research Laboratory Groundbreaking." September 8. Ames, IA, Iowa State University.

Jacobson, Sid, and Ernie Colón. 2006. *The 9/11 Report: A Graphic Adaptation.* Hill and Wang.

Jasinski, James. 2001. "Amplification." In *Sourcebook on Rhetoric: Key Concepts in Contemporary Rhetorical Studies*, by James Jasinski, 12–13. Sage.

Jasinski, James. 2001. "Epideictic Discourse." In *Sourcebook on Rhetoric: Key Concepts in Contemporary Rhetorical Studies*, by James Jasinski, 209–15. Sage.

Jasinski, James. 2001. "Ethos." In *Sourcebook on Rhetoric: Key Concepts in Contemporary Rhetorical Studies*, by James Jasinski, 229–34. Sage.

Jirsa, Pye. 2024. "Ghosting." Photography Terms and Definitions. SLR Lounge. www.slrlounge.com/motion-blur-vs-ghosting-the-difference-between-these-2-artifacts/.

John Deere Company. 2025. "A History of Resounding Impact: The Evolution of Our Logo." www.deere.com/en/our-company/history/.

Kaplan, Fred. 1987. *Sacred Tears: Sentimentality in Victorian Literature*. Princeton University Press.

Katula, Richard A. 2003. "Quintilian on the Art of Emotional Appeal." *Rhetoric Review* 22, no. 1: 5–15. https://doi.org/10.1207/S15327981RR2201_1.

Kayla F. 2020. "The Complete History of the John Deere Logo." *Logo Design Magazine*. https://www.logodesign.org/the-complete-history-of-the-john-deere-logo/.

Kimball, Miles A. 2006. "London Through Rose-Colored Graphics: Visual Rhetoric and Information Graphic Design in Charles Booth's Maps of London Poverty." *Journal of Technical Writing and Communication* 36, no. 4 (October): 353–81. https://journals.sagepub.com/doi/10.2190/K561-40P2-5422-PTG2.

Kimball, Miles A. 2016. "Mountains of Wealth, Rivers of Commerce: Michael G. Mulhall's Graphics and the Imperial Gaze." In *Visible Numbers: Essays on the History of Statistical Graphics*, edited by Miles A. Kimball and Charles Kostelnick, 127–52. Ashgate.

Kimball, Miles A., and Charles Kostelnick, eds. 2016. *Visible Numbers: Essays on the History of Statistical Graphics*. Ashgate.

Kinneavy, James L. 2002. "*Kairos* in Classical and Modern Rhetorical Theory." In *Rhetoric and Kairos: Essays in History, Theory, and Praxis*, edited by Phillip Sipiora and James S. Baumlin, 58–76. State University of New York Press. State University of New York Press. Project Muse. https://muse.jhu.edu/book/4489.

Kinross, Robin. 1985. "The Rhetoric of Neutrality." *Design Issues* 2, no. 2 (Autumn): 18–30. https://doi.org/10.2307/1511415.

Kosara, Robert, and Jock Mackinlay. 2013. "Storytelling: The Next Step for Visualization." *Computer* 46, no. 5 (May): 44–50. https://doi.org/10.1109/MC.2013.36.

Kostelnick, Charles. 1994. "From Pen to Print: The New Visual Landscape of Professional Communication." In "Social and Historical Perspectives on Business and Technical Communication," special issue, *Journal of Business and Technical Communication* 8, no. 1 (January): 91–117. https://doi.org/10.1177/1050651994008001004.

Kostelnick, Charles. 2004. "Melting-Pot Ideology, Modernist Aesthetics, and the Emergence of Graphical Conventions: The Statistical Atlases of the United States, 1874–1925." In *Defining Visual Rhetorics*, edited by Charles A. Hill and Marguerite Helmers, 215–42. Lawrence Erlbaum.

Kostelnick, Charles. 2016. "Mosaics, Culture, and Rhetorical Resiliency: The Convoluted Genealogy of a Data Display Genre." In *Visible Numbers: Essays on the History of Statistical Graphics*, edited by Miles A. Kimball and Charles Kostelnick, 177–206. Ashgate.

Kostelnick, Charles. 2016. "The Re-Emergence of Emotional Appeals in Interactive Data Visualization." *Technical Communication* 63, no. 2 (May): 116–35. http://www.jstor.org/stable/44809638.

Kostelnick, Charles. 2019. *Humanizing Visual Design: The Rhetoric of Human Forms in Practical Communication*. Routledge.

Kostelnick, Charles. 2019. "Pervasive and Perplexing Pies: Our Evolving Relationship with a Data Display Genre." *Information Design Journal* 25, no. 2 (December): 192–213. https://doi.org/10.1075/idj.25.2.04kos.

Kostelnick, Charles, and Michael Hassett. 2003. *Shaping Information: The Rhetoric of Visual Conventions*. Southern Illinois University Press.

Kostelnick, Charles, and John Kostelnick. 2016. "Online Visualizations of Natural Disasters and Hazards: The Rhetorical Dynamics of Charting Risk." In *Science and the Internet: Communicating Knowledge in a Digital Age*, edited by Alan G. Gross and Jonathan Buehl, 157–90. Baywood Publishing.

Kostelnick, Charles, and David D. Roberts. 2011. *Designing Visual Language: Strategies for Professional Communicators*. 2nd ed. Longman/Pearson.

Kress, Gunther, and Theo van Leeuwen. 1996. *Reading Images: The Grammar of Visual Design*. Routledge.

Kurlinkus, William C. 2018. *Nostalgic Design: Rhetoric, Memory, and Democratizing Technology*. University of Pittsburg Press. Project Muse. https://muse.jhu.edu/book/63553.

Kurlinkus, William C. 2021. "Nostalgic Design: Making Memories in the Rhetoric Classroom." *Rhetoric Society Quarterly* 51, no. 5: 422–38. https://doi.org/10.1080/02773945.2021.1972133.

Langley, Batty. 1728. *New Principles of Gardening: or, The Laying out and Planting Parterres, Groves, Wildernesses, Labyrinths, Avenues, Parks, &c. After a More Grand and Rural Manner, Than Has Been Done Before*. London, A. Bettesworth and J. Batley.

Laocoön and His Sons. ca. 40–30 BC. Marble sculpture. Vatican Museums, Vatican City.

Lauer, Claire, and Christopher A. Sanchez. 2023. "What Eye Tracking Can Show Us About How People Are Influenced by Deceptive Tactics in Line Graphs." *IEEE Transactions on Professional Communication* 66, no. 3 (September): 220–35. https://doi.org/10.1109/TPC.2023.3290948.

Lazar, Nomi Claire. 2019. *Out of Joint: Power, Crisis, and the Rhetoric of Time*. Yale University Press.

Lerner, Gerda. 1997. *Why History Matters: Life and Thought*. Oxford University Press.

Lessing, Gotthold Ephraim. 1766. *Laocoön: An Essay upon the Limits of Painting and Poetry*. Translated by Ellen Frothingham. Boston, Roberts Brothers, 1874. HathiTrust. https://hdl.handle.net/2027/hvd.32044020521290.

Letter Friend. 2025. "Letter Friend: Handwritten Direct Mail Services." Letter Friend. letterfriend.com/.

Leutze, Emanuel. 1861–1862. *Westward the Course of Empire Takes Its Way*. Mural. U.S. Capitol Building, Washington, DC. Architect of the Capitol. www.aoc.gov/explore-capitol-campus/art/westward-course-empire-takes-its-way.

Li, Li. 2020. "Visualizing Chinese Immigrants in the *U.S. Statistical Atlases*: A Case Study in Charting and Mapping the Other(s)." *Technical Communication Quarterly* 29, no. 1: 1–17. https://doi.org/10.1080/10572252.2019.1690695.

Lima, Manuel. 2011. *Visual Complexity: Mapping Patterns of Information*. Princeton Architectural Press.

Lincoln, Abraham. 1864. "Letter from Abraham Lincoln to His Cabinet Members, August 23, 1864." Library of Congress, Manuscript Division. https://www.loc.gov/resource/msspin.pin2201/?sp=1&st=image.

Logie, John. 2022. *Writing in the Clouds: Inventing and Composing in Internetworked Writing Spaces*. Parlor Press.

Lovejoy, Arthur O. 1948. "The Chinese Origin of a Romanticism." In *Essays in the History of Ideas*, 99–135. John Hopkins University Press,.

Ludlow, James Meeker. 1885. *Ludlow's Concentric Chart of History: Giving at a Glance the Separate and Contemporaneous History of Each Century*. University of Oregon Libraries Time OnLine, 2016. timeonline.uoregon.edu/ludlow/index.php.

Lupton, Ellen. 1986. "Reading Isotype." *Design Issues* 3, no. 2 (Autumn): 47–58. https://doi.org/10.2307/1511484.

Lupton, Ellen. 2010. *Thinking with Type: A Critical Guide for Designers, Writers, Editors & Students*. 2nd ed. Princeton Architectural Press.

Lyotard, Jean-François. 1984. *The Postmodern Condition: A Report on Knowledge*. Translated by Geoff Bennington and Brian Massumi. University of Minnesota Press.

MacDonald-Ross, Michael. 1977. "How Numbers Are Shown: A Review of Research on the Presentation of Quantitative Data in Texts." *AV Communication Review* 25, no. 4 (Winter): 359–409. http://www.jstor.org/stable/30217944.

Mackenzie, Henry. 1771. *The Man of Feeling*. London, T. Cadell.

Major, Richard, C., engraver. ca. 1848. *Moorhead's Improved Graduated Magnetic Machines*. New York, Richard C. Major. Library of Congress, Prints and Photographs Division. hdl.loc.gov/loc.pnp/pga.02095.

Mann, Michael E., Raymond S. Bradley, and Malcolm K. Hughes. 1999. "Northern Hemisphere Temperatures During the Past Millennium: Inferences, Uncertainties, and Limitations." *Geophysical Research Letters* 26, no. 6 (March): 759–62. https://doi.org/10.1029/1999GL900070.

Marey, Etienne-Jules. 1878. *La Méthode Graphique dans les Sciences Expérimentales et Principalement en Physiologie et en Médicine*. Paris, G. Masson.

Margolin, Victor, ed. 1989. *Design Discourse: History, Theory, Criticism*. University of Chicago Press.

Marinetti, Filippo Tomaso. 1909. "The Founding and Manifesto of Futurism." In *Marinetti; Selected Writings*, edited by R. W. Flint, translated by R. W. Flint and Arthur A. Coppotelli, 39–44. Farrar, Straus and Giroux, 1972.

Massachusetts Bay Transportation Authority. 2025. "Schedules and Maps." Boston, MBTA Schedules and Maps. https://www.mbta.com/schedules.

Massey-Harris Company. 1897. *Massey-Harris Company Catalog*. St. John, New Brunswick, Massey-Harris Company. Iowa State University Library Digital Collections, Lawrence H. Skromme Agricultural Machinery Literature Collection. https://n2t.net/ark:/87292/w9pg1hs7m.

Mast, Foos & Co. N.d. "Buckeye Agricultural Works Advertisement Cards." Springfield, OH, Mast, Foos & Co. Iowa State University Library Digital Collections, Lawrence H. Skromme Agricultural Machinery Literature Collection. https://n2t.net/ark:/87292/w9nz80t90.

Mayo Foundation for Medical Education and Research. 2024. Sales letter for the *Mayo Clinic Health Letter*. June. Rochester, MN, Mayo Clinic Press.

McCauley, Elizabeth Anne. 1985. *A. A. E. Disdéri and the Carte de Visite Portrait Photograph*. Yale University Press.

McCloud, Scott. 1993. *Understanding Comics: The Invisible Art*. Kitchen Sink Press.

McCloud, Scott, and the Google Chrome Team. 2009. *Google Chrome*. Online comic. June. www.google.com/googlebooks/chrome/.

McCormick Harvesting Machine Company. 1885. *Fifty Fourth Illustrated Annual Catalogue, McCormick Machines*. Chicago, McCormick Harvesting Machine Company. Iowa State University Library Digital Collections, Lawrence H. Skromme Agricultural Machinery Literature Collection. https://n2t.net/ark:/87292/w9697026q.

McCormick Harvesting Machine Company. 1886. "McCormick Harvesting Machine Company Advertisement." Chicago, McCormick Harvesting Machine Company; Buffalo and Chicago, Cosack & Co. Iowa State University Library Digital Collections, Lawrence H. Skromme Agricultural Machinery Literature Collection. https://n2t.net/ark:/87292/w9jm23m0j.

McCormick Harvesting Machine Company. 1888. *57th Annual Catalogue of the McCormick Machines*. Chicago, McCormick Harvesting Machine Company. Iowa State University Library Digital Collections, Lawrence H. Skromme Agricultural Machinery Literature Collection. https://n2t.net/ark:/87292/w9xw4813c.

McCormick Harvesting Machine Company. 1893. *McCormick Harvesting Machine Company Catalog*. Chicago, McCormick Harvesting Machine Company. Iowa State University Library Digital Collections, Lawrence H. Skromme

Agricultural Machinery Literature Collection. https://n2t.net/ark:/87292/w9c24qs1d.

McCormick Harvesting Machine Company. 1901. *McCormick Harvesting Machine Company 1901 Catalog*. Chicago, McCormick Harvesting Machine Company. Iowa State University Library Digital Collections, Lawrence H. Skromme Agricultural Machinery Literature Collection. https://n2t.net/ark:/87292/w9ft8dp50.

McKeever, Amy. 2021. "What Can the Faces on Its Currency Tell Us About a Country?" *National Geographic History & Culture*, March 15. National Geographic Society. https://www.nationalgeographic.com/history/article/what-can-faces-on-currency-tell-us-about-country.

McNeil, Paul. 2017. *The Visual History of Type*. Laurence King.

Melville, Herman. 1853. *Bartleby, the Scrivener: A Story of Wall-Street*. Project Gutenberg, 2024. https://www.gutenberg.org/ebooks/11231.

Meriam & Morgan Paraffine Company. 1880. "Paragon Axle Grease." Cleveland, Meriam & Morgan Parrafine Company. Library of Congress, Prints and Photographs Division. http://hdl.loc.gov/loc.pnp/pga.03987.

Metra, Commuter Rail Division of the Regional Transportation Authority and the Northeast Illinois Regional Commuter Rail Corporation. 2024. "Milwaukee North Line: Weekday Schedule." Chicago, Metra, July 15. https://schedules.metrarail.com/pdf/MD-N.pdf. ridertools.metrarail.com/maps-schedules/train-lines/MD-N.

Michelangelo. 1501–1504. *David*. Marble sculpture. Galleria dell'Accademia, Florence.

Minard, Charles Joseph. 1878. *Carte Figurative des Pertes Successives en Hommes de l'Armée Française dans la Campagne de Russie 1812–1813*. Map and graph. In *La Méthode Graphique dans les Sciences Expérimentales et Principalement en Physiologie et en Médicine* by Étienne-Jules Marey, p. 72, figure 37. Paris, G. Masson.

Ministère des Travaux Publics. 1878–1896. *Albums de Statistique Graphique*. Paris, Imprimerie Nationale.

Minneapolis Harvester Works. ca. 1882. "Minneapolis Twine Binder: Minnie Looks After the Sheaves." Chicago, Shober and Carqueville. Iowa State University Library Digital Collections, Lawrence H. Skromme Agricultural Machinery Literature Collection. https://n2t.net/ark:/87292/w9p55dm81.

Moras, Ferdinand, printer. ca. 1867–1869. "F. Moras' Lithographic Establishment, Philadelphia, 610 Jayne St., Plain & Fancy Printing." Philadelphia, Moras's Lithographic Establishment. Library of Congress, Prints and Photographs Division. http://hdl.loc.gov/loc.pnp/ppmsca.24839.

Morrill, Justin S. N.d. "Carte de Visite: Justin S. Morrill." Photograph, Library of Congress, Manuscript Division. https://www.loc.gov/item/mss4429700125/.

Morris, Charles N., engraver. ca. 1873. "St. Nicholas Restaurant. Shell Oysters Received Daily by Express." Poster. Cincinnati, Charles N. Morris. Library of Congress, Prints and Photographs Division. http://hdl.loc.gov/loc.pnp/pga.02243.

Munch, Edvard. 1893. *The Scream*. Tempera painting. National Museum, Oslo, Norway.

Muybridge, Eadweard. 1887. *The Human Figure in Motion*. Dover, 1955.

National Aeronautics and Space Administration (NASA) and the Jet Propulsion Laboratory. 2020. "Visions of the Future: Relax on Kepler-16b." Illustrated by Joby Harris. Visions of the Future Poster Series. NASA and the Jet Propulsion Laboratory, December 24. https://www.jpl.nasa.gov/images/kepler-16b-jpl-travel-poster/.

National Business Show Company. 1906. "National Business Show." Poster. Chicago, National Printing Co. Library of Congress, Prints and Photographs Division. http://hdl.loc.gov/loc.pnp/cph.3b48121.

Navarro, Maria del Mar. 2022. "Isotype of the Conquest: Pictographic Numeracy in Sixteenth-Century Colonial México." *Information Design Journal* 27, no. 1 (December): 35–51. https://doi.org/10.1075/idj.22012.nav.

Neurath, Otto. 1936. *International Picture Language: The First Rules of Isotype*. Kegan, Paul, Trench, Trubner.

New York Central and Lake Shore and Michigan Southern Railroad. 1857. "Change of Time. Summer Arrangement. The New York Central and Lake Shore and Michigan Southern Rail Road Line Forms the Shortest and Most Expeditious Route via Niagara Falls of Buffalo, to Cleveland, Columbus, Cincinnati, Toledo, Chicago." Poster. Whitehall. Library of Congress, Rare Book and Special Collections Division. http://hdl.loc.gov/loc.rbc/rbpe.1220200a.

New York State Department of Health. 2023. "First Aid for Choking." Poster (0603). https://www.health.ny.gov/publications/0603.pdf.

Nightingale, Florence. 1859. *A Contribution to the Sanitary History of the British Army During the Late War with Russia*. London, John W. Parker. Harvard Library, CURIOSity Digital Collections, Contagion: Historical Views of Diseases and Epidemics. https://curiosity.lib.harvard.edu/contagion/catalog/36-990101646750203941.

Norman, Donald A. 2005. *Emotional Design: Why We Love (or Hate) Everyday Things*. Basic Books.

Norman, Donald A. 2013. *The Design of Everyday Things*. Rev. ed. Basic Books.

Nye, David E. 1994. *American Technological Sublime*. MIT Press.

O'Brien, April L., and Josephine Walwema. 2022. "Countering Dominant Narratives in Public Memory." *Technical Communication* 69, no. 3 (August): 40–55. https://doi.org/10.55177/tc985417.

O'Neal, Jacob. 2024. "Animagraffs." Animagraffs.com. https://animagraffs.com/.

Ong, Walter J. 1982. *Orality and Literacy: The Technologizing of the Word*. Methuen.

Ovink, Gerrit Willem. 1938. *Legibility, Atmosphere-Value and Forms of Printing Types*. A. W. Sijthoff.

Palmer, A. N. 1894. *Palmer's Guide to Business Writing*. Cedar Rapids, IA, Western Penman Publishing. Library of Congress. https://www.loc.gov/item/11026563.

Palmer, O. R. 1893. *Type-Writing and Business Correspondence: A Manual of Instruction, Practice Exercises, and Business Forms and Expressions, for Short-Hand Students and Type-writer Operators*. Philadelphia, J. B. Lippincott. Google Books. https://www.google.com/books/edition/Type_writing_and_Business_Correspondence/_HdQAAAAYAAJ?hl=en&gbpv=1.

Patterson, F. B. 1904. "'Stand Pat!' Under Present Prosperity." Broadside. New York. Library of Congress, Rare Book and Special Collections Division. http://hdl.loc.gov/loc.rbc/rbpe.1310110b.

Peabody, Elizabeth Palmer. 1856. *Chronological History of the United States, Arranged with Plates on Bem's Principle*. New York, Sheldon, Blakeman. Library of Congress. https://www.loc.gov/item/ltf90039979/.

Peale, Richard S., ed. 1886. *The Home Library of Useful Knowledge*. Chicago, Home Library Association.

Perelman, Chaïm, and Lucie Olbrechts-Tyteca. 1969. *The New Rhetoric: A Treatise on Argumentation*. Translated by John Wilkinson and Purcell Weaver. University of Notre Dame Press.

Perkins, Jane, and Nancy Blyler. 1999. "Introduction: Taking a Narrative Turn in Professional Communication." In *Narrative and Professional Communication*, edited by Jane M. Perkins and Nancy Blyler, 1–34. Ablex.

Pernot, Laurent. 2015. *Epideictic Rhetoric: Questioning the Stakes of Ancient Praise*. University of Texas Press. Project Muse. https://muse-jhu-edu.eu1.proxy.openathens.net/book/40127.

Pevsner, Nikolaus. 2004. *Pioneers of Modern Design: From William Morris to Walter Gropius*. 4th ed. Yale University Press.

Phillips, Kendall R., ed. 2004. *Framing Public Memory*. University of Alabama Press.

Piranesi, Giovanni Battista. 1748. *Arco di Pola in Istria*. National Gallery of Art, Washington, DC. www.nga.gov/collection/art-object-page.125683.html.

Pitman, Isaac, & Sons. 1893. *A Manual of the Typewriter: A Practical Guide to Commercial, Literary, Legal, Dramatic and All Classes of Typewriting Work*. London, Isaac Pitman and Sons. Google Books. https://www.google.com/books/edition/A_Manual_of_the_Typewriter/xadbAAAAQAAJ?hl=en.

Plato. 1901. *The Republic of Plato*. Translated by Benjamin Jowett. Rev. ed. F. P. Collier & Son. Internet Archive. https://archive.org/details/republic00plat_5/mode/2up?view=theater.

Playfair, William. 1801. *The Commercial and Political Atlas, Representing, by Means of Stained Copper-Plate Charts, the Progress of the Commerce, Revenues, Expenditure, and Debts of England, during the Whole of the Eighteenth Century*. 3rd ed. London, T. Burton. David Rumsey Map Collection, David

Rumsey Map Center, Stanford Libraries. https://www.davidrumsey.com/luna/servlet/detail/RUMSEY~8~1~354689~90121665:Chart-Shewing-the-Amount-of-the-Exp?sort=pub_list_no_initialsort%2Cpub_list_no_initialsort%2Cpub_date%2Cpub_date&qvq=q:William%20Playfair;sort:pub_list_no_initialsort%2Cpub_list_no_initialsort%2Cpub_date%2Cpub_date;lc:RUMSEY~8~1&mi=24&trs=148#.

Pollak, Martha D. 1991. *Military Architecture, Cartography and the Representation of the Early Modern European City: A Checklist of Treatises on Fortification in The Newberry Library*. The Newberry Library.

Polt, Richard. 2015. *The Typewriter Revolution: A Typist's Companion for the 21st Century*. Countryman Press.

Polt, Richard. 2019. "The Life, Death, and Rebirth of the Typewriter." In *Routledge Companion to Media Technology and Obsolescence*, edited by Mark J. P. Wolf. Routledge.

Postaglia Ink. 2015. "Handwritten Mail . . ." https://www.postalgia.ink/.

Prelli, Lawrence J., ed. 2006. *Rhetorics of Display*. University of South Carolina Press.

Price, Uvedale. 1796–1798. *An Essay on the Picturesque: As Compared with the Sublime and the Beautiful*. 2 vols. London, J. Robson.

Priestley, Joseph. 1765. *A Chart of Biography*. London.

Priestley, Joseph. 1769. *A New Chart of History*. London, Joseph Johnson. David Rumsey Map Collection, David Rumsey Map Center, Stanford Libraries. www.davidrumsey.com/luna/servlet/detail/RUMSEY~8~1~254828~5519513:A-new-chart-of-history—J—Priestle#.

Principal Financial Services, Inc. 2025. "History of Principal Financial Group." Interactive chart. Des Moines, IA. https://www.principal.com/about-us/history-principal-financial-group.

Pritchard, Shannon. 2015. "Guido Reni, *Aurora*." Smarthistory, July 19. https://smarthistory.org/reni-aurora/.

Pryibil, P. 1880. "New Double-Feed Planing Machine." *American Machinist* 3, no. 16 (April 17), p. 1.

Pugin, A. Welby. 1843. *An Apology for the Revival of Christian Architecture in England*. London, John Weale.

Queen, James Fuller. 1864. "Great Central Fair for the Sanitary Commission." Pictorial letter sheet. Philadelphia, P.S. Duval & Son. Library of Congress, Prints and Photographs Division. https://loc.gov/pictures/resource/ds.00296/.

QuidelOrtho Corporation. 2021. "User Instructions: QuickVue At-Home OTC COVID-19 Test." Document 1479701 (05/21). San Diego, CA, QuidelOrtho Corporation, May 2021.

Quintilian, Marcus Fabius. 1920–1922. *The Institutio Oratoria of Quintilian*. Translated by H. E. Butler. 4 vols. Loeb-Harvard University Press, 1959–1963.

Radcliffe, Ann. 1794. *The Mysteries of Udolpho*. 4 vols. London, G. G. and J. Robinson.

Ramelli, Agostino. 1588. *Le Diverse et Artificiose Machine del Capitano Agostino Ramelli dal Ponte della Tresia.* Parigi, Italy.

Rawlins, Jacob D., and Greg D. Wilson. 2014. "Agency and Interactive Data Displays: Internet Graphics as Co-created Rhetorical Spaces." *Technical Communication Quarterly* 23, no. 4 (September): 303–22. https://doi.org/10.1080/10572252.2014.942468.

RDG Planning & Design. 2020. Rendering of a Parish Hall for Sacred Heart Church, Boone, IA. Randall L. Milbrath, lead architect. Omaha, NE, RDG Planning & Design.

Reni, Guido. 1613–1614. *Aurora.* Ceiling fresco. Casino dell'Aurora, Rome.

Repton, Humphrey, 1806. *An Enquiry into the Changes of Taste in Landscape Gardening.* London, J. Taylor.

Richardson, Samuel. 1748. *Clarissa; or, the History of a Young Lady.* London.

Robert, Hubert. 1796. *Imaginary View of the Grande Galerie of the Louve in Ruins.* Oil on canvas. Musée du Louvre, Paris, France.

Roberts, Elizabeth. ca. 1950. "A Cook's Tour of Your New Gas Range." Publication no. 25091–25. Montgomery Ward.

Robles, Vincent D. 2018. "Visualizing Certainty: What the Cultural History of the Gantt Chart Teaches Technical and Professional Communicators about Management." *Technical Communication Quarterly* 27, no. 4: 300–321. https://doi.org/10.1080/10572252.2018.1520025.

Rodin, Auguste. 1903. *The Thinker.* Bronze sculpture. Musée Rodin, Paris, France.

Rosenberg, Daniel, and Anthony Grafton. 2010. *Cartographies of Time: A History of the Timeline.* Princeton Architectural Press.

Rosenberg, Daniel, and University of Oregon Libraries. 2016. Time OnLine. University of Oregon Libraries. timeonline.uoregon.edu/.

Rosling, Ola, Anna Rosling Rönnlund, and Hans Rosling. 2025. Gapminder. https://www.gapminder.org/.

Rosling, Ola, Anna Rosling Rönnlund, and Hans Rosling. 2025. "CO2 Emissions per Capita/GDP per Capita." Bubble chart. Gapminder. https://www.gapminder.org/tools/#$model$markers$bubble$encoding$y$data$concept=co2_pcap_cons&space@=geo&=time;;&scale$domain:null&zoomed:null&type=log;;&-frame$value=2022;;;;;&chart-type=bubbles&url=v2.

Ross, Derek G. 2008. "Dam Visuals: The Changing Visual Argument for the Glen Canyon Dam." *Journal of Technical Writing and Communication* 38, no. 1 (January): 75–94. https://doi.org/10.2190/TW.38.1.e.

Rousseau, Jean-Jacques. 1761. *Julie, ou La Nouvelle Hélöise.* Amsterdam, Marc Michel Rey.

Ruskin, John. 1867. "The Nature of Gothic." In *The Stones of Venice*, 2nd ed., 2:151–231. London, Smith, Elder.

Sandburg, Carl. 1950. *Complete Poems: Carl Sandburg.* Harcourt, Brace, & World.

Sandby, Paul. 1758. *Ruins with a Seated Man, and a Covered Wagon in the Distance*. Etching. National Gallery of Art, Washington, DC. www.nga.gov/collection/art-object-page.54467.html.

Sarony, Napoleon, lithographer. 1851. "Mrs. Emma Willard's Chronographer of Ancient History." Lithograph. New York, A. S. Barnes. Library of Congress, Prints and Photographs Division. http://hdl.loc.gov/loc.pnp/ppmsca.09327.

Schiller, Friedrich. 1795–1796. *On the Naive and Sentimental in Literature*. Translated by Helen Watanabe-O'Kelly. Carcanet New Press, 1981.

Schriver, Karen A. 1997. *Dynamics in Document Design: Creating Texts for Readers*. John Wiley & Sons.

Schulten, Susan. 2012. *Mapping the Nation: History and Cartography in Nineteenth-Century America*. University of Chicago Press.

Scribeless. 2025. "Make Business Feel More Personal." www.scribeless.co/.

Shearman & Hart. N.d. "Beautifully Illuminated Show Cards for Every Business." Library of Congress, Prints and Photographs Division. http://hdl.loc.gov/loc.pnp/pga.04144.

Sikoryak, R. 2017. *Terms and Conditions*. Drawn and Quarterly

Simply Noted. 2025. "Send Real Handwritten Notes to Your Clients." www.simplynoted.com/.

Sloane, Charles S., and the U.S. Bureau of the Census. 1914. *Statistical Atlas of the United States*. U.S. Government Printing Office.

Sloane, Charles S., and the U.S. Bureau of the Census. 1925. *Statistical Atlas of the United States*. U.S. Government Printing Office.

Smith, Adam. 1759. *The Theory of Moral Sentiments*. 4th ed. London, W. Strahan, et al., 1774. HathiTrust. https://hdl.handle.net/2027/nyp.33433070252915.

Smith, Adam. 1776. *An Inquiry into the Nature and Causes of the Wealth of Nations*. 2 vols. London, W. Strahan and T. Caddell. Internet Archive. https://archive.org/details/inquiryintonatur01smit_0/page/n3/mode/2up.

Smith, John E. 1969. "Time, Times, and the 'Right Time'; *Chronos* and *Kairos*." *Monist* 53, no. 1 (January): 1–13. https://www.jstor.org/stable/27902109.

Smith, John E. 2002. "Time and Qualitative Time." *Rhetoric and Kairos: Essays in History, Theory, and Praxis*, edited by Phillip Sipiora and James S. Baumlin, 46–57. State University of New York Press. Project Muse. https://muse.jhu.edu/book/4489.

Snow, John. 1855. *On the Mode of Communication of Cholera*. 2nd ed. London, John Churchill.

Spafford, Horatio Gates, H. O. McDaid, and John P. Wilson. ca. 1875. Business card of attorneys. Chicago. Library of Congress, Manuscript Division. https://hdl.loc.gov/loc.mss/ms010123.mamcol.007.

Spencer, Platt R. 1866. *Spencerian Key to Practical Penmanship*. New York, Ivison, Phinney, & Blakeman Co. Hathi Trust. https://babel.hathitrust.org/cgi/pt?id=uc1.c046819857&seq=9.

Stacy, James G. 1880. "Jas. G. Stacy, Photographer, Rahway, N.J." Business card. Boston, J. H. Bufford. Library of Congress, Prints and Photographs Division. http://hdl.loc.gov/loc.pnp/ppmsca.51345.

Standard Wax Company. ca. 1880s. "Wax Strings." Sales illustration. Middletown, OH, Standard Wax Co. Library of Congress, Prints and Photographs Division. http://hdl.loc.gov/loc.pnp/pga.11390.

Steiner, Wendy. 1988. *Pictures of Romance: Form Against Context in Painting and Literature*. University of Chicago Press.

Steiner, Wendy. 2004. "Pictorial Narrativity." In *Narrative Across Media: The Languages of Storytelling*, edited by Marie-Laure Ryan, 145–77. University of Nebraska Press.

Sterne, Laurence. 1768. *A Sentimental Journey Through France and Italy*. 2 vols. London, T. Becket and P. A. De Hondt.

StockCharts.com. 2025. "Dow Jones Industrial Chart: Daily View." Interactive chart. March 7. StockCharts.com. https://stockcharts.com/freecharts/gallery.html?$INDU.

StockCharts.com. 2025. "MarketCarpet of S&P 500." Interactive chart. March 6–7. StockCharts.com. https://stockcharts.com/freecharts/carpet.html?[SEC].

Stowe, Harriet Beecher. 1852. *Uncle Tom's Cabin; Or, Life Among the Lowly*. Boston, John P. Jewett.

Swales, John M. 1990. *Genre Analysis: English in Academic and Research Settings*. Cambridge University Press.

Szczepaniak, Phil. 2025. "Portfolio: 3D Animation in Action." 3Deeit.com. https://3deeit.com/portfolio/.

Tange, Andrea Kaston. 2004. "Envisioning Domesticity, Locating Identity: Constructing the Victorian Middle Class Through Images of Home." In *Defining Visual Rhetorics*, edited by Charles A. Hill and Marguerite Helmers, 277–301. Lawrence Erlbaum.

Tebeaux, Elizabeth. 1991. "Visual Language: The Development of Format and Page Design in English Renaissance Technical Writing." *Journal of Business and Technical Communication* 5, no. 3 (July) 246–74. https://doi.org/10.1177/1050651991005003002.

Tebeaux, Elizabeth. 1993. "From Orality to Textuality in English Accounting and Its Books, 1553–1680: The Power of Visual Presentation." *Journal of Business and Technical Communication* 7, no. 3 (July): 322–59. https://doi.org/10.1177/1050651993007003003.

Tebeaux, Elizabeth. 1997. *The Emergence of a Tradition: Technical Writing in the English Renaissance, 1475–1640*. Baywood Publishing.

Tebeaux, Elizabeth. 2010. "Safety Warnings in Tractor Operation Manuals, 1920–1980: Manuals and Warnings Don't Always Work." *Journal of Technical Writing and Communication* 40, no. 1 (January): 3–28. https://doi.org/10.2190/TW.40.1.b.

Tebeaux, Elizabeth. 2014. *The Flowering of a Tradition: Technical Writing in England, 1641–1700.* Routledge.

Thornton, Tamara Plakins. 1996. *Handwriting in America: A Cultural History.* Yale University Press.

Tirrell, Jeremy. 2015. "Latourian *Memoria*." In *Thinking with Bruno Latour in Rhetoric and Composition*, edited by Paul Lynch and Nathaniel Rivers, 165–81. Southern Illinois University Press.

Tominski, Christian, Wolfgang Aigner, Silvia Miksch, and Heidrum Schumann. 2017. "Images of Time: Visual Representations of Time-Oriented Data." In *Information Design: Research and Practice*, edited by Alison Black, Paul Luna, Ole Lund, and Sue Walker, 23–42. Routledge.

Toyota Motor Corporation. 2024. "History: 75 Years of Toyota." In *Integrated Report 2023*, 102. Integrated Report / Annual Report Archives. https://global.toyota/pages/global_toyota/ir/library/annual/2023_001_integrated_en.pdf.

Tufte, Edward R. 1983. *The Visual Display of Quantitative Information*. Graphics Press.

Tufte, Edward R. 1990. *Envisioning Information*. Graphics Press.

Turner, Joseph Mallord William. 1794. *Tintern Abbey: The Crossing and Chancel, Looking Towards the East Window*. Graphite and watercolor on paper. Tate Gallery, London.

Turner, Joseph Mallord William. 1812. *River Wye*. Etching. National Gallery of Art, Washington, DC. www.nga.gov/collection/art-object-page.10672.html.

Typewriting Instructor and Stenographer's Hand-Book. 1892. Practical Text-Book Company. HathiTrust. https://hdl.handle.net/2027/hvd.hn5znu.

U.S. Department of Agriculture. 1941. "12 Tricks in Saving Electricity." *Consumers' Guide* 8, no. 1 (November 1): 5–6.

U.S. Department of the Army. 1951–1971. *PS Magazine: The Preventive Maintenance Monthly*. Illustrated by Will Eisner. Virginia Commonwealth University Libraries Scholars Compass. https://scholarscompass.vcu.edu/psm/.

U.S. Department of the Army. 1954. "Joe Dope: How Free-Turn-In Works." Illustrated by Will Eisner. *PS Magazine: The Preventive Maintenance Monthly*, no. 17: 781–88. Virginia Commonwealth University Libraries Scholars Compass. https://scholarscompass.vcu.edu/psm/17/.

U.S. Department of Commerce, U.S. Census Bureau. 2019. "The 15 Fastest-Growing Cities: By Percent Change Between July 1, 2017, and July 1, 2018." Infographic, May 23. U.S. Census Bureau Infographics & Visualizations. www.census.gov/library/visualizations/2019/comm/15-fastest-growing-cities.html.

U.S. Department of Commerce, U.S. Census Bureau, Geography Division. 2022. "Mean Center of Population for the United States: 1790–2010." Data map, rev. January 14. United States Census Bureau Infographics & Visualizations. www.census.gov/library/visualizations/2010/geo/center-of-population-1790-2010.html.

U.S. Department of Commerce, National Oceanic and Atmospheric Administration (NOAA). 2025. Sea Level Rise Viewer. Map and picture. coast.noaa.gov/slr/#/layer/slr/6/-9227628.982527956/2969936.6913797804/8/satellite/50/0.8/2050/interHigh/midAccretion.

U.S. Department of Commerce, National Weather Service, National Oceanic and Atmospheric Administration (NOAA). 2025. "NWS Organizational Structure." National Weather Service. https://www.weather.gov/organization/.

U.S. Department of Commerce, National Weather Service, National Oceanic and Atmospheric Administration (NOAA). 2025. Weather map. March 4. National Weather Service. https://www.weather.gov/.

U.S. Department of Health and Human Services, Centers for Disease Control and Prevention. 2021. "COVID View Summary Ending January 2, 2021." Centers for Disease Control and Prevention. https://archive.cdc.gov/#/details?archive_url=https://archive.cdc.gov/www_cdc_gov/coronavirus/2019-ncov/covid-data/pdf/covidview-01-08-2021.pdf.

U.S. Department of the Interior. 2008. Organization chart. University of North Texas Library End of Term Web Archive. webarchive.library.unt.edu/eot2008/20080916004331/http://interior.gov/secretary/officials_orgchart.html.

U.S. Department of Labor, Bureau of Labor Statistics. 1946. "Bureau Studies Why Women Are Paid Less." *Labor Information Bulletin* 13, no. 12 (December 1946): 3.

U.S. Department of Labor, Bureau of Labor Statistics. 1947. "Key Steps in Conciliation." *Labor Information Bulletin* 14, no. 4 (April 1947): 20.

U.S. Department of Transportation, Federal Highway Administration. 2023. *Manual on Uniform Traffic Control Devices for Streets and Highways.* 11th ed. https://mutcd.fhwa.dot.gov/pdfs/11th_Edition/mutcd11thedition.pdf.

U.S. Federal Emergency Relief Administration, Division of Research, Statistics and Finance. 1935. *On Relief: General Relief Program.* Prepared by the Graphic Unit of the Research Section. U.S. Federal Emergency Relief Administration.

U.S. Geological Survey Earthquake Hazards Program. 2022. "Frequency of Damaging Earthquake Shaking Around the U.S." Data map, March 2. U.S. Geological Survey. https://www.usgs.gov/media/images/frequency-damaging-earthquake-shaking-around-us.

Van Winkle, Kevin. 2022. "Above All Made by Themselves: The Visual Rhetoric of W. E. B. Du Bois's Data Visualizations," *Technical Communication Quarterly* 31, no. 1: 17–32. https://doi.org/10.1080/10572252.2021.1906450.

Veltsos, Jennifer R. 2009. "More Than Decoration: An Investigation into the Role of Visual Rhetoric and Ethos in Corporate Visual Identity." PhD diss., Iowa State University.

Venditti, Simona, Francesca Piredda, and Walter Mattana. 2017. "Micronarratives as a Form of Contemporary Communication." *Design Journal* 20, supplement 1 (April): S273-S282. https://doi.org/10.1080/14606925.2017.1352804.

Venturi, Robert. 1966. *Complexity and Contradiction in Architecture*. Museum of Modern Art.

VGraphs. 2025. Interactive rank charts. YouTube. https://www.youtube.com/@vgraphs.

Virgil. 2016. *The Aeneid*. Translated by Barry B. Powell. Oxford University Press.

Wainer, Howard. 1997. *Visual Revelations: Graphical Tales of Fate and Deception from Napolean Bonaparte to Ross Perot*. Copernicus.

Wainer, Howard. 2005. *Graphic Discovery: A Trout in the Milk and Other Visual Adventures*. Princeton University Press.

Wainer, Howard, Polina Harik, and John Neter. 2013. "Stigler's Law of Eponymy and Marey's Train Schedule: Did Serjev Do It Before Ibry, and What About Jules Petiet?" *Chance* 26, no. 1: 53–56. https://doi.org/10.1080/09332480.2013.772394.

Walker, Francis A., and the U.S. Census Office. 1874. *The Statistical Atlas of the United States Based on the Results of the Ninth Census 1870*. Julius Bien. Library of Congress, Geography and Map Division. https://www.loc.gov/resource/g3701gm.gct00008/?sp=1&st=image.

Walker, George H. ca. 1860–1880. "Portland Star Match Co., Portland, ME." Bird's-eye view lithograph. Boston, George H. Walker & Co. Lith. Library of Congress, Prints and Photographs Division. http://hdl.loc.gov/loc.pnp/pga.04222.

Walker, Sue. 1984. "How Typewriters Changed Correspondence: An Analysis of Prescription and Practice. *Visible Language*, 18, no. 2 (Spring): 102–17.

Walker, Sue. 2003. "The Manners of the Page: Prescription and Practice in the Visual Organization of Correspondence." In "Studies in the Cultural History of Letter Writing," special issue, *Huntington Library Quarterly* 66, nos. 3–4: 307–29. http://www.jstor.org/stable/3818085.

Walker, Sue. 2014. *Typography & Language in Everyday Life: Prescriptions and Practices*. Routledge.

Walker, Sue. 2018. "Modernity, Method and Minimal Means: Typewriters, Typing Manuals and Document Design." *Journal of Design History* 31, no. 2 (May): 138–53. https://doi.org/10.1093/jdh/epx018.

Ware, Isaac, and Inigo Jones. 1756. *A Complete Body of Architecture, Adorned with Plans and Elevations, from Original Designs*. London, T. Osborne and J. Shipton.

Watkins, Robert, and Tom Lindsley. 2020. "Sequential Mapping: Using Sequential Rhetoric and Comics Production to Understand UX Design." *Technical Communication Quarterly* 29, no. 3: 304–18. https://doi.org/10.1080/10572252.2020.1768292.

Weaver, Richard. 1970. *Language Is Sermonic: Richard M. Weaver on the Nature of Rhetoric*, edited by Richard L. Johannesen and Rennard Strickland. Louisiana State University Press.

Wei, Hong-Kang. 2004. "Rhetoric as Collective Ethos: From Chinese Classical Texts to Postmodern Corporate Images." PhD diss., Iowa State University.

Wei, Hong-Kang. 2008. "Ethos on the Web: A Cross-Cultural Approach." In *Writing the Visual: A Practical Guide for Teachers of Composition and Communication*, edited by Carol David and Ann R. Richards, 146–67. Parlor Press.

West, Russell W. 1936–1938. "Posters: Federal Art Gallery." Poster. Boston, Federal Art Project. Library of Congress, Prints and Photographs Division. http://hdl.loc.gov/loc.pnp/cph.3b48875.

Westermann, Capitain Ludewig Martin. 1768. Meeting announcement. Hamburg. Library of Congress, Prints and Photographs Division. http://hdl.loc.gov/loc.pnp/pga.11903.

Wharton, Thomas. 1747. "The Pleasures of Melancholy: A Poem." London, R. Dodsley. Internet Archive. https://archive.org/details/bim_eighteenth-century_the-pleasures-of-melanch_warton-thomas_1747/mode/2up.

Whately, Thomas. 1770. *Observations on Modern Gardening*. 2nd ed. London, T. Payne.

Wheelwright, Philip. 1959. *Heraclitus*. Princeton University Press.

Wiebenson, Dora. 1978. *The Picturesque Garden in France*. Princeton University Press.

Wilkie, Brian. 1967. "What Is Sentimentality?" *College English* 28, no. 8 (May): 564–75. https://doi.org/10.2307/374718.

Willard, Emma. 1839. *Atlas, to Accompany a System of Universal History*. New York, F. J. Huntington. Library of Congress, Geography and Map Division. http://hdl.loc.gov/loc.gmd/g3201sm.gct00409.

Williams, William Carlos. 1986. *The Collected Poems of William Carlos Williams*, edited by A. Walton Litz and Christopher MacGowan. Vol. 1, 1909–1939. New Directions.

Williamson, Jack H. 1986. "The Grid: History, Use, and Meaning." *Design Issues* 3, no. 2 (Autumn): 15–30. https://doi.org/10.2307/1511481.

Winter's Art Lithography Company. ca. 1892. "Electrical Building—World's Columbian Exposition." Illustrated by C. Graham. Lithograph. Chicago, Winter's Art Lithography Company. Library of Congress, Prints and Photographs Division. http://hdl.loc.gov/loc.pnp/pga.03063.

Wisconsin Historical Society. 2005. "Letterheads—Image Gallery Essay: Promotions on Paper, 1850s–1975." Wisconsin Historical Society. https://www.wisconsinhistory.org/Records/Article/CS3591.

Wolfe, Tom. 1981. *From Bauhaus to Our House*. Farrar, Straus and Giroux.

Wood, Grant. 1930. *American Gothic*. Oil on wood. Art Institute of Chicago, Chicago.

Wordsworth, William. 1965. *William Wordsworth: Selected Poems and Prefaces*, edited by Jack Stillinger. Houghton Mifflin.

Works Progress Administration (WPA). 1936/1937. "Know the World You Live in, Free Informal Study Groups: Workers Education Project." Poster. New

York: Federal Art Project. Library of Congress, Prints and Photographs Division. http://hdl.loc.gov/loc.pnp/cph.3b48854.

World Food Prize Foundation. 2023. "Heidi Kühn 2023 Laureate: Planting the Roots of Peace." World Food Prize Foundation. https://www.worldfoodprize.org/documents/filelibrary/Heidi_Kuhn_Laureate_Story_Booklet_1803609CD4C45.pdf.

Wright, Orville. 1903. "Telegram from Orville Wright to Bishop Milton Wright Announcing the First Successful Powered Flight." December 17, Kitty Hawk, NC. Library of Congress, Manuscript Division. https://www.loc.gov/item/mcc.061/.

Wyatt, Christopher Scott, and Dànielle Nicole DeVoss, eds. 2018. *Type Matters: The Rhetoricity of Letterforms*. Parlor Press.

Yates, JoAnne. 1985. "Graphs as a Managerial Tool: A Case Study of Du Pont's Use of Graphs in the Early Twentieth Century." *International Journal of Business Communication* 22, no. 1 (Winter): 5–33. https://doi.org/10.1177/002194368502200101.

Yates, JoAnne. 1989. *Control Through Communication: The Rise of System in American Management*. Johns Hopkins University Press.

Yau, Nathan. 2013. *Data Points: Visualization That Means Something*. John Wiley & Sons.

Yu, Han. 2015. *The Other Kind of Funnies: Comics in Technical Communication*. Baywood Publishing.

Zhang, Yuejiao. 2016. "Illustrating Beauty and Utility: Visual Rhetoric in Two Medical Texts Written in China's Northern Song Dynasty, 960–1127." *Journal of Technical Writing and Communication* 46, no. 2 (April): 172–205. https://doi.org/10.1177/0047281616633599.

Zöllner, Frank, and Johannes Nathan. 2003. *Leonardo da Vinci, 1452–1519: The Complete Paintings and Drawings*. Taschen.

Index

Note: A page number in *italics* indicates a figure on the corresponding page.